THE DEFOLIATION OF AMERICA

NE**X**US

NEW HISTORIES OF SCIENCE, TECHNOLOGY, THE ENVIRONMENT, AGRICULTURE & MEDICINE

NEXUS is a book series devoted to the publication of high-quality scholarship in the history of the sciences and allied fields. Its broad reach encompasses science, technology, the environment, agriculture, and medicine, but also includes intersections with other types of knowledge, such as music, urban planning, or educational policy. Its essential concern is with the interface of nature and culture, broadly conceived, and it embraces an emerging intellectual constellation of new syntheses, methods, and approaches in the study of people and nature through time.

THE DEFOLIATION
OF AMERICA

✳———————————————————————✳

Agent Orange Chemicals, Citizens, and Protests

AMY M. HAY

THE UNIVERSITY OF ALABAMA PRESS   TUSCALOOSA

The University of Alabama Press
Tuscaloosa, Alabama 35487-0380
uapress.ua.edu

Typeface: Scala Pro

Cover image: Agent Orange demonstration in Milwaukee, 1981;
courtesy of Vietnam Veterans Against the War

Cover design: Michele Myatt Quinn

Cataloging-in-Publication data is available from the Library of
Congress.
ISBN: 978-0-8173-2108-6
E-ISBN: 978-0-8173-9379-3

For Dolores Huston Hay, Jo Ann Carrigan, and Anthony Carrigan

Contents

Figures

Foreword

Few chemicals have captured the public imagination in quite the same way as Agent Orange. Its use during the Vietnam War—to define boundaries, uncover North Vietnamese Army routes, limit attacks by removing cover, and target enemy food supplies—rendered it a synecdoche for the American war effort as a whole and a synonym for phenoxy herbicides writ large. Its lingering effects on the landscape of Vietnam and its consequences for those servicemen and Vietnamese civilians exposed to it have attracted the attention of scholars and made their way into film documentaries and undergraduate lectures. But for all its familiarity, Agent Orange remains two-dimensional in most accounts, which treat it as a chemical that proved germane to the conflict far away in Vietnam and which affected the home front only through returning soldiers. The domestic use of Agent Orange chemicals and the grassroots opposition they inspired have remained largely obscured—until now.

In this important addition to the Nexus series, Amy Hay moves ordinary citizens to the center of the story. In so doing she significantly reframes our understanding of a story we thought we knew well and in the process introduces us to what is arguably the most important anti-toxics movement of the postwar era. Beginning with a world in which most citizens welcomed the application of phenoxy herbicides, Hay shows how the unintended consequences of the chemicals inspired critiques from a growing number of quarters, as religious organizations, university students, and scientists, both domestic and international, expressed a growing sense of alarm. Expecting politicians to protect the citizens they represented rather than corporate profits, and believing that corporations ought to take responsibility for the products they manufactured and marketed, activists grew increasingly frustrated with what they saw as an inadequate, even dismissive, governmental response to their concerns.

Led by overlooked figures like Billee Shoecraft, Ida Honorof, and Carol Van Strum, these activists worked to inform the public of the hidden risks of the chemicals' application, and over time they inspired a diverse and widespread movement that sought to halt the domestic application of Agent Orange chemicals. Hay's accounts of their efforts call attention to the manifold ways in which phenoxy herbicides were applied in the United States—from increasing water runoff to meet the needs of an expanding population in Phoenix, to controlling brush in California, to facilitating timber companies' access to forests in Oregon. Significantly, they also point to the outsized role of women in the efforts to check the use of the chemicals, an aspect of the movement that previous studies have ignored. Moreover, Hay casts new light on the men who have played the key roles in narratives about opposition to Agent Orange to date. Where previous studies of the veterans' protests have centered on the veterans themselves, Hay adroitly demonstrates how the veterans often focused on the chemicals' effects on their families and children. In tracing the threads that connected the efforts of women like Shoecraft, Honorof, and Van Strum to those of the veterans, Hay highlights the degree to which an emphasis on gender can continue to reframe ostensibly familiar stories.

By calling attention to the ways in which the activists, male and female alike, challenged state and scientific authority, Hay offers fresh insights into the place of citizen science in the second half of the twentieth century. Hay makes a compelling case that by operating on the assumption that phenoxy herbicides were safe in the absence of clear evidence to the contrary, companies and government agencies effectively shunted the burden of proof to those who endured the effects of the chemicals. Later studies would largely support the activists' claims about the perils posed by Agent Orange chemicals, but Hay is less interested in assessing the veracity of their claims than in exploring the significance of their engagement in the *process* of knowledge making, in their claims that ordinary citizens have a right to have a say about the consequences of actions that have real bearing on their lives, and especially their health. Thus, in tracing the divergent motivations of her activists and the uneven results with which they met, Hay offers a sympathetic and convincing account of the ways in which their efforts laid a profoundly democratic foundation for citizen engagement with science.

Writing with a clear moral compass, Hay has produced a definitive study that fundamentally reshapes the contours of scholarship on Agent Orange chemicals. Situated at the nexus of the history of science, environmental

history, and the history of health, it is an ideal fit for this series, not only because it makes significant contributions to those fields but because it is ambitious in its scope. Historians of toxics and toxicants, of citizen science, of
technology, of gender, of political activism in postwar America, and of modern American environmentalism will all find much of value in Hay's analysis. Finally, and perhaps most importantly, it is a study with profound implications for our world today. It is a study that deserves a wide readership in
the academy and beyond, and we are pleased to see its publication as part of
the Nexus series.

MARK D. HERSEY, on behalf of the Nexus series editors

Acknowledgments

Aside from being done, the gratitude is the best part. First, my thanks to Claire Lewis Evans, my editor at the University of Alabama Press, for her knowledge and insight, patience, and timely support. I am very pleased to be published as a part of the press's NEXUS series and want to especially thank series editor Mark Hersey, who expressed interest in my work and took the time at an international conference to speak with me about publishing with the press. I would also like to thank the anonymous reviewers who gave such thoughtful, helpful, and encouraging feedback about the book. It is better for their care.

The University of Texas Pan American (now University of Texas Rio Grande Valley) supported this work with a College of Arts and Humanities grant that allowed me to research at the National Agricultural Library in Beltsville, Maryland. Thanks to the helpful archivists there. Archivists and staff at the Texas Tech University Vietnam Archive in Lubbock, Texas, the Chemical Heritage Foundation in Philadelphia, Pennsylvania, and the Forest History Society in Durham, North Carolina, also aided me in negotiating their rich collections. At the Forest History Society, Jaime Lewis and Steve Anderson were great hosts and made my research trip productive and pleasant. I should also thank the various archives, groups, and individuals who provided images for this book.

Two residential writing fellowships supported this project. The Chemical Heritage Foundation (now the Science History Institute) gave me money, space, and a home in the summer of 2009. I had a chance to research the Dow Company Collection, think about Agent Orange, and write about some of my activists. Jody Roberts helped both intellectually and practically, offering insights about the chemical industry and finding me a place to live while in Philadelphia. His support and encouragement of my research were just a happy bonus.

The Rachel Carson Center for Environment and Society, located in Munich, Germany, funded me to come and live and participate with a brilliant group of environmental humanities scholars in 2012. My fellowship changed my research and my life in important and wonderful ways. Directors Christof Mauch and Helmuth Trischler created what one friend described as an academic utopia (and she was right). My officemates, my Carson Fellow cohort, and all of the thoughtful, gifted, and supportive people I met while there and afterward deserve recognition and thanks; and I cannot begin the list, because it would be a very, very long one indeed. You know who you are. (I have a whole new level of respect for awardees who must thank a host of deserving people in thirty seconds or less.) I should also say thank you to the Federal Republic of Germany for its funding of the Carson Center.

I came back from a conference once and realized something: I was a well-trained gender historian. I owe so very much to my mentors and am so glad to be able to publicly acknowledge the wisdom, patience, and good humor they displayed while guiding me. With an undergraduate degree in pharmacy, I presented some interesting challenges. Jo Ann Carrigan taught me that I could research and write the history of medicine, with help from Marty Pernick. Lisa Fine and Darlene Clark Hine were crucial to my understandings of gender, class, and race (what one of my professors called the historical Holy Trinity). Kirsten Fermaglich, you know everything you have done. Thank you all. I continue to grow as a scholar and a person because of the foundations, and friendships, you all have given me.

Like almost any academic, I see many of my friends at academic conferences. If we eat together, your name should be in this list (and sorry if I forgot). From the Carson Center and ASEH, love to Arielle Helmick, Erika Bsumek, Katie Ritson, Melanie Arndt, Michelle Mart, Ellen Arnold, Ruth Morgan, Sherry Johnson, Fritz Davis, David Vail, Jim Rice, and Rob Gioielli. I am very lucky to have friends like Teresa Sabol Spezio, Amy Cummins, Shawn Thomson, Dora Garcia Saavedra, Russ Skowronek, Peg Graham, Anne and Dave Gossage, Pamela Herring, and Pamela Anderson-Meijas, all of whom have generously (both in quality and quantity) cared about me and my work over many years. Y'all kept me laughing and sane. My tribe, connected to me by blood and/or choice, includes some awesome people. Thank you to my sister Cris Hay Merchant and my brothers Bill, Scott, and Jim Hay, along with their families; to Susan Kroeg, Donna Kroeg, Brad Wood, and Clare Wood;

Grace and the extended Legaspi family; Kristine Wirts; Chris Miller; Tamer Balci, Tania Han, and Ayda and Ayla Han Balci.

Finally, I want to acknowledge and thank the activists whose stories appear here, and the people I met while finding them. I regret that Ida Honorof and Maude DeVictor died while I was working on this project and before I got to meet them. I would like to thank Carol Van Strum, Faye Honorof, Nga Le, Susan Hammond, and Heather Morris Bowser for meeting with me and for all their help. Activists shed blood, sweat, and tears—all while remembering to laugh—in their efforts to make a better world. I am grateful beyond words that their vision, determination, and action give us hope, and eventually, make change for a better world.

THE DEFOLIATION OF AMERICA

Introduction

The Defoliation of America:
Chemical Uses and Protests in Post-1945 America

The original scientific investigations of the chemical compounds 2,4-D and 2,4,5-T emphasized their ability to accelerate plant growth. Scientists initially hoped that the quantity and quality of fruits and vegetables might be improved. In an apocalyptic twist, researchers realized the chemicals' real value lay with their ability to accelerate growth and kill plant life in the process. It would be these chemicals that were used in a 50/50 mixture to form one of most infamous weapons of war, Agent Orange. These phenoxy herbicides, as they were commonly known, became an essential component of Cold War counterinsurgent response, as they offered one means to environmentally expose the insurgent forces invading South Vietnam. The wartime defoliation missions raised the public's awareness of these chemicals and increased the number of people concerned about their use.

From the beginning, using the chemicals provoked controversy and protest. This book investigates those protests. It uses the phenoxy herbicides as a lens to consider citizen resistance in post-1945 America by exploring the ways different groups challenged the domestic and international use of these chemicals on various grounds. Examining these challenges tells us much about citizen understandings of state authority and obligation in the postwar period as the United States assumed a new global stature and achieved significant economic prosperity. As science and technology gained prominence in both military and civilian society, scientists became key players in the emerging technocratic state. In the process, government authorities assigned responsibility for mediating conflict connected to the emerging "risk society" wherein industrialization and capitalism had created new hazards, ones often unseen and dangerous in minute quantities. In this new landscape of risk, the scientists' authority did not go unchallenged, as ordinary citizens worried about their fate and fought for protection.[1]

In the aftermath of World War II, Americans participated in several grassroots—mostly progressive—social movements. These movements included civil rights, self-determination (power movements), student democracy, environmentalism, and women's liberation. One grassroots movement that has gone mostly unrecognized is the antichemical/anti-toxics.[2] Anti-toxics activists worked at the intersection of human and ecological health. These campaigns were often overlooked in subsequent understandings of environmentalism, despite the iconic place of Rachel Carson's *Silent Spring* (1962) in conventional narratives of the environmental movement.[3] Several factors may have led to the invisibility of these decades-long movements, including its major actors (women), geographies (dispersed and local), and contested nature (scientific uncertainty). To be sure, anti-toxic activism never fully cohered in its membership, ideology, or tactics. Carson's 1964 death deprived the movement of a visible, acclaimed, and intuitive leader. Yet, despite the diversity of anti-toxic campaigns, most challenged the Cold War political consensus in much the same way as other progressive movements. While this book tells the history of a broader set of protests over the use of the phenoxy herbicides, it is the use of these herbicides in South Vietnam that elevated the issue and amplified activists' concerns. Even the name of the chemicals took on the branding of the orange-striped barrels used to transport the chemical defoliant during the war. For this reason, I often refer to the phenoxy herbicides 2,4-D; 2,4,5-T; and the other "rainbow" defoliants that contained one or the other of the phenoxies as "Agent Orange herbicides."[4]

The science of postwar America provides the scaffolding that my stories of protest rest upon. The first addresses the perennial dilemma in scientific studies of chemical toxicity. Chemical manufacturers almost always claim safety in the absence of concrete harm and almost always contest charges of illness or demands that the chemicals be treated as hazardous to human and environmental health. In the case of the phenoxy herbicides, the chemical industry was not alone in playing this game of toxic uncertainty. Powerful allies in other business sectors like agriculture and timber, and governmental agencies like the United States Department of Agriculture, all reinforced narratives of safety and economic benefit. In the absence of definitive scientific studies, the argument went, the chemicals must be safe. Individuals and groups concerned about potential harm bore the burden of proof in showing ill effects. In this manner chemical manufacturers successfully muddied the regulatory waters. One way this line of thought succeeded lay in its dismissal of animal studies as not applicable to human beings. As National Institute of Environmental Health Sciences director Linda Birnbaum explained in 2004, while individual studies cannot show definitive causal links, virtually all the studies show the adverse effects of dioxins. The chemical, a contaminant present in the manufacturing of 2,4,5-T, disrupts basic biological processes, and thus biological systems. These include the immune, reproductive, cardiovascular, nervous, and endocrine systems.[5]

This continued absence of definitive scientific and medical proof that the phenoxy herbicides caused ill health provided another challenge in writing the story of these chemicals. For the most part, activists' claims of concern (and harm) are treated as legitimate. Unfounded charges and those instances of when individuals and groups made questionable (and specific) scientific assertions are noted. Activists' concerns in questioning the safety of these chemicals are considered legitimate. I came to this work after researching and writing about another case of hazardous chemicals. In the 1970s residents of Niagara Falls' LaFalce neighborhood challenged public health and elected officials regarding the safety of their community, which had been built over an uncontained buried waste site filled with more than 21,000 tons of toxic chemicals. The episode became known as the Love Canal chemical disaster. Here too toxic uncertainties and politics permeated the situation, and while the chemical disaster resulted in some positive legislative outcomes—most significantly the passage of the 1980 legislation known as Superfund—it also revealed major fault lines in postwar American society. The use of Agent Orange herbicides represents a different kind of chemical disaster, what scholars have recognized as "slow violence." This incremental disaster proves equally revealing about not just the state of America but its place within the global community. As with my research on Love Canal, here I am more interested in the *process* of how these competing paradigms of hazard play out than in trying to determine a *singular truth* about chemical toxicity.[6]

Science in the form of US chemical policies and regulations offers a compelling lens through which to examine post-1945 grassroots social movements. The chemicals were used as a part of a broader Cold War project that sanctioned a relationship between government and industry and used chemicals like DDT or the phenoxy herbicides to achieve state-sanctioned outcomes, such as their use in South Vietnam. The concluding chapters show a different way the use of the chemicals reflected a Cold War mind-set that demanded citizen compliance. Like other groups that protested the political consensus that arose in the post-1945 period, what historian Arthur Schlesinger called "the vital center," the individuals and groups protesting the use of the phenoxy herbicides offered a critique of postwar governmental policies and actions.[7]

My own perspective tends toward sympathy and support of citizens asking that they be a part of decisions made that they think affect them. I also think science has yet to catch up with respect to detection and understanding despite the new research field of epigenetics, an area of research that examines how phenotypic characteristics can be inherited without changing DNA. This research also suggests that male genetic material may be as vulnerable to external factors as female genetic material. In the story of the phenoxy herbicides, the emergence of new scientific paradigms and understandings has generally supported activists' concerns about harm, and not industry and government claims of safety.

The history of humans, chemicals, health, and the environment cuts across a broad range of disciplinary areas. These include the histories of agriculture, science, the environment, and health. Many of these works were published concurrently and, in the early decades, spoke to different audiences and often not to each other. This brief review, arranged chronologically within fields, outlines the most significant research and explains what *The Defoliation of America* contributes to the scholarship. Agricultural history was one of the first fields to examine chemicals generally, and the phenoxy herbicides specifically, over an extended period. Gale Peterson's 1967 essay examines the discovery of 2,4-D. Like Peterson, Nicolas Rasmussen focused specifically on the phenoxy herbicides and their development domestically as a means of controlling weeds, with the understanding they could be potential wartime weapons. Pete Daniel's *Toxic Drift* considers the aerial spraying of agricultural chemicals undertaken as a partnership between the United States Department of Agriculture (USDA) and the chemical industry and the health consequences for crop dusters and field workers. J. L. Anderson focuses on the use of chemicals in modernizing the Corn Belt, where the use of chemicals and machines transformed the postwar Midwest.[8]

In the history of science, studies of chemicals, chemical policy and regulation, and chemical hazards proliferate. The influence of Carson on the field could be seen from the start. In one of the first histories of chemical policy that appeared after the publication of Carson's *Silent Spring*, James

Whorton investigated US chemical policy in the decades prior to 1962. Like Whorton, Thomas Dunlap concentrated on chemical policy, specifically examining the controversies over DDT and its eventual banning in 1972. The United States government has long sponsored the development and use of hazardous chemicals. Edmund Russell examines this relationship in his examination of industry, hazardous chemicals, and their use in agriculture and warfare. Joshua Buhs's work complements Russell's, albeit in a much more narrowly focused study of the USDA's massive campaign to eradicate the fire ant through the use of DDT. Several works examine questions of chemical toxicity, safety, and risk. Sarah Vogel considers the safety of bisphenol A (BPA), challenging traditional understandings of toxicity in the process. Frederick Rowe Davis examines the process by which measurements and standards of toxicity were developed by agencies like the Food and Drug Administration. Michelle Mart's cultural look at Americans' relationship with pesticides tries to answer why people continued to use chemicals they understood to be unsafe. The use of chemicals and their health consequences represents an overlapping area of study.[9]

Many of the initial studies on chemicals and health examined them in occupational settings, focusing on lead, asbestos, and other toxic chemicals. David Rosner and Gerald Markowitz have published extensively on occupational health, lead, and industrial pollution. Christopher Sellers has examined occupational health in the United States and worldwide, tracing its evolution from a focus on industrial disease to one on health. Christian Warren examines the history of lead poisoning over the twentieth century and the two primary routes of exposure: leaded gasoline and paint. Diethylstilbestrol (DES) was used for a variety of reasons over the course of the twentieth century—fed to cattle to increase weight gain, young girls to decrease their height, and eventually women to forestall morning sickness and menopause. Nancy Langston examines the problems these various uses posed as the scientific and medical communities recognized DES as an endocrine disruptor causing health problems across generations. David Kinkela considers the dilemma of using DDT in various global public health campaigns as officials and physicians weighed the risks and benefits. Similar to Kinkela's reframing of DDT studies, Jennifer Thomson recovers environmental activists' using the idea of health as a fundamental way to frame their ecological understandings and interventions.[10]

Vulnerable communities experienced toxic contamination as one of the major parts of environmental racism. Sociologist Robert Bullard's studies of the prevalence of waste sites in southern African American communities appeared just five years after the landmark 1985 United Church of Christ report on environmental racism. Andrew Hurley used Gary, Indiana, as a case study to examine the ways poor minority communities bore the brunt of industrial pollution in that midwestern steel and auto town. Eileen McGurty examines the case of Warren County, North Carolina, where county residents protested the state's plan to locate thousands of pounds of soil resulting from illegal dumping. In two works on toxic contamination, race plays a more subtle and complicated role. Many environmentalists consider the 1978 Love Canal chemical disaster, located in Niagara Falls, New York, a pivotal moment in environmental activism. The success achieved by Lois Gibbs and the activist homemakers she led in protesting the hazardous chemicals buried under the community became a national story. Elizabeth Blum examines the more-than-two-year conflict that eventually saw over nine hundred families relocated from their homes surrounding the dump site, and the tensions between white and Black residents. Kate Brown compares two cities, "plutopias," essential in producing the radioactive materials needed for atomic weapons. Richmond, Washington, home to the Hanford nuclear plant, and Ozersk, Russia, the Russian production center for plutonium, both traded health and civil liberties for comfortable consumerism. Ellen Griffith Spears's study of Anniston, Alabama, and its contamination with PCBs produced by the local Monsanto chemical plant combines many of the themes with this broader body of scholarship—vulnerable minority communities, often located in the South, situated in the midst of industrial pollution produced by international companies (or their own government). The risk presented by the phenoxy herbicides themselves and their use in Vietnam has produced a voluminous body of research as well.[11]

Studies of Agent Orange appeared soon after the formal end of the Vietnam War in 1975 and have continued to be published. Fred Wilcox's account of the veteran protests over the ill health they thought was caused by exposure to Agent Orange and the failure of government actors to respond provides a scathing critique of a government waiting for its soldiers to die. Peter Schuck does a close examination of the 1978 lawsuit pursued by veterans trying to hold chemical companies and the US military and government to account. To the veterans' misfortune, the latent nature of the illnesses caused by exposure to the phenoxy herbicides hurt their attempts to be compensated for their health problems. Wilbur Scott examines the cases of Agent Orange and the Post-Traumatic Stress Disorder (PTSD) suffered by US veterans. Several books complement these earlier investigations in examining a broader array of actors involved with Agent Orange. David Zierler looks at the scientists who framed the herbicide-spraying missions as "ecocide"—the willful destruction of an ecosystem. Edwin Martini scrutinizes the continued uncertainty surrounding exposure to Agent Orange and claims of its ill effects within both the US and Australian militaries. Peter Sills reconsiders the legal and scientific uncertainty that surrounded use of the herbicides.[12]

The Defoliation of America complements and expands these previous studies in several ways, primarily in its focus on dissent. It recovers the different incarnations of antichemical protest by focusing on groups of citizens: religious leaders and laity, students, scientists, veterans, labor, and environmental activists. It uses the phenoxy herbicides to examine US chemical policies, and the state's relationships with various industries, a reinsertion of Dwight D. Eisenhower's "military-industrial complex." It continues the studies of large-scale government projects—their planning, implementation, and outcomes. It also revises our understanding of the postwar American environmental movement, the politics of chemical uncertainty, and the erosion of scientific authority. In one sense it carries on Rachel Carson's "gentle subversiveness" in recovering the citizen challenges to human hubris in their interactions with nature.[13] The various groups of actors structure the book. Each chapter examining these groups advances the story chronologically, although parts do overlap.

Part 1 (1940–70) identifies the creation of the phenoxy herbicides, their assorted uses, and the first protests over their use. Chapter 1 considers the uses of the phenoxy herbicides prior to and during the Vietnam War. It argues that while domestic and international applications of the phenoxy herbicides were initially received with enthusiasm and widespread use, problems emerged that led to protests, which are examined in subsequent chapters. Chapter 2 follows three different groups of US citizens—religious communities, university students, and biological scientists—who integrated critiques of herbicide defoliation in their broader protests of the US involvement in Southeast Asia. Their protests set the stage for international critiques. Chapter 3 considers two different spaces of condemnation—scientific testimony at various international tribunals, and in the negative assessments made by two scientists, a North Vietnamese surgeon and an East German botanist, in scientific and protest literature.

Part 2 (1965–80) focuses on the activism of three Western women challenging herbicide spraying in response to various environmental challenges.

These chapters shift from the transnational development, use, and protest over the Agent Orange herbicides during war to uses of the chemicals on public lands. There are several reasons for this regional case study approach. Carson examined the phenoxy herbicides in *Silent Spring*, highlighting the extensive use of the chemicals on rangelands in the West, where the chemicals like 2,4-D were used for brush control and timber management. While there were many other anti-toxic protests, some over the phenoxy herbicides, an examination of these three women/states provides a means to consider all the common domestic uses of the herbicides. A case study approach allows the examination of what part the phenoxy herbicides played in large-scale government projects and private enterprises to increase water supplies, control fires, and increase agricultural and timber yields. Influenced by Carson, print media played an important part in each woman's activism as they wrote books and newsletters to alert the public. The women also formed the backbone of a loose environmental activist network motivated specifically by chemical contamination. In the process they challenged the US government's chemical regulatory policies, practices, agencies, and authority.

Chapter 4 looks at Arizona and the constant need for water in the western United States. The US Forest Service used the phenoxy herbicides to increase water runoff in Arizona's Tonto National Forest, in part to service the water needs of the expanding Phoenix metropolitan area. One local woman, Billee Shoecraft, was exposed to the chemical spray, along with other community members, and led community organizing and protests. Chapter 5 examines the state of California, where the phenoxy herbicides were used for a variety of purposes, including brush control in Southern California, agricultural use in the Central Valley, and forest access in Northern California. Longtime activist, consumer reporter Ida Honorof used her weekly radio broadcast on progressive station KPFK 90.7 and a monthly newsletter to inform the public about the chemical hazards present throughout the California landscape. Chapter 6 shifts to Oregon, where the phenoxy herbicides were used by road crews to make forests accessible, and by the timber industry to increase more profitable kinds of timber stock. Here, Carol Van Strum and her family, who had moved to western Oregon as a part of the postwar back-to-the-land movement, were directly exposed to the herbicides. The Van Strums helped organize a community response to the routine roadside and aerial spraying. In the process of challenging the chemical and timber industries and Forest Service, Van Strum argued for a new approach to chemical regulation.

Part 3 (1970–95) looks at the phenoxy herbicides' continuing toxic legacies.

The concluding chapters examine the protests of new groups of citizens dealing with the toxic legacies of the phenoxy herbicides. President Richard Nixon, responding to fears of addicted soldiers returning from Vietnam, declared a "war on drugs" in 1971, a war dependent on chemical herbicides. These same veterans protested their exposure to those same chemicals, anxious about the potential harm done to Vietnamese and American children. The concluding chapters consider the role the phenoxy herbicides played in controlling counterculture protests, deepening veterans' disillusionment, and the growing awareness that the routine use of chemicals harmed vulnerable populations. In the process, these protests challenged official narratives of chemical safety.

Chapter 7 examines the use of the chemicals to attack the counterculture, one site of resistance to the political consensus in general, and Nixon's policies and activities in specific. What has been called the "herbicide wars" arose when residents who embodied a new environmental consciousness moved into longtime farming and ranching communities in Northern California and began challenging local practices though greater environmental regulation. Chapter 8 focuses on another group that began challenging the safety of the phenoxy herbicides in the late 1970s, American Vietnam veterans. The claim that Agent Orange had hurt veterans' health launched a class action suit and empowered veterans' groups. Chapter 9 considers the "unexpected casualties" of chemical use in warfare and peacetime—children. The conviction that exposure to the phenoxy herbicides had caused the birth defects and illnesses experienced by their children haunted veterans, farmworkers, and Vietnamese families and motivated their efforts to stop the continued use of the chemicals. These citizens challenged industry and government interests and sought restitution for the consequences of environmental degradation and harm to human health.

Before phenoxy herbicides' widespread domestic, and later wartime, use, scientists had approached them with caution. A 1942 *Science* news update discussing a newly identified group of chemicals known as plant hormones noted the need for further research. The question was whether this new category of chemicals, these growth regulators, would harm crops, or if their actions might promote growth and even be added to fertilizers. The news brief ended with the observation, "The idea would be good, but the results might be disastrous."[14] Uncertainty abounds in science and is what scientists attempt to resolve. Chemists conduct their science primarily within a laboratory setting and usually under carefully controlled field tests. Biologists, especially ecologists, practice under much more chaotic circumstances as they try to understand nature's "unruly complexity."[15] Like many of the activists who appear within these pages, this book was inspired by Carson's love and advocacy for the natural world, and the challenges Carson's work offered to ideas of science, knowledge, decision making, informed consent, and the relationship between human beings and the natural world. This book recovers the stories of successful, failed, and uneven outcomes of citizen activism and engagement in the democratic process. I hope these stories about the struggles to hold government and business responsible for people's health and well-being remind us of past wrongs and suggest ways to better address them in the future.

PART ONE

The Origins of the Phenoxy Herbicide
Uses and Protests, 1940–70

Part 1 examines the initial development and use of, and eventual protests over, a new category of chemicals that were found to selectively inhibit the growth of plants, many of which were considered weeds. The phenoxy herbicides —2,4-D and 2,4,5-T—started out as potential chemical agents to be sprayed on enemy crops in World War II. The commercial manufacture and sale of the herbicides began in 1946, and the herbicides saw widespread use across the United States in deserted urban lots, in agricultural fields, and eventually for brush control in fires and to improve timber stock. The enthusiastic responses to the new chemical herbicides did prompt some concerns, particularly among farmers who dealt with chemical drift and the appearance of resistant weeds. It was the use of the herbicides in South Vietnam that raised their profile and mobilized diverse groups, both domestically and internationally, to protest their use.

CHAPTER ONE

Controlling Jungle Lawns and Jungle Wars

Domestic and International Uses of the Phenoxy Herbicides

Interviewed about the student demonstrations over Dow's production of na-palm, a form of jellied gasoline being used in South Vietnam, Dow Chemi-cal president H. D. "Ted" Doan took pride in Dow's patriotism. He noted that while he supported the right to peacefully protest, Dow would "continue to produce napalm and other materials as long as they are needed by our Gov-ernment."[1] The phenoxy herbicides, better known as Agent Orange, were in-cluded among the "other materials" Dow Chemical and other chemical com-panies provided the US military during the Vietnam War. Prior to their use to defoliate tropical jungles, these chemical herbicides had widespread domestic applications, and were used to clean up urban landscapes, modernize Ameri-can agriculture, and control suburban lawns. The chemicals were used inter-nationally as defoliating agents in South Vietnam, where the United States' presence increased after the expulsion of the French and the end of their co-lonial rule. The wartime uses of the various herbicides increased the chemi-cals' visibility and intensified the various concerns regarding chemical regu-latory policy within the US.[2]

The 1950s and 1960s saw extensive spraying of the miracle weed killers in deserted city lots, in rural fields and pastures, and on suburban lawns. Chemical manufacturing represented a major American industry, and a suc-cessful one. Dow Chemical epitomized the classic story of American busi-ness achievement, with its founder, Henry Dow, starting the company in Mid-land, Michigan, in the 1900s and the company's achievement in becoming an international conglomerate in the postwar period. Despite the successes of the chemical industry, Americans showed increasing uncertainty about the presence of plastics and other chemicals in their lives.

Chemical companies like DuPont, Dow, Monsanto, American Cyanamid, Union Carbide, and Allied saw significant expansion, over twice as fast as the gross national product (GNP), in the two decades after the war. Characterized as a "growth industry," chemical manufacturing shaped Americans' everyday lives, economically, culturally, and even politically. A 1952 financial journal promoted the current state of the chemical industry and its bright future for profits. The article listed the wide range of chemical products available to the investing public. An impressive number of these goods were being developed, including man-made fibers, dyestuffs, petrochemicals, synthetic resins, and agricultural chemicals. Americans had also begun to rebel, however, as the chemical industry increasingly attracted the attention of cultural critics, labor, and civil rights, health, and environmental activists concerned about chemical pollution and possible toxicity. Prior to, during, and after WWII, the chemical industry experienced tremendous growth and success.[3]

Building on its advancements during and after the First World War, the American chemical industry emerged from WWII with increased capacity and knowledge. Isolation during WWI forced chemical companies to build up manufacturing capacity, identify core products, and modernize management. Dye manufacturing represented a good example of interwar product expansion. Cut off from German dye supplies, American companies produced 12 million pounds by 1925, an amount that had increased to 100 million pounds by 1935 and showed "the vigor of a chemical industry."[4] During the 1930s, Dow Chemical produced all the United States' magnesium, around 3,300 pounds yearly. During WWII, Dow helped ensure that wartime needs were met, with national production capacity reaching 291,000 pounds, an almost 100 percent increase. In 1942 the company produced 84 percent of the magnesium used in the war.[5]

Commenting on the postwar industry outlook, one marketing journal noted, "The tremendous abundance of some materials and our advanced knowledge of many more will lead to the redesign of countless products, and the introduction of countless new products."[6] These manufacturing advancements helped the industry to surpass its European competition in postwar product development. By 1946 Monsanto celebrated record corporate earnings of $10 million, and sales of $100 million. Wartime needs accelerated chemical manufacturers' innovations and creation of new products. Chemical companies like Dow and Monsanto spent much of the postwar period promoting their reputations along with those products.[7]

The American chemical industry worked hard in the postwar decades to continue expanding while selling the nation's consumers goods and questionable assurances of chemical safety. New products quickly appeared everywhere, made from materials not in existence before the war and representing 40 percent or more of companies' sales.[8] Continuing their wartime research, the chemical industry created an impressive number of petrochemical merchandises. Plastic parts began appearing in homes and cars, clothing, and even recreational items, like audiocassettes and fishing lines. Working with DuPont materials, entrepreneur Earl S. Tupper transformed kitchens using injection molding to make his polyethylene line of plastic cups and containers and other household items.

The American chemical industry faced a major public relations problem, however, beyond the frequently poor quality of plastic materials. The atomic bomb and the harmful effects of radiation raised Americans' awareness of chemical toxicity. In 1948 a major industrial accident threatened the town of Donora, Pennsylvania, which justified such concerns for many. The Manufacturing Chemists' Association (MCA), the industry trade association, carefully monitored air pollution legislation prompted by anxiety over chemical pollution. Even as the industry offered more helpful things for Americans, it worried about its image and potential governmental oversight. Chemical manufacturers used their military and agricultural connections to influence regulation and legislation so that it would be more favorable to industry. They also emphasized the tremendous social good chemicals achieved, such as eradicating malaria or helping American farmers feed the starving world. In this respect the agricultural chemical lobby emphasized what historians have labeled the "politics of growth" during the 1940s and 1950s.[9]

Dow Chemical exemplified the prominence and success of the chemical industry. Dow saw itself as a technological innovator, an economic powerhouse, and a good corporate citizen throughout its existence in the twentieth century. As one corporate biographer described it, "Dow straddled the globe . . . spending significant sums in a daily, continuing search for better products, better ways to make them, and a better understanding of the science that makes them possible."[10] More than any of the other major American chemical companies, Dow Chemical owed its postwar success to petrochemicals and the everyday household products developed from them. One measure of this success could be seen in the significant expansion in sales, from $16 million in 1930 to $171 million in 1949.

It could also be seen in the scientific honors bestowed upon Dow employees. Dr. Hans Grebe, head of Dow's Physics Lab, won the Chemical Industry Medal for his work in plastics in 1943. E. C. Britton won the Perkin Medal in 1955, the highest honor awarded by the Society of Industrial Chemistry.[11] If visibility represented a company's achievement, then the pervasive presence of Dow's products signified the chemical industry's triumph. Polystyrene, one of Dow's new plastics, appeared in the form of "refresher boxes" in refrigerators across the country. As a *New York Times* feature observed, housewives appreciated the popular storage containers because they kept food fresh, did not retain odors, and could be washed in hot water.[12] Dow Chemical took pride in providing a range of materials and products for wartime and peacetime needs, including agricultural chemicals.

Just as new plastic products became ubiquitous in suburban homes, other chemical products fashioned rural, urban, and suburban landscapes. One of the new products that would be sold by chemical companies like Dow and Monsanto would be the plant growth hormones, selective weed killers developed in the decades before World War II and considered as a possible chemical weapon by American military strategists. These phenoxy herbicides worked by accelerating the plant growth process to the point of death. Here too chemical companies provided an important product, one that helped modernize American agriculture while at the same time taming lawns in the exploding suburban housing developments surrounding cities. These weed killers offered a simple solution to the cost and labor demands experienced by American farmers, even as they helped homeowners realize the American dream: a suburban home surrounded by a well-manicured lawn. These routine, useful, supposedly safe chemicals would also eventually become an essential component of American military involvement in Southeast Asia, clearing the jungle for strategic and ideological purposes.[13]

Research on plant hormones, primarily on the auxin-like compounds, began during the interwar period, although who discovered the herbicidal effects of the compounds appeared to be a contested point, at least at the immediate conclusion of World War II.[14] Many saw the origins of the plant growth regulators dating back as far as Darwin, in his experiments with plant movement toward light. Advancements in the plant growth regulators proceeded throughout the early twentieth century, with efforts intensifying in the 1930s.[15] Scientists identified chemical compounds that mimicked plant hormones that regulated plant growth. Prior to World War II, the research occurred simultaneously in Great Britain and the United States. Although British research efforts predated American ones, American scientists published their results first.[16] The advent of the war escalated research efforts but also ended publication of research results, as both governments classified the growth regulators as potentially important war work.[17] The Chemical Warfare Service conducted experiments on plant growth regulators at Fort Detrick, Maryland. George Merck, of Merck and Company and director of biological and chemical research during the war, claimed that only the cessation of the war prevented the use of the plant growth regulators against Axis enemies.[18]

This wartime designation also allowed a fertile cross-pollination between academic, industry, and military scientists. Major research was conducted at Rothamsted under the auspices of the Agricultural Research Council and Imperial Chemical Industries' Jealott Hill Research Station in Great Britain. American research teams came from the Boyce Thompson Institute in New York state, Hull Botanical Laboratory at the University of Chicago, Cornell University's New York State Agricultural Laboratory, the United States Department of Agriculture's Beltsville, Maryland, research station, and Fort Detrick under the auspices of the US Department of Defense.[19] A key conceptual shift occurred during this intense time of experimentation when researchers began thinking of the chemicals in terms of killing plant growth rather than using them to promote growth such as for ripening fruits, promoting root growth, and in developing seedless fruits.[20]

The substituted phenoxy compounds 2,4-D and 2,4,5-T emerged as two of the most effective chemicals in experiments in killing plants, in part because the chemicals affected plant parts beyond just the point of application.[21] British researchers described another important advantage: "It [2,4-D] is not readily leached from soil but yet possesses the advantage of ultimately losing its toxicity; thus it would be unlikely to poison the land."[22] Further experiments discovered which plants were the most sensitive, with the good news that common weeds like bindweed, dandelions, marigold, buttercups, thistle, and sweet pea appeared to be particularly vulnerable. Even more important were the plants that appeared to be resistant to the chemicals, cereal and rice crops.[23] The phenoxy herbicides represented a significant breakthrough in weed control, and coverage in the popular media meant that chemical companies could not wait to begin satisfying the newly created need.

Although WWII ended before the phenoxy herbicides could be used, the herbicides still became weapons of war, a domestic war on weeds. For public health officials, these chemicals offered a new weapon in a very old battle against noxious plants that triggered allergic reactions like hay fever. By the 1930s the number of hay fever sufferers had risen to 3 percent of the United States' total population, with increasing numbers of city dwellers afflicted.[24] Ragweed, king of urban vacant lots, was public enemy number one. Earlier efforts had focused on manual removal of weeds from public lands, such as New York City's Works Progress Administration unit that cleared 112,000 city lots in 1934 and 1935.[25]

Critics like Chicago botanist Paul Standby expressed doubts that ragweed could be eradicated no matter what the manpower dedicated to removal. He joined other scientists who questioned the feasibility of ragweed crusades.[26] After the war, chemical innovations offered a new weapon in the arsenal, and new hope. Early gardening columns like the one entitled "Death to Weeds! Powerful Weapon in an Endless Battle Are the New Synthetic Plant Hormones" noted the wide number of 2,4-D commercial products.[27] The article also discussed their effectiveness in eradicating bindweed, poison ivy, and ragweed while leaving lawn grasses alone. Inexpensive cost and ease of application made the chemicals attractive not only to the home gardener but for public health campaigns as well.

The New York City Public Health Department under the leadership of Dr. Israel Weinstein led the most visible campaign to remove ragweed from urban spaces. Dubbed "Operation Ragweed," in 1946 the health department launched an ambitious program to remove weeds, with 850,000 gallons of 2,4-D sprayed on three thousand acres of public spaces (roads, lots, sidewalks) and private land. The health department enlisted the aid of the police, who reported ragweed infestations identified on their patrols. Boy Scouts helped educate the general public about their civic and legal obligations to remove ragweed from their properties.[28] By the 1950s New York City's ragweed removal programs had spread to over 150 New Jersey towns; other big cities like Chicago, Detroit, and Washington, DC, and countless small towns were committed to fighting ragweed.[29] In what historian Zachary Falck calls the "ragweed wars," the New York City Health Department focused on destroying ragweed but neglected to replace the empty lots with some other kind of plant growth. It failed to create a new, or to adjust the old, urban ecology. Simply removing ragweed from lots did not prevent ragweed growth the next year.[30] Yet Americans continued to see the phenoxy herbicides 2,4-D and 2,4,5-T as important weapons in maintaining their fields and lawns.

At the same time public health departments embraced miracle weed killers in their war on ragweed, American farmers welcomed a new means of controlling crops: 2,4-D was tested in fields across the country at agricultural extension stations. The United States Department of Agriculture (USDA) demanded that toxicity studies be done, which when completed indicated no problems.[31] American Chemical produced the first commercial weed killer as Weedone in 1945, with products from Dow Chemical and Sherwin-Williams soon following.[32] One measure of how quickly industry responded can be seen when examining production numbers. "In 1945, the first year of public testing when only limited amounts of 2,4-D were available, total production climbed to 5,466,000 pounds—an increase of nearly 500 percent. By 1950 annual production exceeded 14,000,000 pounds."[33]

Another means of measurement would be the total number of farm acres sprayed. By 1960 "more than 50 million acres of agricultural land were treated chemically for weed control. . . . The cost of herbicide application was nearly $137 million."[34] These numbers obscured a social revolution, however, as farmers' attitudes shifted from concern about chemical poisons to acceptance of their usefulness and necessity.[35] Chemical companies offered effective herbicides and, with DDT, pesticides that helped farmers to increase crop production and modernize farm operations. Farmers increasingly relied on herbicides to replace the more labor-intensive practice of cultivation, even as criticism mounted in the late 1950s and into the 1960s.

The case of American farmers provides the most extreme use of the phenoxy herbicides as a means of modernizing agricultural operations. The effectiveness of the phenoxy herbicides, especially 2,4-D's less potent action, and preferential plant eradication made them important contributions in the chemical weed-control arsenal. Previously farmers had used cultural practices to control weed growth. These practices included spring and fall plowing, and routine cultivation during the growing season, all time- and labor-intensive practices during peak work periods. Herbicides cost very little, about $10 a gallon in 1948, and approximately half that in the decades after 1950.[36] Using the herbicides made money in two ways: it cut expenses (fuel and labor) even as it increased crop production. The neglect of fields during the war meant farmers were especially desperate to achieve some control over the weeds choking their crops. As one county extension skit written in the mid-1940s suggested, 2,4-D represented the hero against "Weedy the Thief" in saving Verda Land.[37] Unfortunately, while county and state extension agents urged a mix of weed eradication practices, using cultivation and chemicals, American farmers enthusiastically and single-mindedly embraced chemical herbicides. Even as the amounts of herbicides increased, though, so too did criticism and concern over chemical weed control.

Applying the phenoxy herbicides resulted in three major problems that caused tensions among farmers and between farmers and the public. Farmers wrestled with awareness that chemicals could harm their fields but recognized the need to control damaging insects and weeds. The first issue, the drift of herbicide sprays onto other crops susceptible to the chemicals, such as grapes and cotton, resisted any kind of easy solution. "A three-mile-per-hour breeze could carry a droplet eight miles when applied from ten feet above the field."[38] Farmers began complaining about crops killed by herbicide drift as early as 1948 to both state and federal agricultural officials. Kansas farmer Ralph McGinty expressed a common sentiment when venting his frustration over an illegal spraying that affected his crops and shade trees. "How long will we tolerate destruction to drift with the wind?"[39] The second dilemma appeared by 1960 when farmers saw new, resistant weeds appearing in their fields. This problem was addressed by the development and use of more chemicals.[40] The apparent solution, however, led to the third difficulty confronting farmers and chemical suppliers: the growing concerns over chemical/herbicide safety and environmental degradation. This last concern also arose in the other major site of phenoxy herbicide use: suburban lawns.

Perfect lawns once symbolized the wealth and status of seventeenth- and eighteenth-century European aristocracy, but by the mid-twentieth century immaculate lawns represented American democracy. Lawn care became the other major market for the phenoxy herbicides, as homeowners tried to achieve the perfect lawn—smooth, unblemished grass vistas duplicating those of previous generations' elites. The USDA played an important role in home lawn care, just as in agriculture, in creating a desired product with intensive chemical intervention. Just as farmers got greater crop yields with less labor, suburban homeowners sought to achieve the perfect lawn aesthetic.[41] Ecological imbalance like what happened in the appearance of new kinds of weeds farmers saw in their fields also resulted with the creation of the monoculture lawn.[42] While 2,4-D worked well on dandelions, it allowed the other bane of homeowners, crab grass, to flourish.[43]

The US postwar housing boom resulted in an especially strong market for lawn care herbicides.[44] "By 1960 the nearly thirty million home lawns in the United States were being added to at the rate of almost half a million a year. . . .

Thirty million lawns in 1960 added up to more than thirty-six hundred square miles of turf."[45] So, in the immediate post-1945 period, the United States led the world in its use of the phenoxy herbicides for agricultural and domestic homeowner use. Chemical lawn care formulations helped reduce the time and labor needed for lawn maintenance, just as it did for fields. "Weed and Feed helped boost 2,4-D production from fourteen million pounds in 1950 to thirty-six million in 1960 to fifty-three million just four years later [1964]."[46] Suburban homes, and the lawns that came with them, gave proof of American affluence and the virtues of democracy.

In his 1959 "kitchen debate" with the Soviet Union's Nikita Khrushchev, Vice President Richard M. Nixon asserted that the comfort and technology contained in the suburban home represented America's superiority. This affluence acted as one element in containing the Communist threat, as all Americans had the opportunity to buy their own home. The suburban home represented America's economic prosperity, becoming the first line of defense against Communism in the process. Women, too, would thrive as Americans sought a return to normalcy, one in which wives and mothers worked to make the home a refuge for their families.[47]

The American home, as embodied in bomb shelters built in basements and backyards, also offered the idea of protection amidst fears about nuclear war. Despite the fact that only 3 percent of Americans built bomb shelters, numerous presidential administrations promoted the shelters as a means of allaying citizens' fears and preserving American society. While the idea of a bomb shelter represented a space of shelter, the well-tended front lawn signified one of bravery, as homeowners brought order to the lawns—and their lives—through the eradication of noxious weeds. And the constant care the lawn required kept Americans out of trouble—a sentiment expressed by William J. Levitt, the visionary builder who had built the Levittown housing development. "No man who owns his own house can be a Communist. He has too much to do."[48] The proper containment of American workers, women, and lawns, whether it was ideological or literal, realized the proper commitment to American values and patriotism.[49]

One of the more revealing novels of the 1950s, *The Man in the Gray Flannel Suit*, opened with evidence of its protagonists' unfitness, the breach their domestic malaise caused. This failing appeared clear to the entire neighborhood. "The ragged lawn and weed-filled garden proclaimed to passers-by and the neighbors that Thomas R. Rath and his family disliked 'working around the place' and couldn't afford to pay someone else to do it."[50] The novel reflected Americans' unease about the postwar landscape. For many, investing in their homes offered one way to cope with the uprootedness, the uncertainty of the nuclear age. Key to that investment was creating and maintaining proper lawns.

This emphasis on the importance of pristine lawns as a means of soothing Cold War anxieties could be seen in a variety of forms. *Time* magazine's August 2, 1954, cover showed an eight-armed man, dressed casually in a checked shirt atop his lawn mower happily performing a multitude of tasks, which included painting, mowing, and spraying for weeds. Ostensibly about the "do-it-yourself" industry, the cover showed lawn care as a major component of suburban life. In his study of the new "organization men," sociologist William Whyte Jr. noted that homeowners spent most of their time on lawn maintenance, and that "the sharing of tools, energies, and advice that the lawns provoke tend[ed] to make family friendships go along and across the street rather than over the back yards."[51] People bonded over their lawns. "Charlie used to make fun of us for spending so much time planting and mowing and weeding. . . . You should see him now. He's got sprays and everything."[52] Whyte argued that the new middle class saw their suburban homes as a way to "develop a new kind of roots to replace what [they] left behind."[53] While Sloan Wilson's novel captured the ambivalence of suburbia and its discontents, many others saw the nuclear family and their home as the first line of defense.[54]

A 1962 *Life* humor piece made an even more explicit connection between lawns and the Cold War. In it the author recounted his "bitter neighbourhood war with Fred Morgrew," wherein battling lawns represented the opposing political positions of the United States and Soviet Union. Morgrew had replaced the lush, abundant plant growth of his new home with crabgrass, to be achieved through a "Five-Year Plan of Landscape Reform." Morgrew planned a total crab-grass domination of the neighborhood. Like the superpowers, the neighbors had their own arms race, represented by the Power Lawn Mower phase, followed by Morgrew's construction of a "spite" fence vaguely reminiscent of the Berlin Wall. Morgrew's envy of lawn and lawn ornaments made the "cold war between us grow warmer." The American protagonist finally went to a garden store and asked for the best chemical remedy for crab grass. His plans to use the herbicide were thwarted when Morgrew too acquired the chemical weapon. The men lived in uneasy standoff until Morgrew abandoned his hideous, if cheap, crab grass lawn and embraced the virtues of proper Merion blue grass. Even though the ultimate herbicidal weapon could not have been the phenoxy herbicides, as they are ineffective against crab grass, clearly the American lawn represented Cold War struggles, even in jest.[55]

While these lawn chemicals represented modern convenience and appeared as an everyday part of life in cities, suburbs, and rural areas, various Americans began questioning their effects on human beings and the natural world. But just as other progressive social movements had begun to challenge the Cold War political consensus, the various constituencies of the nascent environmental movement of the 1950s became increasingly aware of the environmental degradation occurring around them. Several conditions helped promote the emerging environmental movement: Americans' postwar affluence, increased awareness of new hazards after the use of the atomic bomb, and new concepts in ecological science. These individuals questioned the widespread use of chemicals and their supposed safety.[56]

As the suburbs were considered domestic space, women took up such causes as preserving open spaces, stopping pollution, and reducing the use of pesticides. Women experienced the environmental havoc caused by rapid suburbanization, with problems of transportation, sanitation, and overbuilding complicating the ideal of the home with a white picket fence. One striking example of urban and suburban women's environmental activism would be the League of Women Voters' focus on clean-water issues. League members were encouraged to research the problem in their local areas, disseminate information to their community, and form alliances with public officials. Another visible women's group protested nuclear pollution. Concerned about the contamination of milk by strontium 90, radioactive chemical fallout from nuclear weapons testing, Dagmar Wilson and Bella Abzug created Women Strike for Peace (WSP) in 1961. Wilson described the sentiments motivating the group's formation: "There are times, it seems to me, when the only thing to do is let out a loud scream. . . . Just women raising a hue and cry against nuclear weapons for all of them to cut it out."[57] Like other middle-class women, WSP combined their environmental concerns with broader social and political challenges to Cold War militarism.[58]

Citizen challenges to DDT provide a good example of Americans' growing unhappiness with the routine use of what they increasingly feared were harmful chemicals. State and local agency campaigns against two persistent pests—the gypsy moth and fire ants—in the 1950s and 1960s provoked citizens to act. Two court cases, both based on DDT spraying on Long Island, bridged the emergence of what historian Christopher Sellers calls the "environmental imaginary," what he argues was the first political identity in the

United States directly connected to environmental concerns. That they occurred before and after the 1962 publication of *Silent Spring* was not accidental. The first court case took place in 1958 when fifteen Long Island residents sued to halt DDT spray campaigns to eradicate the gypsy moth. The residents based their lawsuit on long-standing public nuisance laws and their status as property owners. They presented evidence suggesting the potential harm DDT posed, but much of their testimony was anecdotal or suggestive rather than definitive. While the plaintiffs lost their suit against the USDA and local agencies, the lawsuit inspired others. They marked the beginning of a changed understanding of chemical pesticides from miracle chemicals to ones of threat. The first Long Island case directly influenced Carson and the writing of *Silent Spring*, and it helped the emerging grassroots protests to coalesce and strengthen.[59]

Some significant differences can be seen in the second trial. Prompted by a massive fish kill at a local lake, lawyer Victor Yannacone and his wife, Carol, brought a class action lawsuit against the Long Island Mosquito Commission. With Carol Yannacone as the sole plaintiff, Victor Yannacone sued to protect local natural resources for the people of Suffolk County. Although its protagonists were considered middle-class, both had strong and recent working-class roots. The case focused on DDT's harm to local wildlife, rather than on human health. It used legal tactics pioneered by labor and civil rights activists; these tactics became the origins of environmental law. It also spurred the creation of a new environmental group, the Environmental Defense Fund (EDF). The EDF went on to successfully use these same tactics in its challenges to DDT spraying in Wisconsin.[60]

The concerns identified by liberal advisers, middle-class women, and young Americans led to what historian Maril Hazlett framed as the "ecological turn in American health."[61] Here, ordinary Americans increasingly challenged scientific claims of pesticide safety and hubris in controlling the natural world. Middle-class, white women living in the suburbs led this ecological turn, as they experienced harm through their bodily senses. This awareness of bodily harm also suffused the 1958 Long Island DDT case. Two of the plaintiffs, Marjorie Spock and her housemate Mary Richards, had begun to practice organic gardening in response to Richards's chemical sensitivities. They had tested their food and were dismayed to find out it showed high levels of DDT. Even Rachel Carson demonstrated an awareness of the human body (as a part of the ecological system) as threatened versus threatening. This perception contrasted with other scientists who considered the increasing numbers of human beings to be the problem, a view expressed in Paul Ehrlich's 1968 *The Population Bomb*.[62] The uncertainties concerning chemical utility and safety continued as the United States initiated a new use of chemicals in its Southeast Asian conflict. Here the chemicals were used by masculine institutions in conducting a new kind of war, one technological and irregular in its weapons, and halfway around the world.

⁂

Chemical herbicides preceded American soldiers in South Vietnam. Offering a low-tech solution to enable more high-tech military technology, the phenoxy herbicides 2,4-D and 2,4,5-T were used to destroy the jungles of South Vietnam. The very same chemicals that had proved effective in destroying lawn weeds could be used to destroy tropical jungles. President John F. Kennedy, determined to prevent the expansion of Communism in Asia, asked his administration advisors to assess ways to aid the Ngo Dinh Diem government. Military advisors, frustrated by ill-defined boundaries and an inability to assess ground conditions, were amenable to using chemical herbicides to clear jungle growth. As historian Bui Thi Phuong-Lan described, a policy of "extreme visibility" was implemented. Along with such strategic military considerations, though, were centuries of imperialist understandings of, if not the dangers of jungles, at the least the problem of jungles.[63]

Certain natural environments, like forests, swamps, and jungles, have long been used as spaces of resistance. In *Landscape and Memory* historian Simon Schama examines the power of the natural landscape on myth, memory, and ideology. Dating back to Tacitus's description of German forests as explanatory of Germanic success in resisting Roman conquest, such understandings of the ways natural environments have shaped national identities have been intertwined with ideas of conquest and empire.[64] Tropical forests, or jungles, figured as an essential element in imperial conflicts during the eighteenth and nineteenth centuries as European nations, the "West," scrambled to establish global empires in India, Africa, and Southeast Asia, often represented as the inscrutable "East." Western attitudes toward the classical tropical environment, wet and hot, betrayed a deep ambivalence, seeing tropics as both enticing and dangerous. Colonial administrators constructed a new interpretation of the tropical environment: what they called tropicality. This idea embodied an understanding of the jungle as uncivilized, as producing illness, as a place of degeneration, and a site of resistance. Such spaces existed as much conceptually as they did physically and represented an alien "other" both politically and culturally as well as environmentally.[65]

In their initial encounters with tropical forests, Westerners saw these environments as a kind of bountiful Eden, but they also quickly associated them with, as David Arnold noted, "a negative otherness."[66] Greg Bankoff's study of horse breeding in the nineteenth-century Philippines provides an example of these attitudes and the newly emerging justification for imperialism to continue. In attempting to improve the Philippines' horse stock, Spanish colonial administrators worried about the unsuitable nature of the tropics for white colonialists. But their biological breeding endeavors also exposed the need for Western intervention in reversing the degeneration of animals and human beings in the tropical environment.[67] The end of the nineteenth century showed the dilemma of empire as literally embodied in tropical jungles, which harbored disease, barbarianism, and resistance. This understanding continued into the twentieth century through WWII, and afterward, as jungles became commonplace arenas of conflict, especially in the proxy conflicts of the Cold War.

Herbicides played a crucial role in a new American military strategy. American imperialism had encountered jungles before US intervention in Southeast Asia. Exerting influence initially in the Caribbean, and Central and South America, Americans displayed the same ambivalence as other colonial enterprises in seeing tropical landscapes as Edenic and threatening at the same time. Both understandings justified imperial intervention.[68] By the time American military advisors began assessing the challenge of South Vietnam's jungle landscape, they had been dealing with tropical environments for decades. As Kennedy began engaging in Cold War conflicts, South Vietnam became the perfect laboratory for testing the administration's theories on counterinsurgency, which composed a significant part of their anti-Communist strategies. As jungles represented the paramount site of counterinsurgent resistance, the use of herbicides became a fundamental part of American involvement in South Vietnam. The defoliation program would be focused on four military goals: expose troops on the borders of Cambodia and Laos; expose guerilla troops in the Mekong Delta; destroy manioc and rice crops to deprive Viet Cong troops of food; and destroy mangrove swamps to expose Viet Cong retreats.[69] These military concerns came with previous understandings of tropical environments as lush and beautiful while at the same time remaining impenetrable and resistant.

Defoliation advanced these military goals indirectly. Historian David Briggs identified one goal in spraying the Mekong Delta region as the removal of a resistant landscape that had hidden insurgents for literally centuries.[70] Another aim appeared to be psychological. The Diem government viewed chemical defoliation as a means of terrifying unruly peasants (who often feared that the fog raining down was poisonous gas), which helped relocate peasants into government-constructed hamlets, or villages.[71] Diem's earlier attempts to create such settlements, called "Agrovilles," failed, but it did not stop him from urging Americans to carry out a similar policy. Known as the Strategic Hamlet Initiative and influenced by Kennedy advisor Walt Rostow's macroeconomic theories of modernization, the program sought to relocate rural peasants into government-constructed villages. It was argued that these fortified compounds protected villagers from attacks by National Liberation Front (NLF) fighters. It also attempted, as Bui noted, to "separate the people from the foliage so that control of both the people and the land could be achieved."[72] Chemical defoliation altered the physical environment, and at the same time it caused psychological fear. While the program failed at modernizing South Vietnamese society, it succeeded in alienating South Vietnamese peasants.[73]

American military advisers overlooked the role the jungle played in South Vietnam's agricultural economy. From their perspective, relocating peasants into fortified villages identified friendly and unfriendly forces, which in turn justified bombing civilians who refused to relocate to the strategic hamlets. In the process of attacking supposedly empty jungles, though, the US military ignored long-standing traditions of field-to-forest agriculture. In South Vietnam subsistence slash-and-burn agriculture, also called forest agriculture, was still practiced in the central highlands by a minority ethnic group, the Montagnards. Here, farmers planted a variety of crops—manioc, rice, melons—within the forest itself.

While other parts of South Vietnam practiced more settled agricultural cultivation, peasant society still centered on the village. Governed by elderly notables, and supported by intensive rice cultivation, small crafts, and subsistence hunting and gathering, rural Vietnamese peasants grew up in villages located near waterways, convenient for rice cultivation, and forests filled with edible plants and animals. Vietnam's peasant economy proved remarkably resistant to change, surviving until the end of the twentieth century. This kind of agricultural economy also gave rebellious peasants the ability to elude governmental scrutiny and control. Villages could be moved, disappear into another part of the jungle if necessary. Given this, these premodern forms of agriculture also represented impediments to the modernization of South Vietnamese society, and defoliation could be justified inasmuch as it disrupted or damaged such practices. Chemical defoliants became a crucial part of counterinsurgency efforts, necessary because of the understanding of jungles, "as dangerous places peopled with suspect populations, particularly near international borders."[74]

American air force personnel displayed the same uneasy understandings of jungles. The defoliation campaign began as a joint venture between the United States and the Diem regime but quickly became a solely American operation. This meant that American pilots would fly missions to spray defoliants along the roads, waterways, and croplands of South Vietnam. Over a nine-year period, from 1962 to 1971, approximately twenty-two thousand gallons of chemical herbicides were sprayed; 2,4-D and 2,4,5-T were used in varying concentrations and combinations as Agents Pink, Green, Purple, White, and Orange (and an even more potent version known as Orange II). One Ranch Hand pilot described the countryside as beautiful but admitted it was less so after spraying.[75] Charlie Hubbs, another pilot, also testified to the destruction of the jungle. "We could go back six weeks after we'd sprayed a particular area and it was barren." Talking about the purpose of the missions, Hubbs claimed success. "They had trails that they pulled over the top and made tunnels so you never would, never have seen from the air. They were just like highways."[76] Another pilot emphasized the need for accurate maps, for the "extreme visibility" necessary for military planning.[77]

Pilot interviews and other observers' perspectives also revealed cultural assumptions that justified destroying forests. These imperialist understandings of the jungle viewed occupied or unoccupied space as equally dangerous. As one Ranch Hand pilot described it: "I don't know if anybody was living in them or not. But at least, that would be weird. Usually it was just straight jungle."[78] Yet several other pilots demonstrated an awareness of spraying chemicals on North Liberation Front troops. Ralph Dresser admitted that while pilots avoided spraying peasants deemed friendly, herbicides might be used occasionally against enemy combatants.[79] Robert Turk declared that of course people were sprayed: "We didn't stop."[80]

As early as 1963, South Vietnamese leaders were defending spraying enemy troops. Responding to philosopher Bertrand Russell's charges that the United States was using chemical weapons, Madame Ngo Dinh Nhu, President Diem's sister-in-law, asked "So what?" She continued, "If Communists do not like it, why do they stay in OUR jungles, breathing it? Have we not the

Figure 1. Jay Cravens led the US Agency for International Development team sent to South Vietnam because of his previous experiences in the Forest Service and with herbicides. Courtesy of Forest Service Museum.

right to defoliate our OWN leaves whenever we like, and the more so in wartime?"[81] Another pilot noted that Ranch Hand operatives were told that the herbicides were safe for humans and pets, which made it okay to spray over the jungle and the human beings and animals that lived within it.[82]

In 1968 forestry consultant Jay Cravens described the problems of rubber plantations, which were off-limits to defoliation and bombing missions. "These were dark and spooky places, full of VC and sympathizers. . . . The trees kept the area in the dark, deep shade, ideal for concealment and movement of enemy troops."[83] Cravens also, however, expressed his frustration and concern about what defoliation had done to Vietnam's valuable hardwood timber stands. He questioned the feasibility of removing the top story of jungle forests, which left the understory brush growth intact and still effective cover for enemy troops and supplies to move through. In a May 27, 1967, letter home, Cravens described the countryside he was flying over, nothing that the dense understory growth remained, where the "VC, NVA and Chinese armies could hide out." He lamented the destruction of the jungle. "I am supposed to be here to help the Vietnamese manage and utilize their timber resources and the spray planes and bombers are doing everything within their power to destroy a valuable resource."[84] A National Liberation Front fighter described defoliation's effect from within the forest in a 1966 interview. "All the paddy and manioc were dead. Many big trees in the nearby forest were also killed. They were stripped of leaves and the trunks were black."[85]

While the reasons offered for defoliation emphasized the military necessity of visibility for tactical decisions and protection of the troops, a more amorphous reason appears to underlie the concerted attack on the South Vietnamese jungle. US military advisers and officials were convinced that South Vietnam's jungles, and the guerilla fighters they sheltered, posed a significant threat to military action in Southeast Asia. More than that military consideration, however, shaped their understanding and opposition to the jungle, which over centuries of imperialist encounters became the embodiment not only of environmental obstacle, but a cultural one as well. Fighting against an alien "other" could be carried out against the natural environment, a longstanding understanding of the resistant jungle. As the war intensified, human beings also became more resistant as they challenged American imperialism and, for some, focused on the military-industrial complex. The first signs of resistance appeared almost immediately and united several disparate groups in their opposition to Agent Orange herbicides. The next chapter examines the actions of religious groups, students, and scientists in protesting the use of these chemical herbicides.

CHAPTER TWO

The Quickening Conscience

Seminarians, Students, and Scientists
Protest the Phenoxy Herbicides

The nine antiwar protestors splattered blood on the walls, threw company records out the window, overturned office furniture, and wreaked havoc at the Dow Chemical Company's Washington, DC, offices. Group members went peacefully when police officers arrived to arrest them. Eight of the demonstrators identified themselves as Roman Catholic. Five were clergy, one was a nun, one a former nun, and one a Catholic layperson. The protest appeared to be carefully planned, as reporters and photographers met beforehand with one of the protesters, who took them to a corner, advised the "action" would begin shortly, and left them.[1] In their "Open Letter to American Corporations," the group condemned Dow and other American companies for their exploitation of Third World nations, claimed their products were death and their markets war. "Your offices have lost the right to exist. It is a blow for justice we strike today."[2] News accounts noticed the similarity of the action to draft protests conducted earlier in Baltimore, Catonsville, and Milwaukee by Catholic activists.[3] Biographies of the group, called the D.C. Nine, expressed their antiwar sentiments. Member Mike Doherty, a "missionary for the peace movement," urged individuals to "confront their oppressors."[4] Dow's oppression took the form of the chemical weapons of war they manufactured, which included napalm, nerve gases, and herbicidal defoliants.

Concerns about military ideology motivated a surprisingly diverse group of Americans to protest the manufacture and use of napalm and Agent Orange herbicides. Many might consider this opening episode a minor skirmish

in the Catholic Left's campaign against what they characterized as an imperialist and immoral war in Southeast Asia, an intensification of the Church's
grassroots pacifism. It demonstrates, however, the challenges ordinary citizens made to the military-industrial complex, identified by former president
Dwight D. Eisenhower, as a part of their efforts to hold corporations and the
state culpable. Here, war protesters connected the seemingly innocuous weed
killers with the chemical biological weapons (CBW) being used to fight the
war in Vietnam, recognizing the potential harm the US defoliation campaign
posed for the South Vietnamese people and countryside. Beginning with the
Catholic Left and other religious protesters, who linked the imperial war in
Vietnam with broader global injustices, to uncovering student demonstrations against herbicides as well as napalm, to the antiherbicide scientists who
pressured President Richard Nixon's administration to rethink the systematic destruction of South Vietnamese forests, Americans questioned the ethics and effectiveness of chemical defoliation when used as a weapon of war.

The Catholic Left and other religious denominations undertook direct actions or protests that targeted industrial war collaborators, most specifically
Dow Chemical. Unlike the student or scientific antiwar protests, which appeared in the post–WWII context, radical pacifism shows a longer history in
America, dating back to the beginning of the twentieth century, such as the
Women's International League for Peace and Freedom, which was established
in 1915. For Catholics, ideas about peace emerged during the 1930s. As embodied by Dorothy Day, one of the leading American pacifists, the pacifist
movement became linked to the radical Left of the Catholic Church. Day espoused antiwar beliefs before she converted to Catholicism, in the decades
prior to World War I when she wrote as a journalist for the socialist press.
Day's conversion to Catholicism united her profound commitment to social
justice with a new commitment to nonviolence, a mature pacifism. As one
of the founders of the Catholic Worker movement, which sought to restore
spirituality to socialist causes, Day provided a thread of continuity that other
prominent Catholic thinkers would join as the Church became increasingly
critical of the Vietnam War even as the Church became more critical of war
in general.[5] Voices like Day's and theologian Thomas Merton's represented
a new intellectual current in American Catholicism, which, unlike Protestant denominations, had little history of pacifism. This influential minority
of American Catholics questioned whether the Church's principle of "just
war" could be applied to the Vietnam conflict and began articulating a cohe-
sive pacifist rationale.[6]

Initially American bishops either supported or remained silent on American involvement in Vietnam. As the war progressed, a growing minority of bishops began speaking out. This shift in support paralleled the changed attitudes of the American Catholic laity and press, who strongly supported the war in the early 1960s, in part because of a strong opposition to Communism, but increasingly grew to question it. Formed in 1964, the Catholic Peace Fellowship, affiliated with the Fellowship of Reconciliation, showed antiwar influence in its sponsors and inclinations. The group offered counseling on conscientious objection, and Catholics showed the greatest increase in individuals claiming this status.[7]

The antiwar movement of the 1950s and 1960s, which united liberal internationalists and radical pacifists, became increasingly concerned with the perceived gap between democratic decision making and citizens. War in the nuclear age appeared futile, and the trial of Nazi henchman Adolph Eichmann, who offered the defense that he simply followed orders, intensified the sense of moral malaise and, increasingly, a sense of moral absurdity. Cold War conflicts, like weapons, proliferated around the globe. Some Americans became increasingly uncomfortable with the way wars were being waged after World War II. Winning justified any measures, and civilian populations were now considered collateral damage. Merton offered an important analogy when he described efforts to challenge governmental actions as "resistance," a term he imported from the various resistance groups involved in subverting the forces of Nazi Germany.[8]

The Catholic Left's contribution to Vietnam War protest came through its "direct actions." Massive protests appeared to have no effect on official US policy and measures, and while the first direct action received little media attention, it pointed the way to a viable protest strategy—interference with the government's ability to conduct the war. On Friday, October 27, 1967, four men entered the Baltimore Selective Services offices and proceeded to pour their own blood (augmented with chicken blood) over the records. The group, composed of two Catholics and two Protestants, inspired the next direct action six months later. On May 17, nine people—two priests, including Rev. Philip Berrigan, who had participated in the Baltimore action, a Christian brother, a former priest and nun, and four lay Catholics—met at the Selective Services offices in Catonsville, Maryland, a Baltimore suburb. Once there the group collected and took draft records to the parking lot, where they burned them with homemade napalm. This action received national attention and created a new protest tactic that emphasized public theater and interference with the draft. Two more direct actions followed, one in Milwaukee in September later that year, and the next in Washington, DC, on March 22, 1969. Unlike previous direct actions, the "D.C. Nine" focused attention on corporate involvement in the war.[9]

Like members of the Catonsville Nine, members of the D.C. Nine came to their activism with a wealth of experiences working domestically and internationally against poverty and for different social justice causes. Arthur Melville returned to Washington, DC, with his brother and sister-in-law, a former priest and nun who had served in Guatemala as Maryknoll missionaries, determined to bring attention to their fears that Guatemala was turning into another Vietnam.[10] Thomas and Marjorie Melville participated in the Catonsville Nine action, while Arthur and his wife, Catherine, also a former Maryknoll missionary, participated in the action against Dow. Other group members included Bernie Mayer and Bob Begin, both members of Cleveland's Thomas Merton community and whose latest protest had involved taking over the city's Catholic cathedral.[11] Two members came from Detroit, Mike Slaski, a young, ex-Catholic draft dodger, and Rev. Denny Moloney. Father Joe O'Rourke, a local Jesuit and friend of the Berrigan brothers, his friend Mike Doherty, and Sister Joann Malone, a nun from St. Louis and friend of one of the Catonsville Nine, rounded out the Catholic members of the group.[12]

The decision to target a corporate rather than government actor demonstrated the more radical stance of several group members. As in previous direct actions, the activists met to discuss possibilities several times. "But as the formation process continued, it became apparent that several of this group's membership adhered to a more militant and leftist ideology than previous action groups," noted a later interviewer. Malone confessed that while she was outraged about the war, she also had other motivations. "By this time my politics were definitely anti-capitalist and anti-imperialist. . . . I came from a working-class background. . . . The decision to act meant a conscious decision to become a dedicated revolutionary."[13] O'Rourke agreed with the shift in focus, from draft boards to corporate offices. "I felt that the war was really a corporate war. Both a defense of corporate expansion and economic hegemony in Southeast Asia by American-based multinationals through Japan and other places."[14] The group's official statement on their action condemned Dow not only for its alleged wartime collaboration with Germany's I. G. Farben but also for its manufacture of the chemical weapons of warfare, chemicals like napalm, defoliants, and nerve gas. The Nine claimed Dow's offices had lost their right to exist. While one member later bemoaned its creation, one of the most interesting documents to come out of the D.C. Nine action was not the Dow files flung out the window, but rather the group's "defense booklet." Activists modified a Dow annual report by incorporating a critique of its products. It was used to rally support for the group and revealed the group's positions and rhetoric at the time.[15]

As with the other direct actions taken by the Catholic Left, the D.C. Nine hoped to bring national attention to US involvement in Southeast Asia. One way of discerning the group's objectives comes from an examination of public relations materials created by the protesters' allies. Crafted by the "D.C. Nine Defense Committee," the group used a radically altered 1968 Dow Annual Report as the basis for a protest pamphlet. The protest document revealed the major complaints the "D[ow] C[hemical] Nine" and others on the Catholic Left had with corporate America. Several pages examined the case against napalm B, noting its horrible effects and Dow's role in manufacturing the jellied gasoline, used in incendiary bombs, allegedly against civilians. In the report's section detailing herbicide sales, the Defense Committee denounced their use for crop destruction, noting that South Vietnam had imported 800,000 tons of rice in 1968, as compared to its status as a rice exporter prior to 1967 in the amount of $134 million.

The annual report's newly added facts also observed the cost of the defoliants and, finally, alerted readers to massive areas of countryside being sprayed with 2,4,5-T and 2,4,-D, the equivalent of the state of Massachusetts, or more than half of South Vietnam's arable land. The altered annual report now included accusations of the chemical defoliants' link to severe birth defects haunting both North and South Vietnamese women, in graphic and shocking detail. The global section of the annual report now connected the sale of "highly poisonous herbicides and insecticides" that were being sold on the open market in Latin America, with minimal cautionary warnings. The new and improved annual report ended by asking "Is Dow Killing You for a Profit?" It was not only "our brothers in Laos, Cambodia, Thailand and Viet Nam" being killed but people in the USA, through chemically induced cancers.[16]

The modified report offered a window into the critiques Catholic peace activists were making against corporations supporting the war and, more broadly, the military-industrial complex. The sources used to modify the annual report included *New York Times* articles on chemical weapons by Seymour Hersh, from the *Wall Street Journal*, and Vietnamese sources. The ominous warnings about herbicides all emphasized their potential harmful effects on human health, or in crop destruction, with brief mention of the country's arable farmland exposure.[17] David Darst, a member of the Catonsville Nine, helped write the D.C. 9 Defense Committee Annual Report, and the address given for the Defense Committee was George Mische's, another Catonsville Nine group member. Activists held Dow responsible even though they were no longer producing napalm for the government. The D.C. Nine hoped to follow in the successful footsteps of other direct action groups brought to trial and challenge the legitimacy of the war, the US military and government, and corporations like Dow.

The trials of the Catonsville Nine and Milwaukee Fourteen provided bully pulpits for antiwar activists to speak out against the war. The Catonsville trial drew hundreds of supporters to Baltimore, many of whom took the opportunity to plan future actions, with two thousand present by October 7, 1968. Supporters flooded the courtroom, giving the defendants a daily standing ovation as they entered. A sympathetic judge allowed group members leeway to explain why they had burned draft cards, which included an "unprecedented forty-minute open discussion with the judge which they concluded by holding hands and, with the astonished court's permission, reciting aloud the Lord's Prayer."[18] The Milwaukee Fourteen achieved similar results when they dismissed their lawyers and represented themselves. The courtroom erupted with the announcement of a guilty verdict, as observers identified themselves as antiwar and declared the verdict as condemning Jesus Christ. As defendant James Harney observed, "Judge, you just lost your authority, whether you know it or not."[19] The trial of the D.C. Nine five months later would not go as well.

The February 1970 trial of the D.C. Nine represented a new response on the part of the judiciary to recover control of its courtrooms. Regaining control of trials appeared even more imperative after the successful disruptions of the 1969 Chicago 8 Trial. Judge John Pratt was determined "to censor any drop of meaning or message from the proceedings."[20] Pratt had attended a seminar focusing on the best way to handle political trials led by the Milwaukee Fourteen judge, and he threatened anyone mentioning the war in Vietnam with contempt of court with no bail.[21] Pratt cited D.C. Nine attorney Phil Hirschkop, a renowned civil rights activist, for contempt, with a thirty-day jail sentence. Going into the trial, liberals expressed the hope that "regardless of the law, what we wanted for the country was better than what the court, the prosecution, and the Establishment had already served them and us."[22] Despite a large crowd, by Washington's standards, protesting the contempt sentences, many supporters struggled with the growing realization that the D.C.

Nine "may have sacrificed themselves to no avail."[23] The trial itself ended in shambles, with two group members pleading no contest, and a fight breaking out in the courtroom. Direct actions had attracted national attention but increasingly were drawing the wrath of local and regional bureaucrats, industry leaders, and law enforcement.

Catholic radicals continued to pressure the system with Selective Service actions, although some groups also challenged the military-industrial complex as embodied in war contractors. On October 31, 1969, members of the direct action group Beaver 55 broke into the Selective Service offices in Indianapolis and destroyed records. A week later, some, possibly all, of the group's members entered Dow Chemical's headquarters in Midlands, Michigan. Here they erased hundreds of computer tapes containing research on chemical weapons, worth up to $100,000. The statement left at the action expressed the group's concerns. "We will no longer tolerate the production of napalm, of nerve gas, of defoliants. We will no longer tolerate the pollution of our air, water and earth. . . . We will no longer tolerate control of subsidiaries in other nations or exploitations of human beings labor and natural resources for profit."[24] Group member Jo Ann Mulert explained that the group had decided to destroy the lethal knowledge encoded in the tapes but spared the machines, the computers, in the hopes that they could be used for good.[25]

The group publicly took credit for the draft board action as well as the one at Dow at a press conference held on November 22 in Washington, DC, acknowledging the ways the D.C. Nine had influenced them. "A growing number of us understand clearly how American foreign policy is used to make possible the rape of human life and natural resources in developing countries."[26] Of the eight group members, three were under twenty-years old and members of the Young Christian Students group. Another three members were in their early twenties. The last two members, Tom Trost (thirty-seven years old) and Jane Kennedy (forty-four years old), were considerably older. Kennedy, a nurse administrator at the University of Chicago hospitals, received the harshest sentence, serving almost three years total in state and federal jails. Her case also attracted national attention as her friends and supporters tried to gain her freedom.[27]

The campaign for Kennedy's release highlighted the sustained opposition to the military contractors. An advocacy group seeking Kennedy's release sent out an explanatory fundraising appeal that noted that Beaver 55's "respect for human life" led them to take dramatic action to protest the wanton death and destruction caused by US intervention in Vietnam.[28] Dow's production of napalm and "poison defoliants" justified the destruction of Dow's research. Kennedy admitted in a later interview that she had no real interest in destroying draft records. She considered "the industrial people, those who made money out of things used to kill other human beings" guilty of supporting and benefiting from the military-industrial complex.[29] Kennedy admitted that the idea that group members were charged with "property crimes" puzzled her. In a conversation with a parole officer investigating to see whether she deserved a reduced sentence, Kennedy expressed her bewilderment with the system. "They charged us with conspiracy to destroy Dow Chemical *property*. Would that we had a society where we could charge Dow Chemical with conspiracy to maliciously destroy and maim human life with its property."[30]

War manufacturers drew the ire of more than just Catholic Left radicals. Other religious protesters, most prominently Clergy and Laity Concerned About Vietnam (CALCAV), an interfaith peace group, joined in the protests over corporate support of the war. CALCAV's origins lay with the response of mainstream denominational and evangelical/Pentecostal groups took in response to the civil rights movement and increased social activism. Broader changes in postwar religious institutions toward ecumenical councils, a growing religious press, and the conditions in Vietnam itself provided the conditions that led to the formation of the group. In addition, the campaign by various government officials to discredit antiwar activists prompted the formation of CALCAV in early 1966.

The organization appeared to be much more traditional. The forty-man national committee collectively tended to be theologically liberal, mostly Protestant (twenty-eight members), with smaller numbers of Jewish and Roman Catholic representatives. Half the committee came from the New York City metropolitan area, three-fourths from the eastern seaboard, with a median age of fifty. Sixty percent of the members were denominational officials or in education. The group saw its mission as primarily education and advocacy, although this changed throughout its existence. One of these transmutations came when the group decided to apply "direct pressure to a defense contractor as a step toward ending the war."[31] Three hundred people showed up outside Dow's corporate headquarters in Midland, Michigan, in 1968, and five CALCAV members participated in the shareholders' meeting. Dow took note of the protest, producing a pamphlet in response to the religious group's concerns.[32]

Just shortly after the D.C. Nine's direct actions against Dow, CALCAV announced it would be attending Dow's annual stockholder's meeting on May 7, 1969, in Midland. A statement the group released announcing their planned attendance aired their specific concerns. They decried Dow's production of napalm B and the phenoxy herbicides. "We are in anguish," the statement read, "that American industry allows its talent and capacity to devise defoliants whose strategic purpose is to deprive innocent civilians, like ourselves, of food."[33] CALCAV's statement repudiated the idea that corporations could allow the military to decide questions of conscience and called upon business executives to show ethical as well as economic leadership. The statement noted that while Dow's production of napalm B had made it the target of antiwar activists, "increasing attention has been given recently to Dow's manufacture of other materials, especially defoliants and instruments of Chemical-Biological Warfare."[34]

Dow's response provides one means of measuring the action's effect. In a September 24, 1969, memo, Dow officials detailed their planned response to the religious community. Item II(e) of the company's Napalm Program for 1969–70 required "establishing channels of communication with the religious community, since much of the attack had originated in that quarter."[35] The public relations office outlined three "major lines of activity," which included the following points: (1) Dow should send materials emphasizing its societal contributions to a mailing list of religious media, with the thought that such materials would positively affect editors even if it was not reflected in print; (2) develop materials suitable for select release to publications specifically for the clergy; and (3) rejected pursuing direct contact with churches, universities, foundations, and other institutions holding Dow stock in an attempt to prevent such institutions from selling Dow stock in protest.[36]

Other CALCAV actions included its participation in efforts to "unsell" the war, a public relations campaign responding to a CBS News program, *The Selling of the Pentagon*. A cooperative effort between Vietnam scholars and Madison Avenue ad men, Help Unsell the War, proceeded to produce a variety of print and media ads costing more than a $1 million. Over 100 television and 450 radio stations ran the advertisements as public service announcements. One of the most effective ads, "Apple Pie," won advertising's Clio Award, the industry equivalent of an Oscar. The piece played upon the connection between apple pie and American identity, but also the idea of a pie that gets doled out to eat. "When they divide up the pie in Washington," the narrator asked, "do you ever wonder who gets the biggest slice?" The voice answers that Americans' taxes buy Vietnam. They buy "bombs instead of schools, defoliation instead of clean air and water, tanks instead of trains, and destroy homes instead of building them." The commercial ended by suggesting America get

out of Vietnam and come back home. Corporations attracted the ire of another group of Americans, university students. Like religious activists, student protests also included criticism of herbicides and defoliation along with
their concerns over napalm.[37]

⚬⚬⚬

The academic wing of the military-industrial complex began with President
Franklin D. Roosevelt's creation of the National Defense Research Committee
(NDRC) in 1940. Under the guidance of Vannevar Bush, the NDRC awarded
155 contracts to forty-one universities. The University of Pennsylvania eventually developed computer and submarine technologies from research funded
by the Defense Department. Most supporters, including Bush, assumed that
wartime academic research labs would be discontinued at the end of the war.
Scientists from the Massachusetts Institute of Technology who had developed
radar technology, for example, mostly returned to pre-war research projects.
But of the government funding awarded to universities, the Department of
Defense (DoD) provided the bulk, as much as 84 percent in 1958. The continued existence of government-university-academic institutions created sta-
ble and productive relationships in creating a national defense system, and
university scientists situating themselves as either teachers or managers producing knowledge and goods. Universities successfully leveraged these relationships with the government and private industry so that academic research
received $401 million in military grants by 1964. Just two years later, Penn
alone had been awarded $25 million in funding, more than a fourth of the
$90 million university budget.[38] The university began specializing in a spe-
cific kind of military research focused on chemical biological warfare (CBW).
    While student protests featured napalm most prominently, defoliants and
chemical and biological weapons were recognized as evidence of corporate
culpability in an unjust war. The opening salvo of campus protests over Agent
Orange herbicides occurred at the University of Pennsylvania in direct response to its CBW projects. In the aftermath of WWI and the banning of poison gas weapons, the Army Chemical Corps carried a tainted reputation, and
the division barely survived in the decades after WWII. Penn's relationship
with the Chemical Corps dated back to 1951, when Penn received its first,
secret, government contract for CBW, codenamed Project Benjamin (short-
ened to Big Ben).[39] The research was funded for seven years under the direc-
tion of chemistry professor Knut Krieger, costing almost $3 million. During

this time, the university created a new entity, the Institute for Cooperative Research (ICR), where interdisciplinary project contributions could be conducted, with an assurance of security and secrecy. Project Summit began in 1955, focusing on "air-delivered chemical and biological munitions for counterinsurgent operations."[40] President John F. Kennedy and his military advisers continued investing in Chemical Corps weapons that promised "war without death."[41] Two more contracts sought information on CBW agents. Project Spicerack followed in 1963, with a contract to research aerial delivery systems for counterinsurgent CBW. The military planned on using the research of both projects for the war in Vietnam.[42]

These various military research projects remained undisclosed until 1965, although once made public they provoked intense passions. An undergraduate student affiliated with a leftist student group, the Trotskyist Young Socialist Alliance (YSA), noticed the ICR's off-campus headquarters and used his position in the bookstore to identify the research materials ordered by the institute. Robin Maisel, the student, contacted his history professor, Gabriel Kolko, who verified the information and proceeded to rouse faculty attention. In the summer of 1966, graduate student Jules Benjamin, a member of the Committee to End the War in Vietnam (CEWV), sent information on Projects Summit and Spicerack to two antiwar magazines, *Ramparts* and *Viet-Report*, revealing the programs to a national audience. In response to a CEWV letter requesting termination of the contracts, university president Gaylord Harnwell admitted that both programs were conducting research on aerial spraying systems for defoliants.

In an October 1966 follow-up article, "The Weed Killers—A Final Word," author Carol Brightman noted the disappearance of the issue in the American press. Brightman, a graduate student at New York University, had started *Viet-Report* with a $3,000 grant. *Viet-Report* gathered and printed information on the war in Vietnam, designed to allow people interested in the war to stay informed.[43] The piece questioned how effective a recent protest by twenty-two scientists against CBW had been. The article revealed two major concerns about current CBW projects and research like the kind being done at Penn. The first concerned the harm being done to South Vietnam's rural populations as defoliants destroyed the human habitat; and the ways the "success of 'nonlethal' chemical warfare in Vietnam" led to "stepped up research, development and production of more esoteric 'disorientation' agents (BZ), along with deadly nerve gases."[44]

The situation at Penn heated up as the projects' lead researcher and university president sent confused messages about the proper role of classified research at academic institutions and the status of the programs and their future. CBW research caused as much grief to military officials as to university administrators. A 1967 *Science* brief quoted Pentagon officials who expressed dismay that such weapons were the "most easily misunderstood . . . because [such research] provokes the most emotional distress and moral turbulence."[45] These concerns led the military to cloak CBW in secrecy, with procurement figures classified by 1967. It also generated controversies like the one happening on the University of Pennsylvania campus. The protesters included radical students, faculty who condemned the war as immoral, and faculty who denounced the classified nature of the research. Some faculty supported CBW, and Penn's military research programs, as a necessary part of national defense. The anti-CBW factions won with respect to mobilizing the faculty, if not convincing the broader student community, in large part because the majority of faculty united to reject CBW research programs.

The student newspaper, the *Daily Pennsylvanian*, from the beginning questioned the administration's actions. Along with critical editorials, it included visual satire as well. In its February 1967 issue, the paper "advertised" a new product. Called "It's a GAS," the chemistry set was developed by Krieger, the chemistry professor doing research for Project Spicerack. The ad showed a picture of a young boy and a popular home chemistry set. Bullets highlighted the features of "It's a GAS," which included creating your own ICR, and the enticement "Be the first kid on your block to develop complete dispersal equipment for mustard gas and crop defoliants." Once made, the equipment could be used to spray friends or graduate assistants.[46]

The Penn controversy over CBW contracts continued for several years as Harnwell found himself in trouble because of either poor communication or deceptive intent regarding the status of the program. His unwillingness to terminate the DoD contract increasingly angered students and faculty. The situation grew progressively volatile, with local, regional, and national media covering events on the Penn campus, students organizing and demonstrating, and faculty passing resolutions to end the defense contracts. The crisis was resolved on May 4, 1967, when the board of trustees voted thirty-nine to one to terminate the Summit and Spicerack contracts. Penn's student radicals had initiated the challenges to the Cold War political and military consensus in their protests against CBW, which included chemical defoliants; other students across the nation followed.[47]

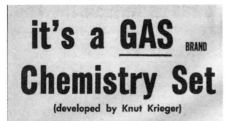

it's a **GAS** BRAND
Chemistry Set
(developed by Knut Krieger)

- Build your own ICR

- Be the first kid on your block to develop complete dispersal equipment for mustard gas and crop defoliants

- Use it on your friends or your graduate instructors

YOU KNOW IT'S MATELL . . . IT KILLS REAL WELL!

Figure 2. The *Daily Pennsylvanian*, the student newspaper, mocked the University of Pennsylvania's secret defense contracts with this satirical cartoon offering a chemistry set that could make mustard gas and chemical defoliants. Courtesy of University Archives, University of Pennsylvania.

The protests started at Wayne State University in Detroit in 1966 and quickly spread to West Coast campuses. Students began demonstrating against Dow Chemical Company recruiters visiting campus to find bright young (mostly) men to work in the chemical industry. Like Catholic Left and other religious activists, following the lead of students at the University of Pennsylvania, bright, intelligent, well-to-do students began protesting university policies that allowed the US military, the CIA, and war contractors like Dow to come to college campuses and recruit students. They also protested university relationships with the military establishment and their private industry counterparts. Dow remained on campus even when the military and CIA left. Students increasingly focused their concerns, frustrations, and ire against the company. By the end of 1967, there had been over 133 campus demonstrations, with violent confrontations at San Jose State College (where the Dow recruiter escaped from an angry mob out the back window) and the University of Wisconsin. The Harvard recruiter failed to escape his captivity, and was imprisoned for eight hours, the longest thus far. Demonstrations began to increase in tone and frequency, and Dow found itself mired in a public relations quandary.

Dow took the campus agitations seriously, monitoring them and issuing an in-house "Napalm Report Bulletin." Historian David Maraniss describes Dow's reaction as aggressive, as company leaders framed the company's actions as courageous and patriotic. Most mainstream media supported the company, such as the November 24 Detroit *Daily Express* editorial, which Michigan congressman John Dingell had read into the *Congressional Record*. The editorial chastised deserting sailors and students at San Jose State, declaring both groups wrong to disobey the law in protesting the war. The company had also received thousands of letters condemning Dow's production of napalm for the war. Dow commissioned a public opinion poll in 1966 that conducted 4,500 interviews of telephone subscribers three times over the course of the year. "Mr. Average Guy" was asked about the pace of the war, and chemical weapons like napalm, defoliants, and tear gas. The poll report noted an increasing dissatisfaction with the war, which had "generated some resentment against those industries which supply war materials."[48] From December 1966 to December 1967, public awareness regarding napalm had increased, while there appeared to be virtually no change in attitudes about defoliants, and an actual decrease in knowledge about tear gas. This report summary specifically noted this lack of change for the defoliants. If Dow executives based their response on this opinion poll, they were handling the demonstrations properly. But campus troubles continued.[49]

Student protests about Dow and defoliants expanded geographically in 1968. One campus protest revealed students' displeasure with their university's continued relationship with a war supplier. In February 1968 Duke University students used a Dow recruiting trip to challenge the company's production of napalm and defoliants, and the university's investment in Dow— represented by five thousand shares of stock. A small number of student activists challenged a Dow representative to debate "Dow's manufacture of napalm and defoliants."[50] They also requested the administration discuss the university's relationship with Dow.

The protest at Duke was written up as an "Organizing Case Study" with a detailed report on what happened. Under "It Starts" the account noted that a small group of three to five people identified the target based on research. At Duke, Dow presented an ideal "target," as it produced war materials and also because Duke was a shareholder. The recruiter coming to campus represented the perfect opportunity. The event chronicle detailed the ways information was disseminated, which included a press release, leafleting, educational materials available to picketers and the general public, and pre-event literature tables with new material and "a lot of free stuff" involving information on different groups and materials handed out to passersby. The report claimed relatively modest success in moving some people from the general campus, to those who might become involved next time, greater commitment by some people who picketed this time, and finally some who participated in the sit-in after picketing last time. The account ended by noting: "We didn't stop the recruiting, didn't make Duke sell its stock. . . . But we now have a lot of new dorm canvassers, a lot more picketers, a few new researchers, and a lot more sympathizers that we didn't even know about."[51] This confidence may have been overly optimistic, but for many the protests against Dow and chemical agents represented their first protest against the war.

Religious and student demonstrations against defoliation took place even as another group began voicing its concerns. Dow Chemical took the student demonstrations very seriously, not just because they disrupted recruitment or potentially attracted negative public attention. In tracking the various student protests and their influence, Dow executives worried about the attitudes of one particularly important group, academic scientists. The Federation of American Scientists began questioning the use of chemical and biological agents in the war as early as 1964. In 1966 twenty-two scientists had urged President Johnson to stop using antipersonnel and anticrop chemical weapons. While neutral on the war itself, the scientists' letter raised concerns about the use of chemical and biological weapons in Vietnam, stating that the use of such weapons "sets a dangerous precedent, with long-term hazards far outweighing the probable short-term military advantages."[52] Dow recognized the danger campus protests represented, and acknowledged the need for positive media coverage of "what 'sane' students and 'sane' faculties must do to preserve their own academic freedom."[53] Dow could not afford for the same fight over chemical weapons to ignite the scientific community, and they took efforts to prevent this.[54]

One measure advocated by Dow Chemical's public relations department dealt with positive press coverage. A major campaign was launched around the use of Silvex (2,4,5-TP, a related chemical compound) to clear jungle growth in Hawaii. An article that appeared in the October 1967 *Hawaii Business and Industry* was reprinted in Dow print media like *Down to Earth*, a review of "agricultural chemical progress." An internal company memo recommended promoting "forest defoliation" as a measure to combat world hunger. Along with using Dow print media, the memo suggested collaboration with the United States Department of Agriculture (USDA) and the World Health Organization (WHO) to promote "the land use concept." The memo counseled working with the Manufacturing Chemical Association and the National Agricultural Chemical Association, organizations that both had "a vested interest in vegetation control."[55]

The company produced a fact sheet on the value of Dow's Tordon, a commercial product with the chemical picloram as its active ingredient. Picloram was combined in a 4:1 ratio with 2,5-D to produce Agent White. Other US companies also addressed their involvement in jungle eradication in Southeast Asia. Dow PR files included a *US News & World Report* news brief on the "jungle-clearing job" being done in South America and Southeast Asia by Rome plows attached to Caterpillar tractors. The plows were specially modified armored bulldozers used in South Vietnam to clear jungle growth. The piece noted that Caterpillar had removed growth around US military bases to "eliminate enemy-sniper havens" but also turned what had been a "wasteland into usable farmland." The company also monitored affairs within the scientific community.[56]

American scientists continued to speak out after the 1966 letter to President Johnson, and Dow Chemical executives worried about the outcomes of such protests. Later that year the American Association for the Advancement of Science (AAAS), an organization of 105,000 scientists, voted 125 to 95 to pass a resolution "expressing concern over long range consequences to the planet by use of chemical and biological agents in Vietnam."[57] While the final resolution omitted specific references to biological and chemical weapons along with any mention of Vietnam, it appeared clear that segments of the scientific community had begun questioning the long-term effect of defoliation on the South Vietnamese ecosystem. The AAAS council, the group's governing body, also set up a committee to investigate the effects of biological and chemical agents on the environment and during times of war. These questions about chemical warfare arose even as academics in other disciplines questioned the Johnson administration's South Vietnamese policies and promotion of the war that many considered undemocratic.

On May 15, 1965, a National Teach-In had taken place in Washington, DC. At the University of Michigan, university professors used teach-ins as a means of protesting the war. Just as faculty and student protests heated up on campus, concerns about herbicides increased within the academic scientific community, and the intensity of activism displayed among student and religious groups accelerated between 1967 and 1970. Unlike these, scientific protest specifically focused on chemical herbicides and defoliation. This focus threatened more serious consequences because of the economic importance the two chemical products represented to the company. Dow produced napalm on a government contract, with almost $5 million in sales in 1967, and up to $6.7 million in 1968. Napalm had limited uses outside of warfare. Herbicides, however, represented an important, and profitable, component of the company's product offerings. In 1961 the average farmer spent almost $500 on 2,4-D annually. By the 1980s Dow's agricultural products brought in $500 *million* a year. Dow feared that scientific criticism of 2,4-D and 2,4,5-T might have much more direct effects on their annual profits, and much further into the future.[58]

Historically, scientists have shown mixed levels of engagement as activists. The scientific ideal emphasizes rationality, neutrality, the impartial nature of scientific endeavors. But the ways modern science was being used also increased pressure on scientists to speak out. As sociologist Kelly Moore has noted, the benefits of state funding came with expectations of scientific silence. The public furor when nuclear physicist Robert Oppenheimer was stripped of his security clearance in 1954 offers one compelling example of what happened to scientists who ran afoul of the national-security state. One striking aspect of postwar scientific activism is the increased numbers of groups specifically challenging understandings of scientists as apolitical, groups such as the Society for Social Responsibility in Science, the Committee for Nuclear Information, and Science for the People. Although they were not formally connected with any of these groups, scientists who questioned the use of chemical warfare expressed the same concerns about the relationship between science, warfare, and the state.[59]

A small group of scientists led the protests opposing chemical warfare in general, especially the use of chemical herbicides. One of the first was Egbert "Bert" Pfeiffer, a zoologist from the University of Montana, who submitted the original draft of the 1966 AAAS resolution. While the group passed a modified version of Pfeiffer's resolution, it rejected, for the time, his call for a committee to investigate the environmental effects of defoliation on the South Vietnamese ecosystem. John Constable, a surgeon at Harvard Medical School, voiced his concerns about the use of herbicides. Yale biologist Arthur Galston contributed another voice criticizing the US military's defoliation activities. Galston had worked on the early studies examining the plant growth chemicals when he was a graduate student at the University of Chicago. Arthur Westing, Galston's graduate student at Yale, also became actively involved in investigating and writing about the environmental damage being done in South Vietnam. Finally, Harvard molecular biologist Matthew Meselson, a longtime chemical and biological weapons activist, completed the core group of the most prominent scientists engaged in challenging chemical herbicides.[60]

These primarily academic scientists, most with expertise in the biological fields, were matched by an assortment of scientists drawn from land grant universities, government agencies, and industry. Dr. Fred Tshirley, a USDA agricultural scientist and later professor at the University of Nebraska, represented a typical "weed scientist" who proclaimed the safety of the defoliating herbicides. Tshirley had visited South Vietnam as part of a DoD study designed to survey herbicidal effects on the country's forests. Boysie Day, a plant physiologist at the University of California, Riverside, was another prominent supporter of the phenoxy herbicides. Dow Chemical's E. C. Britton and Keith Barrons had researched, published, and defended the phenoxy herbicides. These men and the other members of the powerful agribusiness alliance significantly outnumbered the antiherbicide scientists. Yet this smaller group succeeded in gaining access to scientific organizations like the AAAS; the media in the form of newspapers, journals, and books; and, some have argued, the Johnson and Nixon administrations. For antiherbicide scientists, however, their training in the life sciences allowed them to speak from positions of scientific expertise if not authority. Rather than rehash these works, this section considers the ways these scientists identified their intervention(s) as challenges to the political consensus and its military-industrial complex, focusing on scientific visits and appearances, publications, and the final scientific studies produced.[61]

Antiherbicide scientists continued voicing concerns about the military's defoliation campaign after 1966, increasing their efforts to mobilize the broader scientific community. A 1967 petition sent to Johnson's White House

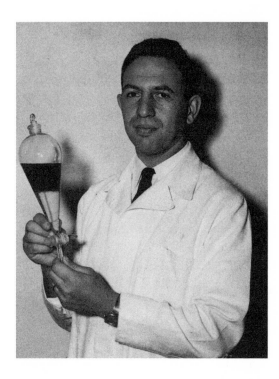

Figure 3. After helping develop the phenoxy herbicides, Yale plant physiologist Arthur Galston spent much of the rest of his life protesting their use. Courtesy of Archives & Special Collections, CalTech.

represented over five thousand independent scientists who expressed concern over the use of CBW and included a call for the cessation of herbicide spraying in South Vietnam. The antiherbicide scientists concentrated their attention on the AAAS, getting some traction at the 1969 annual meeting, already discussed. Earlier that year, in March, Pfeiffer and his friend Gordon Orians had traveled to South Vietnam attempting to evaluate the effects of defoliation on the ecosystem. In assessing their two-week trip sponsored by the Society for Social Responsibility in Science (SSRS), Pfeiffer offered a scathing judgment on the harm being done to South Vietnamese animals and plant life. He also expressed his belief that the military needed to defoliate the jungle growth to protect troops from ambush, although not everyone agreed with the success of the herbicides in thwarting sniper attacks. Despite Pfeiffer's mixed message, his Vietnam trip helped bring broader public attention to the questions surrounding the use of herbicides in South Vietnam.

At the 1969 AAAS meeting, a special session focused on herbicide use in war. Several spirited discussions ensued, including military scientist Fred Tshirley's acknowledgment that mangrove forests had been damaged but also his assertion that the forests could recover within twenty years. The AAAS executive committee agreed to send a team of scientists to do their own assessment of ecological conditions in areas sprayed by herbicides, in part because "Department of Defense officials do not have scientific assurance that herbicides cause no seriously adverse ecological consequences in Vietnam." Although it was clear that not all members opposed the war in Vietnam, or even the US herbicide campaign, "most scientists find it difficult to oppose a purely scientific study." The action was viewed by some members as the "quickening conscience of the American scientific community."[62]

Along with the concerns raised by site visits like the Pfeiffer-Orians and Tshirley ones, other information had contributed to the growing concerns over the phenoxy herbicides. A National Cancer Institute study, commonly known as the Bionetics study for the research lab that conducted it, was initiated in 1963. The study was designed to test common chemicals for possible carcinogenic, teratogenic, or mutagenic properties. Repeated tests indicated a strong correlation between exposure to 2,4,5-T and birth defects. Despite strong industry opposition, some study results had leaked out, raising concerns about the safety of the phenoxy herbicides in the process; the study was fully released in 1968. Coupled with the vocal support for an ecological evaluation of conditions in South Vietnam, the AAAS created the Herbicide Assessment Commission (HAC). The HAC, led by Harvard geneticist (and longtime CBW activist) Matthew Meselson, visited South Vietnam in August and September of 1970. With the AAAS sponsorship of the scientific survey, the antiherbicide position was gaining visibility, acceptance, and most importantly, data.[63]

The visit to South Vietnam presented its own set of difficulties, which included limited resources, resistance to the survey, and the difficulties of assessing ecological and health conditions in war zones. Working with a budget of $50,000 from the AAAS, the survey team prepared before leaving for Vietnam with a five-day June conference at Woods Hole, Massachusetts, to review the existing relevant scientific materials, with experts from government, industry, and universities attending. The commission also consulted with French and Vietnamese experts in Paris before their arrival in Saigon. Meselson emphasized professionalism and neutrality to his team members, which included outspoken CBW critic Arthur Westing, a plant physiologist working on forests; John Constable, a French-speaking physician from Boston's Massachusetts General; and Robert Cook, a current graduate student of Yale's Arthur Galston.

Figure 4. Harvard biologist Matthew Meselson poses with a photograph of defoliated South Vietnamese landscape. Meselson led the American Association for the Advancement of Science Herbicide Assessment Committee. Courtesy of Bettman Collection, Getty Images.

Once in South Vietnam, HAC received most of its support from the United States Agency for International Development (USAID), the government bureau where forester Jay Cravens was assigned. The commission could not get access to critical Operation Ranch Hand records, however, which included spray runs, amounts of herbicides sprayed, and the concentration of the herbicide mixtures sprayed. HAC members investigated as many sprayed areas as possible, some for the first time. They "interviewed numerous farmers and village officials for first-hand information on herbicide effects" and "conducted aerial and ground inspections of herbicide treated and untreated areas and conducted studies of possible health and congenital anomaly changes in selected regions."[64] Their findings led to a final report that challenged previous government studies that had downplayed any harm to human beings and the South Vietnamese ecosystem.[65]

The HAC's conclusions drew the attention of many, and the anger of some. In its final assessment, the HAC warned of the significant effect the US military's defoliation program had, noting that over 800,000 people had been deprived of food crops, that South Vietnamese forests had lost hardwood trees worth over $500 million, and 1.2 million acres of mangrove forest had been destroyed. The AAAS team had also found anecdotal evidence of increased miscarriages and birth defects among the South Vietnamese citizens exposed to the defoliating herbicides. Presenting their findings at the 1970 AAAS meeting in Chicago, one dedicated to the environment, Meselson and his team members found themselves in a confrontation with one of the army's highest-ranking officers, Brigadier General William Stone, who argued that the defoliation program was vital to protect American soldiers from jungle ambushes. The HAC investigation should be considered a factor in the Nixon administration's decision to halt defoliation spraying in South Vietnam, and it certainly helped cement the decision. As the war continued, so too did scientists' criticism of defoliation's effects on South Vietnamese people and environment.[66]

The antiherbicide scientists aired their views in a multitude of publications, both scientific and some aimed at a more general audience. Yale botanist Galston was quoted in updates on "the defoliation controversy," wrote brief commentaries in scientific journals, and was an active participant in letter-to-the-editor debates that erupted after essays and articles on the phenoxy herbicides appeared. Remembering his own history with the phenoxy herbicides in a 1971 essay, Galston described the campaign he undertook in bringing the war's defoliation program, and the possible ecological consequences, to the attention of first Johnson and later Nixon administrations. As early as 1967, in a special edition of *Scientist and Citizen* (titled *Environment* after 1969), Galston and others publicly questioned the US policy of crop destruction "as a weapon of starvation."[67]

Galston's correction of a *Science* piece on the possible concentration of 2,4,5-T and subsequent toxicities ignited an exchange of claims with Michael Newton, an Oregon State University professor, and Logan A. Norris from the US Forest Service, contesting the idea that the phenoxy herbicides should be restricted while under further scientific investigation. Galston replied that he disagreed, suggesting that emerging evidence supported a reexamination of the phenoxy herbicides. He listed several questions, including whether the chemicals themselves, not just the dioxin contaminant, were teratogenic (affecting fetal development but not genetic material), could these substances be degraded into dioxin-like compounds, and what were the possibilities these chemicals could contaminate drinking water. He acknowledged restricting the phenoxy herbicides would be inconvenient but asserted this represented "a relatively small price to pay while we [get] the hard data that we need to protect the health of the public."[68] Reflecting in an essay published in the bioethical journal *Hastings Center Report*, Galston admitted that "many scientists are now coming to feel that they cannot surrender control of their findings to businessman, politicians, or others for indiscriminate and unregulated use in social or military conflicts."[69] The loose network of antiherbicide scientists also sought to reach the general public through books and public testimonies.

Galston and other antiherbicide scientists contributed to collections like *Ecocide in Indochina: The Ecology of War* (1970), an edited volume of essays previously published in other venues, edited by Barry Weisberg. As noted in the book's preface, the US military was "fighting the enemy on his own terrain," and one strategy of this war was to destroy that terrain, that the warfare in Indochina had gone beyond deterrence or counterinsurgency to one of total destruction. The book popularized a term credited to Galston, the idea that if genocide was the planned eradication of a people, then *ecocide* was the systematic destruction of a country's ecosystem. *Ecocide* included essays like Jean-Paul Sartre's "On Genocide," Pfeiffer's "From Ecological Effects of the Vietnam War," and Westing's "From Poisoning Plants for Peace." Along with *Harvest of Death: Chemical Warfare in Vietnam and Cambodia* (1972), scientists used these books to raise concerns about the long-term effects of herbicide defoliation and exposure to the South Vietnamese landscape and society.

Beyond influencing the US government to stop defoliation missions in South Vietnam, American scientists offered critiques that formed the nucleus of protests within global scientific and pacifist communities. Here, noted philosophers Bertrand Russell and Jean-Paul Sartre led one of the first international tribunals to decide whether the United States was guilty of war crimes, with American defoliation missions offered as evidence of a wartime atrocity. At the same time, scientists from North Vietnam and East Germany challenged US claims about the safety of the phenoxy herbicides to human health. The next chapter examines the international community's responses to Agent Orange herbicides and their role in US imperialism.

CHAPTER THREE

Ecological Disruption in Vietnam

*International Protests over Crop Destruction,
Defoliation, and Ecological Imperialism*

A weeklong mass demonstration that took place early in 1962 represented one of the first protests over Operation Ranch Hand and its defoliating mission. The protest happened in an area of the South Vietnamese province of Bien Hoa, somewhere by the side of Highway 15, an example of a *dich van*. These popular uprisings, or "action among the enemy," were designed to mobilize the rural countryside, confound the enemy, and culminate in a massive populist uprising to overthrow the Ngo Dinh Diem regime. Although the numbers of those participating were often inflated, the demonstrations revealed an overall strategy and spirit of resistance. They also show that from the beginning of their use in the war, the North Vietnamese government, the National Liberation Front (NLF), and international observers criticized the United States' use of chemical weapons in the Vietnamese conflict. Early critiques charged that the defoliation missions sprayed poison chemicals and destroyed crops. After repeated assertions from the US government and military that the herbicides presented no harm to humans and animals, criticism shifted to the potential health hazards and actual ecological harm being done by the chemicals. Questions about the "ecological consequences" of the war had begun and would continue until well after the war had ended.[1]

As the US military began using the phenoxy herbicides to eradicate the jungle, making boundaries, supply lines, and troops visible in the process, voices in North and South Vietnam began criticizing the defoliation missions. They were joined by a growing body of concerned international observers. One way the NLF attempted to discredit the war appeared within the discourse it generated around defoliation and the US government's defense of the chemical defoliants. These efforts complemented the growing international attention to and condemnation of crop destruction and defoliation. International censure of defoliation can be measured in two different ways. In the first cases examined here, international conferences directly addressed the problem with chemical defoliation with scientific presentations of ongoing research. Another means of gauging international arguments against chemical defoliation can be seen in the critiques offered by two of the more visible individual scientists involved in international protest, one from East Germany and one from North Vietnam. Their arguments offered a broader understanding of American imperialism and ecological and health hazards posed by the chemicals. Both scientists were based in communist countries, but they succeeded in making their concerns, critiques, and challenges about and to Agent Orange herbicides known to the global community. Finally, an international conference held in 1983 demonstrated the ongoing effects of, and concerns about, chemical contamination in Vietnam in the aftermath of the war.[2]

The NFL identified the herbicides as the best-known in the chemical arsenal that they claimed the US was spraying on jungles, fields, and people. As such, reports of poisonous gases and the early conflation of herbicide spraying with gas attacks remain difficult to disentangle. The embedded nature of defoliation protests within broader propaganda against chemical warfare demonstrates the link between defoliation as a tactic of containment—both of unruly jungles and of peasants—as well as an actuality that caused fear, anxiety, crop losses, and societal and ecological disruptions. Some of the earliest Vietnamese protests appeared in the form of *dich van* movements held against defoliation, the strategic hamlet program, and chemical warfare more broadly in 1962, 1963, and 1964.

In his study of the NLF, historian Douglas Pike notes that a *dich van* did not depend on the numbers of people mobilized. He gives the example of one of the most effective *dich vans*, led by a peasant woman who fought against the governor of Ben Tre (Kien Hoa) province. The woman led over 380 demonstrations before she was beaten to death in May 1963. Pike provided an example of one demonstration:

Woman: How many kinds of America are there in the world?

Governor: Only one kind, the United States of America. It has granted aid to Vietnam and belongs to the Free World.

Woman (in a firm voice): In my opinion there are two kinds of America. Disinterested America is good. But the America that has brought bombs and guns to South Vietnam and spread noxious chemicals to massacre the South Vietnamese people is as bad as a mad dog.

Hardly had she finished when other demonstrators shouted "Down with America for spraying poison chemicals."[3]

For many protesting peasants, defoliating herbicides were one of several chemical poisons used by the United States, which was guilty of waging chemical warfare, according to the NLF and its allies.

Early propaganda produced by the NLF characterized defoliation as a part of the US chemical war against militants, attacks that harmed local peoples as well. Some literature charged that the US sprayed other kinds of poisonous chemicals under the guise of its defoliating missions. "Everywhere, after the passage of US planes, the same scene of desolation is seen: rice turning yellow, banana-trees, coco-trees and other fruit-trees withered, poultry, fish dying, women, children, old and sick people affected by colic, diarrhoea, vomiting and often frightful burns."[4] Multiple publications contained the same charges, even using the same text, suggesting that they were a routine part of NLF propaganda.

These charges were echoed in the global communist press, with articles like Wilfred Burnett's "South Viet-Nam: War against the Trees." A Cuban stamp, produced as part of a series called "Genocide in Vietnam," criticized defoliation missions with images of ill and dying Vietnamese. Officials issued condemnations of the "imperialist poison war" as noted by historian David Zierler. Another 1966 "war crime" pamphlet charged, correctly, that the US spraying missions were expanding in geographic area and the amount of chemicals being sprayed, claiming that concentrated solutions of 2,4-D and 2,4,5-T "have a harmful effect on human beings, animals, and vegetation."[5] Quoted in same pamphlet, American journalist Robert Smight supported the position that these herbicides represented a health hazard. "If we try to make ourselves believe that weed-killer, in a quantity sufficient to kill jungle growth, is not going to harm a living soul, then we are just as stupid as our hired propagandists believe us to be."[6] In response, the US military asserted the safety of the chemicals, producing propaganda materials of their own.[7]

The US military and the Diem government worked to convince rural Vietnamese regarding the safety of the herbicides and that they would be paid for their crops accidentally damaged. Historian Edwin Martini describes the extensive efforts made by the Army of the Republic of Vietnam (ARVN) and the US military to educate the Vietnamese populace about the herbicide program. These included village visits by special teams where members applied small amounts of Agent Orange to their skin to demonstrate the harmless nature of the chemicals and informed villagers of reparations programs for destroyed crops. The military also produced its own pamphlets featuring "Brother Nam." The booklets addressed a whole host of NLF propaganda, including the safety of defoliation chemicals and assurances that farmers would receive financial compensation for any crops destroyed. Martini notes that NLF propaganda that had emphasized the toxicity of herbicide spraying lost believers as villagers realized that in the short term the chemicals appeared not to cause immediate harm, although later interviews done by RAND researchers implied that rumors of herbicide toxicity persisted among villagers. Later assessment of the crop destruction program done by RAND Corporation consultants also suggested that education campaigns on herbicide safety were not always pursued consistently or effectively.[8]

The damage done to crops, however, remained a contentious issue. Writing in the introduction to an international commission's investigations into US war crimes, former International Voluntary Services director Don Luce described one encounter with defoliation. Luce was working with local farmers on an island in the Mekong River to develop new varieties of watermelon when the entire island was sprayed by an American plane on a defoliation mission. The herbicides destroyed the farmers' watermelon crops, worth at the very least $10,000 and their sole income. When American authorities were approached for reimbursement, Luce and the farmers were told, "The whole damn country is not worth ten thousand dollars."[9] Other Vietnamese planters found themselves in the same situation. Rife with corruption, the purported compensation program for the loss of subsistence and cash crops such as rubber trees won the US no converts. US forester Jay Cravens became the designated USAID herbicide claims investigator late in 1967 and saw firsthand the ill will generated over damaged crops. He observed, "A serious side effect of this destruction was the embitterment of farmers who were loyal to the South Vietnam government. Many became supporters of the Viet Cong."[10]

This bitterness caused a reaction exactly opposite to the US's stated intentions. The defoliation engendered enough controversy that Truong Dinh Dzu, the 1967 civilian candidate for the South Vietnamese presidency, promised to seek the cessation of the defoliation program. Destroyed crops and denuded forests, however, did "encourage" rural peasants to relocate to designated spaces. Although the NLF may have failed to persuade rural villagers about herbicides' health hazards, using the chemicals to destroy food crops in a part of the world that routinely experienced famine drew the censure of the international community.[11]

Mirroring domestic protests in America, international condemnation of the Vietnam War increased after 1966, and chemical warfare, including herbicides, became a routine part of those critiques. Initially, American allies publicly endorsed the war, even as many expressed private concerns about the wisdom of pursuing military action in Southeast Asia. Plans by military leaders and the Lyndon B. Johnson administration to accelerate the war in 1964 were held in check, at least partially, by British prime minister H. Wilson's refusal to endorse escalation of the war. Historian Fredrik Logevall has argued that these considerations resulted in the administration's decision to pursue a policy of "graduated pressure." International criticism of the war fell into two phases. The first phase, from approximately 1967 to 1969, focused chiefly on the herbicides' use in crop destruction. While Agent Orange was used to destroy some crops, it mostly removed jungle cover. Crops were exterminated using Agents Blue and White. Agent Blue's active ingredient, cacodylic acid, provoked sharp criticism as it was an arsenic derivative and posed significant contamination risks. Agent White combined a new herbicide, picloram, and the Agent Orange component 2,4-D in a 1:4 combination. Starting in 1969, international concern had shifted to the massive amounts of forests that had been sprayed. Both included charges of the chemicals' hazardous health effects.[12]

Barred from sending more troops, the Johnson administration applied different kinds of pressure, which may have included increasing the numbers of acres sprayed and killing more crops and forests. When some RAND Corporation research questioned the support of South Vietnamese peasants for the United States or the war, Johnson administration policy makers often ignored the reports. Policy makers and military officials listened only when the researchers' findings supported their own agenda. Findings by some RAND scientists that crop destruction and defoliation hurt the NLF by denying them food and jungle cover encouraged the expansion of the program and justified intensifying other war efforts with the thought that the NLF was tottering. This commitment to the war came despite other interviews that suggested civilians suffered the most from the defoliation program, and that Viet Cong forces continued to find food. The role played by the Rand Corporation merits further discussion.[13]

The RAND Corporation originated in 1945 and became an integral part of the Cold War military research with the recognition that leadership sometimes needed information and analysis best produced by the private sector. Many might consider it an exemplar of the military-industrial complex President Dwight Eisenhower had warned about in his 1957 speech. By the 1960s RAND had become a major consultant to the US military on the war in Vietnam. Analyst Leon Gouré's positive assessment of American air attacks, including crop destruction, and their negative effect on insurgent morale gave military commanders confidence to pursue the war. Other RAND analysts—Russell Betts, Frank Denton, and Anthony Russo—challenged the effectiveness of the crop destruction program as early as 1967, in two separate reports. Betts and Denton's assessment of the crop destruction program "concluded that the program was ineffective," and they argued that the program had a negative effect on peasant attitudes toward the South Vietnamese government (GVN) and the United States. To the civilian population, the spraying program destroyed farmers' livelihood, caused fears of chemical poisoning (based on rumor and firsthand experience), and proved the GVN's disregard for ordinary Vietnamese people. They stopped short of recommending the program's discontinuation. Russo did not. His assessment of the crop destruction project ignited a firestorm.[14]

Russo's experience in the field, and the negative reception of his work, eventually transformed his understandings of the war. Unlike the Betts and Denton study, Russo's assessment of the crop destruction program included more data sets than just interviews with NLF defectors and prisoners of war. His empirical model showed a close correlation between NLF forces and the rice economy, and that the destruction of rice crops had minimal effect on enemy troops. Russo provided one example of one of the most heavily sprayed areas in 1966, with 23 percent of crops eradicated, but only a 5 percent decrease in rice rations. Almost as important, civilians experienced significant and harmful effects from crop destruction, as they lost as much as a hundred pounds of rice for every pound denied to the NLF. Russo ended his report by saying, "The program should be taken under serious review; based on the analysis presented here and on opinions shaped by field experience in South Vietnam, the author's feeling is that the program should be discontinued."[15] Russo later claimed that when Secretary of Defense Robert McNamara was told the report's findings, he asked military command why such an ineffective program had lasted so long. The internal disagreement among RAND researchers mirrored the broader national debate, and the "contrary points of view" supported very different outcomes with respect to the war.[16] The controversy fought within RAND, military, and administrative circles regarding the morality and effectiveness of the crop destruction resonated within the broader international community as well. RAND fired Russo in 1968.[17]

⁂

One of the first international gatherings condemning the United States' actions in South Vietnam cataloged defoliation as a war crime. The atrocities of the twentieth century, which began with the Armenian genocide and ended with the Bosnian, coupled with the failure of national governments to hold themselves responsible, led to international tribunals designed to establish and uphold international laws. An "International Scientific Symposium" that took place in Peking in 1964 played a momentous part in the appearance of one of the most famous of these international bodies. A report given at the end of the gathering, "The Use of Toxic Chemicals as a War Means by the US Imperialists and Their Stooges," indicted American military use of tear gas and defoliants. It also led to the formation of the International Vietnam War Crimes Tribunal. Held under the auspices of the Bertrand Russell Peace Foundation (BRPF) and privately funded (with a large contribution from North Vietnam), the International War Crimes Tribunal took place in 1967 at two Scandinavian locations. It was headed by the philosophers and activists Bertrand Russell and Jean-Paul Sartre. Both Russell and Sartre espoused radical politics, but neither was an unthinking shill for the Soviet Union or communism. Both men vehemently opposed the war in Vietnam and spoke out strongly against American imperialism.[18]

The Russell Tribunal offered supportive space for political and scientific critiques of the US imperialist war. Leftist activists from around the globe gathered to hear evidence of various war crimes committed by the United States in South Vietnam. The tribunal was mostly ignored in the United States, although individual citizens attended; National Security Affairs advisor Walt Rostow represented President Johnson. The tribunal attracted some controversy over its decision to report only non-Communist atrocities, a concern voiced by radical antiwar activist Staughton Lynd and other prominent African leaders who withdrew from the BRPF and the tribunal. The second tribunal session, held in Roskilde (Copenhagen), Denmark, included a series of presentations that featured evidence on chemical warfare, including the US defoliation program. Scientists' tribunal contributions identified crop destruction, and the subsequent starvation, chemicals' toxic effects, and large-scale ecological damage as the major problems associated with the "weed killers" used for defoliation. These testimonies were based mostly upon individuals' and committees' scientific expertise.[19]

Physician and antinuclear activist Abraham Behar addressed the charge given to answer whether "new or forbidden arms [had] been used in Vietnam." The committee had identified three kinds of weapons: incendiary, defoliants, and poison gases. Behar characterized the "weed killers" used in South Vietnam as poisons designed for crop destruction. He acknowledged that ruining crops during war was not banned by international law. He asserted, however, that the chemical herbicides were still toxic substances capable of causing harm. Picking up on this theme, French pediatrician Alexandre Minkowski emphasized the ways that crop destruction and the subsequent food scarcity harmed children the most severely. Along with the plant crops destroyed by spraying missions, Minkowski cited studies that suggested 2,4-D was especially toxic for fish, a significant problem given the ubiquitous presence of that protein in the Vietnamese diet.[20]

The findings of several scientific working groups were also presented at the Copenhagen session. Chemist Edgar Lederer gave the subcommittee report on chemical warfare in Vietnam, compiled by his French colleagues, a group of biochemists and nutritionists. The section devoted to defoliation and crop destruction began by noting a 1965 article published in *Le Monde* that use of chemical agents in Vietnam had attracted the censure of US scientists. Lederer's testimony covered the military rationale for defoliation, the kinds of chemical agents used, the need for greater quantities of defoliants (both because of poor jungle growth eradication and because of expansion of the program), the problems with famine and crop compensation, and the toxicity of the chemicals. Based on Vietnamese sources, the report estimated 320,000 acres of crops had been destroyed in 1963 and had more than doubled to 700,000 acres by 1965. The subcommittee also discussed the ways chemical defoliation harmed the equilibrium of the human environment. Farmers hesitated to plant crops, even when reassured they would receive compensation for accidental exposures. Drifting chemical sprays had put rubber tree plantations at serious risk. The destruction of forests and crops harmed humans through famine. The report made a more tenuous connection between environmental destruction and increases in the already endemic diseases of plague and malaria. The report also challenged US State Department claims of chemical safety, providing information from Dow Chemical's package instructions, which listed several cautions for users. Lederer concluded his testimony by noting "there can be no doubt that defoliation of the forests, the jungle and the bush are already having dangerous repercussions on the condition of the human environment. . . . The use of chemical warfare is in danger of provoking wholly unsuspected biological devastation in the very near future."[21]

The second testimony presented excerpts taken from the two different Japanese Scientific Committees' assessment of the effects of agricultural chemicals in Vietnam, which primarily detailed the damage done to vegetables. Japanese scientists used the observations of Vietnamese civilians, as the report acknowledged the difficulty of obtaining samples and verifying the aftereffects of the chemicals. Among the food crops studied were tapioca, taros, papayas, bananas, water plants (used for pig fodder), pumpkins and snake gourds, various spice crops, and sweet potatoes. In 1967 Japanese researchers performed an experiment done with "excessive" farm chemicals in Japan. When a 1/100th solution of 2,4-D was applied to sweet potatoes, they withered in a manner consistent with reports from Vietnamese sources. The Japanese scientists did other experiments on soybean crops, rice plants, and weeds. All died, although the withering process varied according to species. The Japanese report discussed the harmful effects of arsenic extensively (Agent Blue, which contained no phenoxy herbicides). Professor Yoichi Fukushima, a member of the Japan Science Council, noted the amounts of chemicals sprayed amounted to ten times the quantity used in Japanese agriculture.[22]

The second session summary condemned the use of defoliants for crop destruction, noting that US Army manuals enjoined such action, especially using chemical agents—even those theoretically nonharmful to man—any crops that are not intended to be used exclusively for the food of the armed forces. The Russell Tribunal offered one of the first international forums where defoliation and crop destruction were condemned. The testimonies on defoliation, along with those on civilian bombings and forced relocation (aided by defoliation), all proved compelling enough that tribunal members judged the US guilty of carrying out genocide. Students for a Democratic Society president Carl Oglesby described the deliberations that led to the declaration of genocide; these included the charges of defoliation and crop destruction among the criteria by which extermination was determined. The tribunal's judgment would be echoed by contemporary and later critics of the war.[23]

International committees continued to condemn chemical herbicides as a part of their broader critiques of the war. In 1970 the Orsay International Conference, held under the auspices of the World Federation of Scientific

Workers, met and brought together international scientists like Lederer with outspoken American scientists E. W. "Bert" Pfeiffer and Arthur Westing to speak out against the war and the use of herbicides. In a news item highlighting the conference, it was noted that the US had lit an "ecological time fuse" that would "reverberate down the life chain" as plants and animals were destroyed with unforeseeable consequences.[24]

As early as 1966, a Swedish peace group invited international peace organizations concerned about the Vietnam War to meet in Sweden so they could coordinate their efforts. Sweden became home to several leftist groups speaking out against the Vietnam War, with some actively supporting the NLF. Noted Swedish sociologist Gunnar Myrdal led one of the groups, an umbrella organization known as the Swedish Vietnam Committee, considered centrist liberal. The Young Philosophers attracted young Swedes; the New Left, focused on education and discussion. The NLF committees were the most dedicated to the North Vietnamese cause, using public agitation and propaganda tactics, which garnered donations of over $240,000 that were sent to the NLF. It was in this political environment that several meetings took place from 1967 through 1975, later known as the Stockholm Conferences on Vietnam (although three of the eight meetings took place elsewhere). They led in turn to the establishment of the International Commission of Enquiry into US Crimes in Indochina, cochaired by Myrdal.[25]

The International Commission held three consecutive conferences that met in Stockholm in 1970 and 1972, and Oslo in 1971; the conference proceedings of the 1971 and 1972 sessions were published in English. All the meetings provided global forums that included defoliation and crop destruction as attacks on the natural environment, imperialist actions that qualified as war crimes. The conferences were not limited to Scandinavian critics, but like the Russell Tribunal, served as a visible and central location for the international pacifist community to gather. While the last two inquiries occurred after the Nixon administration had officially ended defoliation missions, but before the official end of the war, criticism of the US defoliation program continued. In his introduction to the report of the 1971 commission, Richard Falk indicted the US military's counterinsurgency that resulted in the crime of genocide, warfare that included "crop denial programs" as a part of its arsenal.[26]

Using onsite inspections, written reports, and victims' stories, the 1971 report focused on the effects of chemical warfare in a section "Herbicides." The report described the effects of herbicides on South Vietnam's mangrove and hardwood forests, the forests' bird population, and domestic animals. "Millions of hectares of cropland and forest have been destroyed—in some areas perhaps permanently." Fruit trees like the custard apple and papaya appeared to be grotesquely malformed, and that defoliation proved to be "very effective against subsistence farmer populations of the sort that we have in Indochina."[27] The effects of repeated sprayings, which had occurred in several areas, remained to be assessed. The report included the symptoms exposed individuals experienced, with skin and respiratory irritation appearing first, and then later gastrointestinal effects such as vomiting and diarrhea. Also included was evidence of the chemicals' effects on pregnancy, with three out of four demonstrating abnormal births. Chromosomal tests showed damage, most likely due to the phenoxy herbicide 2,4,5-T, known to affect chromosomes. The investigative team also met with Professor Tùng Tien, a North Vietnamese physician who presented his data on herbicide exposure and its effects on pregnant women.

In the final judgment of the investigative team, the United States had engaged in chemical warfare, spraying "more than six pounds of herbicides per man, woman, and child in South Vietnam." Given the clear evidence of environmental damage, and preliminary evidence of the harmful effects on human health, the Commission condemned the use of the Agent Orange herbicides. "We are of the opinion that the use of herbicides in warfare and on this scale may seriously affect the Vietnamese people, their environment, and the society as a cultural unit for decades to come."[28]

Although the United Nations (UN) Conference on the Human Environment in Stockholm 1972 did not address the Vietnam War, several events held concurrently did. The third international commission held gave the strongest condemnation of defoliation as an attack on South Vietnam's people and environment. The International Commission of Enquiry into US Crimes in Indochina held a special session to meet at the same time as the 1972 UN Conference. The International Commission published its June 2–4 conference proceedings as *The Effects of Modern Weapons on the Human Environment in Indochina*. Of the ten topics presented, over half dealt in part or wholly with defoliation. As American speaker Westing noted, "a growing number of scientists throughout the world, have become appalled by the long-range ecological impact of so-called limited counterinsurgency warfare."[29] Westing lamented the fact that the concurrent UN conference failed to include any examination of the Indochina War and its devastating effects on the South Vietnamese environment. One striking aspect of Westing's accounting of the war's ecological effects was the extensive literature he cited in writing his report, which also included discussions of Roman plows and land cratering. Westing also participated in another event on ecocidal warfare sponsored by Dai Dong, the transnational organization founded by the US pacifist Fellowship of Reconciliation. This one-day symposium also featured Luce, Falk, and other critics of the war's effects on the natural and human landscapes.[30]

Echoing earlier charges, political scientist John E. Fried focused on the ways that modern weapons had enacted an "ecocide" against the South Vietnamese environment. He claimed that the intentional ecological harm was compounded by other nonecological weapons of war as well as the chaotic conditions created during wartime. Fried also wondered what might be the possible legal ramifications of ecocide, which while not yet recognized as a legal concept, might come to be accepted in the same way as genocide.[31]

The role played by information generated by American sources—media, physicians, scientists—appeared prominently in communist critiques offered by North Vietnamese officials, war crime tribunals, and international commissions' public condemnation of American imperialism. These public condemnations included defoliation and herbicides as essential parts of American Cold War imperial efforts. The ecological disruption caused by herbicides significantly contributed to the social disruption of South Vietnam as well. Crop destruction, forest defoliation, and faulty reparations programs pressured ordinary Vietnamese peasants to relocate, to leave mixed agricultural spaces and move to "strategic hamlets." Unlike the charges made regarding poison gas, the harmful effects of the Agent Orange herbicides could be only minimized, not denied. And the work of American scientists gave these critics evidence that they amplified through tribunals, commissions, and validation of Eastern Bloc countries' charges of new warfare tactics that violated international agreements such as the 1925 Geneva Convention. International protests also emerged in the work of Edgar Lederer and his French colleagues and the Japanese scientific committees. Moving beyond international tribunals and commissions, the work and writings of two scientists, one from East Germany and one from North Vietnam, offer a different lens on international protest. Both men gained international visibility in their campaigns against the chemical herbicides used in crop destruction and defoliation missions.

Within the global Cold War conflict, the NLF aligned itself with Communist Bloc nations, which in turn supported the Vietnamese and their protests over the war and, more specifically, defoliation. Although lacking many of the characteristics and accoutrements of statehood, by 1965 the NLF had positioned itself as the legitimate government in Vietnam. It achieved this status, in part, through its relationships with other communist countries and organizations. A random sampling of how many times a country's name appeared in various NLF radio and print media provides one way to evaluate the importance the NLF assigned to other foreign nations. This national ranking shows the German Democratic Republic (GDR) in the fifth position, after Indonesia, Cuba, Japan, and North Korea, and ahead of other European nations like Czechoslovakia and Poland. From the beginning, GDR support for the NLF involved health-care efforts.[32]

Early in the war, the GDR Red Cross donated medical supplies, while the minister of public health met the NLF delegation, headed by Nguyen Van Hieu, which visited East Berlin in August 1962. Over the next few years, various East German groups donated cash and medical supplies to the North Vietnamese government. By 1964 over one million marks in cash and the monetary equivalent of 500,000 marks in medical supplies had been donated, and fifty NLF military personnel had been treated in East Germany. Major contributions had come from the Afro-Asian People's Solidarity Committee of the GDR, considered a Communist-front organization. These committees appeared in countries around the globe and provided significant support for NLF externalization efforts, including hosting visiting diplomatic missions. The GDR Afro-Asian People's Solidarity Committee provided other services in the fight for a reunited Vietnam in the form of publications decrying American imperialism.[33]

In sync with Western scientists who had begun to voice concerns about the ecological harm of 2,4-D and 2,4,5-T, Professor Gerhard Grümmer began speaking and writing about the potential hazards of herbicide spraying by the late 1960s. Born in 1926, Grümmer became a professor of plant physiology in 1958 and taught at the University of Griefswald, and later at the University of Rostuck. Grümmer served in World War II in the German navy until he was captured and became a British prisoner of war. After the war Grümmer attained his PhD at the University of Jena, where he continued to work. Grümmer held a visiting professorship in India and later published on rice cultivation. In 1969 Grümmer's *Herbicides in Vietnam* appeared in English, clearly aimed at an international audience. *Genocide with Herbicides* followed in 1971. Other critiques of herbicide spraying appeared in French and German.

Figure 5. East German plant physiologist Gerhard Grümmer published several booklets protesting the use of defoliants in South Vietnam and America's imperialist war, including *Herbicides in Vietnam* based on his trip to North and South Vietnam. Copy in author's possession.

Interviewed by a reporter from *National-Zeitung* in December 1970, Grümmer discussed his visit to North and South Vietnam. Invited by the Democratic Republic of Vietnam (North Vietnam) Investigating Committee, Grümmer went to "gather information about the criminal use of chemicals, notably herbicides, to introduce measures to protect human, animal and plant life and to inform the world public about the true extent of toxic warfare in that country."[34] Grümmer identified two ways the GDR could help in solving the problems caused by herbicides. Along with "collecting material and facts to indict the barbaric US crimes," Grümmer saw an important role in helping to train agricultural experts to educate Vietnamese farmers in dealing with "the protection of agricultural and forestry production."[35]

Although his politics predisposed Grümmer to view the war as imperialist subjugation, his scientific training focused his attention of the herbicide spraying program and other chemical weapons. In *Herbicides in Vietnam*, Grümmer presented a comprehensive discussion of the plant growth chemicals and arsenical compounds, like cacodylic acid being used for crop destruction and defoliation. He noted the phenoxy herbicides' origins as weed killers, their development during WWII as potential chemical weapons to destroy Japanese rice crops, and their current use in South Vietnam. Grümmer quoted from testimony presented at the Russell War Crimes Tribunal. His accounts documented multiple spraying missions from 1965 through 1967 from both scientists studying the effects and NLF members sharing their experiences. The booklet included pictures of destroyed plants, of troops marching through defoliated terrain, and of barren fields.[36]

Grümmer gave a history of the chemicals' use thus far in the war and their harmful effects on forests, food and commercial crops, and insect and animal life. Of perhaps even greater concern were the chemicals' detrimental effects on South Vietnam's aquatic culture. Fish represented a major source of the country's consumed protein, and studies showed fish were damaged at the concentrations seen in South Vietnamese rivers. Grümmer also examined the relationship between the herbicide spraying program and military-industrial-academic complex, citing the example of activities at the University of Pennsylvania previously discussed, and various research institutes that played a part in producing the technology needed for the spraying program. *Herbicides in Vietnam* demonstrated the ways information about herbicides had spread beyond American sources and permeated the international discourse. It also marked a transition in that discourse from one focused on crop destruction to more general concerns regarding defoliation and its harmful effects on ecosystems.[37]

Writing in *Genocide with Herbicides*, Grümmer specifically addressed the defoliation of the tropical jungle. Grümmer argued that "the forests which are being destroyed now in this manner are lost beyond recall."[38] He cited examples from Java and Cambodia of complete vegetative eradication that led to replacement vegetation and a loss of mineral-rich soil. More than this devastation, Grümmer warned about the ways herbicidal warfare would transform postwar Vietnam. The loss of protective forest cover meant increased flooding, which in turn affected food supplies. Flooding also affected Vietnam's water economy. According to Grümmer, the disrupted ecology led to increased malaria and other contagious diseases. Eradication of forests also destroyed indigenous medicinal plants. Grümmer blamed the increase in mosquitoes (and thus malaria) on the use of herbicides followed by bombs that created trapped pockets of water.[39]

Grümmer's examinations broadened the case against herbicides by extending the scope of the harmful effects on the natural world beyond crops and forests. He made explicit the links between destruction of the natural world and disruption of the human environment. In Grümmer's condemnation of chemical herbicides, his arguments addressed ongoing, current dilemmas rather than speculating on the long-term damage to South Vietnam's forests, an uncertain endeavor and one difficult to prove definitively. Excerpts from *Genocide with Herbicides* appeared in a 1971 essay collection, *The Truth about U.S. Aggression in Vietnam: GDR Authors Unmask Imperialist Crimes*. Most of the essays condemned the war from a legal standpoint. Grümmer's essay was one of two that addressed chemical defoliation, along with a protest declaration of East German medical scientists. A later publication, Grümmer's account of his 1970 visit to North and South Vietnam, brought home his sense of the wrongs crop destruction and defoliation had inflicted on the people of Vietnam.[40]

Initiated by North Vietnamese officials, Grümmer's 1970 visit was envisioned, at least in part, as a mission to "inform the world public about the true extent of toxic warfare in that country."[41] To do this, Grümmer needed to assess conditions on the ground. Grümmer's visit was also planned to recommend relief countermeasures for the herbicide spraying. In the section "Crimes against the Forest," Grümmer focused on the harmful environmental effects of chemical herbicides. Examining an area in the fourth fighting zone, Grümmer lamented the bleak landscape he surveyed. The area had been the target of spraying missions for three years, which had "killed off trees and shrubs, orchids and ferns, herbs, and moss—everything."[42] Grümmer continued his description, noting a mostly barren landscape, with barely enough grass to feed a goat. The hardwood shrubs that had appeared were insufficient to prevent soil erosion, with naked chalk exposed and rich loam washed to the sea. Grümmer described the outcomes of many years of chemical spraying. He noted that the American military had investigated, like farmers, the ideal conditions to kill plants. "This 'experimental activity' earned the American pilots a special nickname—ranch hands. Do they help agriculture? No—destruction is their job, ruthless destruction of huge forest regions and valuable plant life."[43] Grümmer noted that eventually Agent Orange became the chemical weapon of choice.

During his trip Grümmer met with various North Vietnamese experts and officials who presented evidence of the harmful effects of herbicides. Agronomists showed him the defoliated forests, doctors brought him victims of chemical exposure, and party officials showed him the effects of herbicides on sensitive plants. Grümmer made the case that he needed to visit South Vietnam, the real epicenter of herbicide spraying. Grümmer convinced his NLF hosts that to do his work properly, he must take risks: he must evaluate conditions in South Vietnam despite weather and military obstacles. Traveling in a camouflaged car, Grümmer visited a fishing village at the mouth of the Ben Hai. Once there Grümmer observed fishing boats and US naval forces, the Seventh Fleet, off the coast. Here, heavy spraying helped US artillery attacks. Upon his return to Vinh Linh, Grümmer met "two victims of chemical warfare. . . . An old peasant woman and a young man [told] us about their experiences with US herbicides."[44]

Grümmer recounted their stories. The woman, Nguyen Thi Dieu, shared her experience with chemical herbicides. She had been ill for weeks after being sprayed during a defoliation mission in August 1970. The mist had killed breadfruit trees and tea plants, and damaged banana trees in the family garden, suggesting that the spray was chemical herbicides. Grümmer quoted a figure provided by an NLF colonel, claiming that 1.3 million people in North and South Vietnam had suffered serious health problems after being exposed to herbicides. Grümmer, looking at the "haggard" face of the woman before him, saw thousands more represented in her "listless appearance and feeble movements," which spoke far more than words to her afflictions. A woman who had been a "strong countrywoman" had become an invalid, a prisoner of her home. The young man, Nguyen van Chau, worked as a field secretary for a cultivating team. In July of 1969 he "felt a prickling in his eyes, mouth and nose, although he had heard no planes." The young man went into "a kind of coma," experiencing periods of unconsciousness until his friends found him and took him to the doctor. He continued to bleed from his nose and mouth and had been diagnosed with an inflamed liver. Once again, the rolled-up leaves, defoliated trees, and swollen plant stems suggested herbicide exposure. "Only grasses and bamboo were not affected. So it must have been Agent Orange again."[45]

At several points in his account Grümmer acknowledged research on the hazards of Agent Orange. In discussing the young man, Grümmer noted the release of a report by an American institute that had identified liver damage caused not by Agent Orange itself but by a dangerous by-product. Although he did not name the institute, Grümmer successfully had alerted his readers to the Bionetics report and the new evidence indicting 2,4,5-T's dioxin contaminant as a lethal chemical. Grümmer finished his account describing the various physical ailments the young man still suffered, including weakness, an inability to work, trouble eating, and an increased sensitivity to cold and damp. Grümmer notes that his conversation with Nguyen van Chau did not go as smoothly as he had written it. The once vigorous worker now struggled to tell his story. Later, Grümmer described a visit to a field hospital where he was given a tour of the encampment. Grümmer met a seriously injured soldier whose unit had been sprayed with Agent Orange. The men all experienced "inflammation of the skin and rashes, damage to the respiratory organs and temporary disablement."[46] All the soldiers recovered except for this one. As in reports of US workers, he proved especially sensitive to the effects of the phenoxy herbicides.[47]

Unlike his other works on the ill consequences of herbicide spraying in Vietnam, *Accusation from the Jungle* provides much more in the way of personal stories and perspectives, both Grümmer's and those of the people he interacted with and observed. In this way, the book offers an emotional plea along with its scientific information. As one of many communist critics of the war, Grümmer's contribution demonstrates the importance of chemical herbicides in the list of wrongs done in Vietnam, a serious crime committed in America's imperial war. Grümmer's condemnations of herbicides mostly stopped after the 1972 publication of his trip's account, well after the US had ceased its spraying missions. He returned to university life until being forced into an early retirement because of a health crisis. It would be more than a decade later that Grümmer would publish *Giftküchen des Teufels* (1988), "Gifts from the Devil's Kitchen," his reflections on chemical warfare. While Grümmer focused primarily on the ecological effects of herbicides on the tropical jungle, another visible critic of US herbicide spraying focused on the chemicals' potential health hazards.[48]

Skill, initiative, and compassion helped a physician lead the medical critique of Agent Orange's health effects. Before he rose to international prominence, North Vietnamese surgeon Tôn Thất Tùng had fought at the 1954 battle of Dien Bien Phu on the side of the Viet Minh against the French colonizers and served as Ho Chi Minh's personal physician, proof of his commitment to the communist cause. Born in 1912, Tùng came from one of the last imperial families and grew up in the North Vietnamese city of Hue. He received his medical education from the Ecole de Médecine de Hanoi, the first Western (French) medical school established in Southeast Asia, and received his surgical training at Hanoi's Phù Doãn Hospital. His promising career as a liver surgeon was interrupted by the war, which he spent leading medical teams in treating patients at Hanoi's "Viet Nam—German Democratic Republic Hospital." Tùng served as a medical ambassador to other communist countries, visiting China, Korea, and the Soviet Union. He also moved within medical circles in the West, in contrast to scientists restricted by Cold War politics.[49]

Between 1960 and 1973, Tùng served as part of the North Vietnamese government as a vice-minister of public health. It was during this period Tùng attended a massive conference of Vietnamese intellectuals protesting the war. It was also during this time that Tùng met a young French doctor and Communist, Jean-Michel Krivine, who was touring North Vietnam as a part of the Russell War Crimes Tribunal. Tùng's political commitment to Communism, military involvement, personal connections, and work as an internationally recognized liver specialist meant that he held a renowned position in Vietnamese society, part of a small number of "privileged intellectuals" often found within communist countries.

The exigencies of war brought new challenges to the physician. While Tùng initially continued his liver surgery research, the pressing need to respond to Agent Orange exposures shifted his focus. Tùng became one of the most reputable figures in decrying the health hazards imposed on the Vietnamese people through their exposure to Agent Orange herbicides.[50]

As early as 1965, Tùng and his scientific and medical associates criticized the US herbicide program and subsequently began focusing their efforts on documenting the herbicides' harmful effects on human beings. By the early 1970s they suspected that the chemicals had caused an increased number of cancers, miscarriages, and birth defects. Hanoi surgeon Le Cao Dai had established an undercover field hospital in the highlands of Central Vietnam, one of the most heavily sprayed regions during the war. The hospital relocated periodically to avoid detection, particularly in response to repeated herbicide sprayings. Dai recounted a story about Tùng's determination to gather data to prove the harmful effects of Agent Orange, or more specifically, its dioxin contaminant. Tùng wanted Dai to send samples of diseased livers to him in Hanoi, at a considerable distance. "He proposed that we organize a line of runners to carry blood and tissue samples from the forest to Hanoi."[51] Dai's resources proved nowhere near adequate to undertake such a task.

The rest of the world learned about Vietnamese scientists' concerns through various exchanges of people and information, either to or from Vietnam. Tùng presented evidence at the 1970 Orsay conference. He corresponded with scientists in France and overseas Vietnamese scientists. He became the leading expert on dioxin in Vietnam, well aware of the Bionetics study and its findings. Tùng also kept in touch with Krivine, and the men and their families became personal friends. His research focus and results received support from activist scientists in the United States. An infamous visit by a celebrity also brought attention to the issue.

Actor and activist Jane Fonda's visit to Hanoi in July 1972 represents one of the most infamous trips to Hanoi. Opposed to the war, Fonda viewed a display created by the Committee for the Denunciation of US War Crimes in Vietnam. The exhibit included examples and photos of the various weapons used, which included a wide array of the various bombs used. Pictures of defoliated forests also appeared in the glass cases. Fonda then met with Tùng and heard about his research linking Agent Orange exposure to birth defects. "We are seeing more and more of these birth defects," Tùng said, and continued, "Yes, I fear you will soon be seeing these things among your own soldiers."[52] Seeing and talking with North Vietnamese officials, including Tùng, convinced Fonda she needed to speak out. She recorded several radio spots that were broadcast to American troops.

Less controversial but a part of the same antiwar movement, George Perera traveled to Hanoi in 1972 as part of an American Friends Service Committee medical relief mission. Here he met Tùng, who discussed the medical communities' concerns about the herbicides' potentially toxic effects. Tùng told Perera that there had been a fivefold increase in liver cancers, an increase he suspected was the result of exposure to dioxin. In October, Tùng testified in Denmark before a "self-styled International Commission of Enquiry into US Crimes in Indochina" that chemical defoliants had resulted in "a current 'explosion' of cancer of the liver in Vietnam."[53] Sponsored by the American Friends Service Commission, Tùng came to the US in the spring of 1979 on a five-week visit. Tùng toured several university campuses, including the University of Chicago Medical School, the University of Oregon, and the University of California, Berkeley. Speaking in French, Tùng talked about his work as a liver surgeon and as a public health official, and his work examining the possible links between Agent Orange dioxin and a variety of medical conditions, including cancer, miscarriages, and birth defects. (I discuss this aspect of Tùng's work in more detail later.)[54]

While in the country, Tùng met with the Veterans Administration Central Office (VACO) Steering Committee, a unit established to address the deluge of claims submitted after a documentary and death had alerted Vietnam veterans about the possibility of harm from Agent Orange exposure (discussed more fully in chapter 8). VACO committee minutes mention the meeting with Tùng and were quick to dismiss the validity of his scientific claims. Tùng himself acknowledged that conditions in Vietnam during and after the war had made it difficult at best, nearly impossible at worst, to properly set up studies to measure dioxin exposure levels in the population. But he presented strong evidence of harm when he compared the kinds of illnesses being seen in Vietnam and among Vietnam veterans in the US as well as from other countries that sent troops, such as Australia and New Zealand. Despite these dismissals, Tùng's and his colleagues' research received validation, in part because it appeared in so many forums, from French newspapers to the "Letters to the Editor" section of the *New York Times*.

Tùng's position as a respected scientist, his status within Communist circles, and key alliances meant that his opinions and research garnered attention. Interviews with Tùng first appeared within the Communist press. Tùng published an essay, "The Vietnam War and Man's Conscience," in the Vietnamese paper *Nhân Dân* in 1972. The essay denounced the United States' all-out war in Vietnam as having "assumed the position of a giant wrestling with a child. . . . [The US had] mobilized 57 universities and technological institutes to study the problems of destroying man, and it chose Vietnam as its testing-ground."[55] Included in Tùng's litany of US crimes was the destruction of the human environment in Vietnam. Tùng connected exposure to 2,4,5-T with ill health effects, and noted that he and his colleagues had "gathered a group of well-known scientists from many countries to study this harmful effect."[56] A 1974 newspaper article in the Communist *Sun* described his work with liver cancer and quoted Tùng's condemnation of the use of 2,4,5-T in Indochina, especially since it was restricted in the US after 1970. The piece detailed Tùng's work with Harvard biochemist Matthew Meselson in developing a means of measuring minute amounts of dioxin in tissue samples.

A 1975 French report on biological warfare and ecological disruption in Vietnam focused on the condition of Vietnam at the end of the war. It quoted from an October 30, 1975, *Le Monde* article that featured Tùng's claims that liver cancers had increased threefold and told of Tùng's and his colleagues' efforts to document dioxin in the livers. Along with Meselson's help in gathering data, Yale physiologist Arthur Galston publicized Tùng's findings throughout American and Canadian newspapers. In a 1975 essay carried by the United Press International, Galston discussed Tùng's work and showed his support. Even after news reports appeared discrediting Tùng, Galston professed faith in his research. Decades after the war, Galston reminded the public about this important yet ignored body of research that existed, research that might help American veterans, in a letter to the editor of the *New York Times*. Tùng continued to speak out. His research, controversial as it was, represented one of the best indicators of the hazards of Agent Orange exposure, and he was one of the main forces in organizing an international conference to discuss the continuing effects of Agent Orange herbicides.[57]

The Vietnamese government formed the 10-80 Committee (named after the date of its formation in October of 1980) with the intent of evaluating the costs of the Vietnam War. The committee began organizing an international scientific symposium, which was eventually held in Ho Chi Minh City in January 1983. The theme was "Herbicides and Defoliants in War: Long-Term Effects on Man and Nature." Sadly, Tùng did not attend, as he had died the previous year from a heart attack. Reading the conference's final report reveals the major goals and concerns of the Vietnamese scientific community and government. The report's introduction emphasized the 160 scientists and members of international organizations who had attended the conference, and the breadth of representation, from over twenty-one countries. It also took pains to note that the scientists came from the West as well as the East, and that the symposium was a working, dedicated gathering of scientific experts. The conference's stated goals revealed the limited status of information.[58]

The symposium sought to review and evaluate the existing scientific literature on the long-term effects of herbicide ecological and physiological exposure; to identify future research needed; and to establish international scientific cooperation. Of the papers presented, two-thirds focused on the herbicides' continued damage to South Vietnam's environment, like the mangrove forests of the Cau Mau peninsula, one of the most heavily sprayed areas. The conference represented the ongoing international critiques of Agent Orange herbicides and continued issue of the effects of chemical exposure. It sought to bring Vietnamese concerns, and science, to the international community. "Scientists attending the symposium highly valued the contribution made by Vietnamese scientists who, despite the limited facilities and other difficulties during and after the war, were able to overcome these problems and made valuable research contributions."[59]

The US herbicide spraying campaign was meant to expose the movement of supplies and troops, delineate boundaries, and relocate peasant peoples. These goals generated increasing criticism in the various international protests of the Vietnam War. The convening of the Russell Tribunal marked the appearance of a new kind of international activism, providing a forum for the many opponents of the imperialist American war. In the process, herbicide spraying became a common component of a multitude of international commissions and forums. Early voices protested the use of herbicides to destroy crops, with later critiques raising concerns about environmental devastation, and finally the potential human health hazards. Individual scientists built on their colleagues' research and in the process created a network of activism that spanned continents. East German plant biologist Grümmer published books translated in several languages and wrote a moving account of his journey to see the harm done to the jungle and the people that lived there. Tùng, a loyal party member, worked to document the increases in cancers and birth defects he suspected resulted from the exposure to a known teratogen and possible carcinogen.

The international protests occurred during a moment of changing consciousness. *Silent Spring* appeared at the same time the herbicide spraying program began in South Vietnam. The book represented the culmination of new understandings of the natural world, the harm being inflicted by humans, and growing uncertainties about what the relationship between human beings and the natural world should be, even as it provoked new discussions. All of this led ordinary citizens to question the decisions being made by various government agencies.

In his review of *Harvest of Death*, an essay collection about chemical warfare, ecologist Frank Egler observed that the book was already outdated, published after the banning of most of the chemicals discussed. Egler chastised the scientists for their preoccupation with the international use of the phenoxy herbicides and paucity of their scientific research. "I have even a more anguished concern for the peace-time uses of these same chemicals in our homeland, and for the persistent and discreditable blindness of science in not seeing that these two abuses are opposite sides of the same coin."[60] He ended his review extolling an altogether different book by an Arizona woman encountering many of the same obstacles in her quest to find out more about the herbicides being sprayed on her beloved land. Billee Shoecraft's story, and the stories of two other western women activists, tell us about the domestic use of phenoxy herbicides and the protests over the use of such chemicals. Inspired by Rachel Carson, these women questioned the use of chemical herbicides in open ranges, suburban spaces, and forest reserves.

PART TWO

Three Cases in the West
Arizona, California, and Oregon, 1960–80

Part 2 shifts focus and examines three regional case studies in Arizona, California, and Oregon. These case studies allow a consideration of what were some of the typical uses of the phenoxy herbicides, both by government agencies and private industry, and three western women who led protests over the use of the phenoxy herbicides on local, national, and international levels. In Arizona the phenoxy herbicides were used in a project designed to provide more water for the city of Phoenix. In California the chemicals were used for brush and fire control in the south, agriculture centrally, and forest management in the north. In Oregon 2,4-D and 2,4,5-T were sprayed on roadways to ensure access, and on forests to help manage and improve timber stock.

CHAPTER FOUR

Water in the West

Billee Shoecraft and Herbicide Use in Arizona

On a Sunday morning in June 1969, Billee Shoecraft walked out the patio door in her bedroom to investigate the sound of a low-flying aircraft. The helicopter overhead sprayed her, wetting her hair and soaking her pink chiffon nightgown. Shoecraft remembered her first action was to exclaim "Damn!" Shoecraft could not get any answers when she called the local forest ranger, and she set out to talk with the pilot. She got more than she had anticipated when a leaking spray nozzle meant that she and her truck were drenched when she followed the helicopter. Shoecraft recounted the episode in her 1971 book, *Sue the Bastards!* In it she also expressed her love for her home, nestled in the foothills of the Pinal Mountains located in the Tonto National Forest, close to the city of Globe, Arizona. Shoecraft told the story of rebuilding her summer cabin twice after it was burned down by summer forest fires. She described the changing relationship between residents and local forest rangers, bemoaning the increasing distrust of the Forest Service. Once she became aware of her broken trust with one branch of government, Shoecraft found herself questioning other parts of government as well. Her shock at the treatment she and the community received from public officials turned to anger, and then to disillusionment. Shoecraft's new consciousness led her to challenge the herbicide spraying program she experienced on her mountainside.[1]

As Rachel Carson noted, "One of the most tragic examples of our unthinking bludgeoning of the landscape is to be seen in the sagebrush lands of the West, where a vast campaign is on to destroy the sage and to substitute grasslands."[2] Carson goes on to recount the experiences of Supreme Court Justice

Figure 6. Photograph of Billee Shoecraft that appeared on the back cover of her book, *Sue the Bastards!*, which told of her challenges to the US Forest Service spraying program in the Tonto National Forest. Copy in author's possession.

William O. Douglas when he visited Wyoming's Bridger National Park. Here, Douglas viewed lands where sage had been killed, replaced with grasslands. But in the process, the herbicide spraying had killed willow trees and disrupted the ecosystem so that lakes, trout, beaver, and moose had disappeared along with the sage. In his description of his trip and the protests over herbicide sprayings, Douglas defended the rights of citizens to enjoy the natural world. Carson pointed out that along with eradicating sage, utility and timber companies sprayed to remove brush and maintain right-of-way and access roads. She provided the numbers of acres sprayed as well: 50 million by utility companies, 75 million in the Southwest to remove mesquite, an unknown figure but acknowledged millions of acres to increase more desirable timber stock. *Silent Spring* set the agenda for these women as they fought to preserve natural beauty, maintain ecosystems, and minimize potential health hazards in Arizona, California, and Oregon.[3]

Billee Shoecraft's story illuminates several major elements of the domestic use of herbicides as agents of Cold War chemical policy. Shoecraft and other grassroots activists interacted with representatives of the military-industrial complex in the form of the United States Department of Agriculture (USDA); the Forest Service, the subsidiary agency charged with oversight of the country's national forests and grasslands; universities and state agencies; and Dow Chemical, the major supplier of the 2,4-D, 2,4,5-T, and 2,4,5-TP (a chemical analog of 2,4,5-T known by its trade name of Silvex) used in the Forest Service spraying campaign. But as Shoecraft noted in the first chapter of her book, the conflict over chemical policy regulating the herbicide spraying prompted her to question her relationship with the Forest Service, and from there "to more closely inspect many of the other branches of Government also; and in doing so, I have stood a little in shock of those things I have learned."[4] Unlike the war on weeds on suburban lawns, which were an extension of the suburban home and the first line of defense against Communism, the USDA conducted a number of programs to protect American agriculture, which was itself as one historian noted "a Cold War weapon."[5]

A host of chemicals were used to increase agricultural and natural resource production. DDT, one of the best-known chemicals used, was sprayed to increase beef production, eradicate fire ants throughout the South, and kill gypsy moths threatening the nation's hardwood trees and forests. The chemical diethylstilbestrol (DES), a nonsteroidal estrogen hormone, was added to cattle feed to increase growth and produce more beef. The United States exported DDT as a part of the Green Revolution, which along with the Peace Corps, was intended to win the hearts of developing nations and gain them as allies against communism. Herbicides were used in the same way, used to redesign the natural environment to better realize, and profit from, natural resources.[6]

In the West this meant increasing water supplies, decreasing fires, cultivating timber stock, and expanding grasslands. The region only needed water for its agricultural success. Arizona had a long history of land projects intended to improve water supplies across the state. In the absence of that resource, Arizona's wealth mostly came from exporting its mineral resources, especially silver and copper. The Hohokam, a prehistoric people that lived in the region for centuries, built canals, stored water, and farmed in the Salt River Valley, growing to a civilization as large as 4,000 people in 700 AD. The Hohokam built an extensive canal system, and scholars have speculated that changing river channels might have been the reason the canals were eventually abandoned. The Hohokam left the area for unknown reasons, although archaeological evidence suggests that water scarcity may have been a factor. European conquerors and settlers encountered similar problems. The Spanish found the area populated with various Native American tribes in 1539. Acquired in 1848 as part of the Mexican American War, Arizona was initially ignored by Anglo settlers in their westward migrations.[7]

Settlement of the Salt River Valley by Anglo Americans really began with the establishment of Fort McDowell in 1865. The city of Phoenix began as a hay depot. It owed its growth to the series of private canals built between 1867 and 1912. Droughts in the 1870s and 1880s highlighted the need for water storage. By 1889 ambitious local leaders had identified a potential dam site, but to build it would cost between $2 million and $5 million. The business leaders turned to Washington for help. Arizona became the first in an experiment designed to fully realize the promise of western lands through the partnership between state residents and the federal government.[8]

Relief came with the passage of the Reclamation Act in 1902, which offered financial and scientific resources to western settlers so they could recover unused lands with the expansion of water supplies. Under the supervision of the newly created Reclamation Service, federal monies became available for local projects, dependent on two criteria being met: residents had to resolve any conflicting land and water claims, and they would collectively create a landowners' association. Just a year later, a twenty-five-member Salt River Valley Water Users' Association formed to represent almost five thousand individual stakeholders. This marked the first of many water projects conducted in Arizona over the course of the twentieth century. Along with federal-private water projects, the national government exerted its presence in western states in protecting resources through the establishment of national parks. President Theodore Roosevelt established Tonto National Forest in 1905 to protect its watersheds around reservoirs. The creation of the forest followed several decades of boom-and-bust rain and drought cycles that had resulted in 150,000 sheep grazing seasonally at the turn of the century and 85,000 cattle on the range year-round by 1921.[9]

Writing in 1924, noted ecologist Aldo Leopold outlined five obvious characteristics of a large geographic area that encompassed national forests in Prescott, Tonto, Coronado, and Crook and included adjacent lands. Leopold identified links between grass, brush, livestock, and fires. He noted that prior to livestock being brought to the range in the 1880s, lightning and Indians had started fires that periodically cleared out brush. Given the history visible in the natural environment, Leopold argued that conventional wisdom condemning fires was completely wrong. He advocated that the Forest Service needed to play a role in setting policy to reduce erosion and preserve the watershed. In a later, unpublished essay, Leopold again looked to the natural landscape to understand what changes happened with the incursion of Anglo-American cattle ranchers. He sought to help citizens understand the severe erosion conditions present in Arizona and other parts of the Southwest, and the role overgrazing played in producing such conditions. Leopold again challenged the thinking that blamed fires as the cause of erosion. Debates over what policies the Forest Service should implement continued into the postwar decades.[10]

The Forest Service used a variety of approaches to manage the Tonto National Forest watershed over the next several decades. As sociologist Wendy Espeland demonstrated in her study of government agencies and water projects in central Arizona, state bureaucracies often had competing visions of rationality, which meant competing project agendas and strategies. The New Deal provided cheap labor in the form of forty-one thousand Civilian Conservation Corps workers, who constructed half a million temporary dams designed to slow water velocity and stop erosion. Vegetation increased, but not necessarily of the kind cattle would eat. Another approach focused on a reseeding program using different palatable plants in four Tonto environments. Forest Supervisor F. Lee Kirby set a new course for the forest's management, including what was initially called "artificial restoration," later known as reseeding. The projects started in the 1940s used both native and non-native plants and saw a measure of success. Reseeding ranges with vegetation, however, presented challenges, not the least was the time involved. The fact that Tonto encompassed a multitude of different environments showed the complications inherent in restoring grasslands. Plans to increase grasses and grazing on the range remained an ongoing effort, until the growing population of Phoenix forced a reassessment of priorities. The focus shifted from protection of watersheds and promoting grazing to producing more water.[11]

The Forest Service began implementing controlled burns in Arizona in the 1950s, prompted by concerns over water supplies rather than pasturage maintenance. These activities resulted from the agency's involvement in the Arizona Watershed Management Program, which had its origins in a 1955 summer meeting of a number of Arizona ranchers, a Forest Service researcher, and a representative from the Salt River Project. The participants agreed that over the last fifty years foliage and brush coverage had increased, but water yields had decreased. Convinced they were right, the group contacted land management agency administrators, state representatives, members of the media, Congressman Barry Goldwater, and Sen. Carl Hayden to visit a site to demonstrate the theory. This action led to the creation of the Arizona Watershed Management Program, which was charged with testing and evaluating procedures designed to increase water yield and protect watersheds.[12]

The first experiments were conducted in 1957 in Coconino National Forest and the Beaver Creek watershed. Jay Cravens, the US Forest Service forester sent to South Vietnam as a consultant, described the attempts to increase Arizona's water yield as "rational and interdisciplinary."[13] Cravens and other foresters became "experts in applying 2-4-D, 2-4-5-T and other plant killers." He trained others on how to properly handle "these toxic chemicals safely and according to label instructions." It was his experience with the Beaver Creek project that qualified him to be sent to Vietnam. The Salt River watershed became the next site for brush control trials.

The persistent problem of brush, mostly inedible, also affected the watershed negatively and pressured range managers to consider previously shunned techniques. This problem encouraged Tonto National Forest officials to pursue the strategy suggested by Leopold three decades earlier—controlled burns. Setting controlled fires promised to increase grasses, possibly restoring perennial grasses, while at the same time decreasing brush and decreasing fire hazards and fire control costs. Two national forests were selected for the experimental program, Prescott and Tonto, with specific areas designated to test the chemical herbicides. George Glendening, a local man, was picked to oversee the project. Glendening had worked in the Forest Service in the 1940s, leaving the service to join the Copper State Chemical Company, headquartered in Tucson, as an "agricultural technical representative."[14] His government and private sector jobs had given him expertise in using chemicals to control mesquite. A 1958 newspaper article described a Forest Service Program to remove brush using fire and "killing sprays."

The project, located in the Tonto National Forest, sought to find a cost-effective means to remove chaparral types of brush but also to assess if removing such brush cover would increase water and forage for livestock. The project drew upon a host of partners, including researchers from the University of Arizona, personnel from the USDA Agricultural Research Service, and people from state land and fish and wildlife agencies, all of whom cooperated closely in the project. This approach represented a shift in Forest Service range and forest management thinking and actions in their attempts to restore edible grasses, increase water supplies, and reduce soil erosion.[15]

Tonto was one of the first national forests to begin using controlled burns. Within Tonto this meant selecting the area where the fire would be set, typically between 160 and 3,000 acres. Preparation of the area included building firebreaks, setting aside areas of brush for wildlife, and making sure riverbanks were protected. A 1974 Forest Service report noted, "Spraying by aircraft, particularly by helicopter . . . is the only practical or economical method for treating large watershed areas."[16] Given that much of Arizona's most common brush growth, chaparral shrubs, differed in their susceptibility to the phenoxy herbicides, repeated sprayings were often necessary. Despite some concerns over water contamination and their sometimes marginal effectiveness, the phenoxy herbicides were still used because of their low cost and selectivity in sparing grass, where "repeated treatments offer[ed] a method of tipping the ecological balance in favor of grass."[17] Fires were started downslope, burning uphill. In the final part of the burn, helicopters sprayed herbicides, primarily 2,4-D and 2,4,5-T, and used napalm grenades to fully remove the brush. Forest managers created what historian Adam Sowards likened to a "desert Vietnam."[18]

Prescribed burns succeeded in increasing water yield, grazing herds grew from twenty to two hundred cattle, grass cover thrived, and quail and deer numbers multiplied. While the prescribed burns worked, they came with a high economic and political price, a price the Forest Service eventually could not pay. Billee Shoecraft and the group of landowners living around Globe contributed significantly to the political costs of spraying herbicides in the Tonto National Forest.

Billee Shoecraft's artistic sensibilities made her an ardent admirer of her adopted state. A sickly child, Wilma "Billee" White grew up determined to live each day fully, committing herself to finding beauty in everyday life. As a young divorcée, she visited her sister and brother-in-law living in Globe, Arizona, a city of approximately six thousand people in 1940. Here she met Willard Shoecraft, the manager of a local radio station, who began courting her immediately. She moved her family from Indiana and married Willard in 1948. Despite a tumultuous marriage—the couple divorced and remarried twice—the Shoecrafts made a home together and raised three sons. Billee Shoecraft wrote and published poetry and ran an interior decorating business. Shoecraft's planned book projects, *The Misunderstood Male* and *Women Are a Mess*, reflected her traditional understandings of gender roles, and the belief that women needed to learn to take care of themselves and to stop worrying their husbands to death. She also embraced her new surroundings. Living in Arizona allowed Shoecraft freedom and independence. Her love of the outdoors did not mean she considered herself an environmentalist. "I've loved Arizona forever, I guess, and especially her rugged untamed qualities; that special sense she can give you of being alone, but not lonely."[19] Shoecraft enjoyed Arizona for the chance it offered for an active outdoor life—hunting, riding, and even building her own home.[20]

Shoecraft's home, located in the Pinal Mountains on the border of the

Tonto National Forest, grounded her in the Arizona soil. Influenced by Frank Lloyd Wright's philosophy of functional beauty, Shoecraft designed her dream home in the wild, rocky cliffs and canyons of southern Arizona. Here, she and Willard had built "on a long-abandoned pond site with the ancient rock wall of the dam retained as part of the natural contour of the house." The house took eighteen months to build using contract labor and the Shoecrafts' own sweat and ingenuity. Much of the lumber for the home came from local abandoned copper mines and Mexican railroad ties, with one 850-pound living room beam taken from the historic Lyric Theater in Tucson. Within the split-level house, Billee created a home with twenty-five years of auction finds and treasures. Featured in the "Sun Living" section of the *Arizona Republic*, the Shoecraft house embodied the couple's love for and connection to the land. These sentiments would be what fueled Billee Shoecraft's nine-year battle with the US Forest Service; its parent agency, the United States Department of Agriculture (USDA); various local, state, and national elected officials; and scientific and medical experts. Her ordeal started on a Sunday morning in June 1969, just six months after the Shoecrafts' home was featured in the newspaper.[21]

Along with Shoecraft, several other families living in four canyons—Russell, Kellner, Icehouse, and Six-shooter—were also sprayed with herbicides by Forest Service aircraft. Bob McCray and his wife and son were picnicking at the site of their future home, only half-built at the time. Bob McKusick and his family were sprayed while looking at clay deposits McKusick had acquired for his pottery trade. Another Kellner Canyon resident, Pat Medlin, was exposed while sunbathing in her backyard garden. In a 1970 *New York Times* article, McKusick described an increase in birth defects and deaths in his goat herd. Local resident Richard Lewis saw his peach trees killed and then watched as his family suffered a variety of ailments. Even more extreme, Jack Andrews was exposed during the spraying and had lost weight and experienced severe chest pains. When Andrews consulted a Las Vegas specialist, he was diagnosed with herbicide poisoning. The herbicides appeared to have affected female reproductive systems, with increased internal hemorrhaging resulting in hysterectomies. In the interview McKusick admitted tensions were so great that residents were prepared to defend their lands by violence if necessary. Local Forest Service officials denied that the herbicides posed any health hazards, although Tonto National Forest supervisor Robert Courtney decided to suspend spraying operations. His decision came after months of raucous protests, led by Billee Shoecraft.[22]

Shoecraft considered herself to be on good terms with Forest Service personnel despite increasing regulations, but the relationship dramatically worsened. Bemoaning the tangled bureaucratization, Shoecraft wondered when things had changed. What, she asked, motivated the press release announcing the initial herbicide sprayings to include these words: "I also anticipate adverse criticism and harassment from those who devote their lives to criticizing and harassing."[23] The Forest Service at least recognized that some people would be unhappy about the program. Opposition to the program did not coalesce, however, until the June 1969 spraying incident, the one that Bob McKusick described as "when all hell broke loose."[24] In the weeks that followed, the residents sprayed all experienced a host of illnesses and symptoms, including fatigue, headaches, blisters, itchy rashes, shortness of breath, and, in some extreme cases, chest pains.

The June spraying united individuals who had consciously chosen to live in relative isolation, one means of measuring their commitment to individual rights. Approximately eighty residents attended a July 3 city council meeting, where the majority voted on a nine-point proposal crafted by Shoecraft and McKusick. The Forest Service position might not have been helped when during the meeting Ranger William Moehn declared the various claims of harm to plants, wildlife, and humans "malarkey."[25]

Two considerations lay at the heart of the Globe protests: what rights citizens have in challenging federal policies and agencies, and their increasingly anxious health concerns about exposure to potentially toxic chemicals. The plan presented by McKusick and Shoecraft included a variety of demands, primarily insisting the property owners' rights be respected, spraying and other destructive programs cease, and the Forest Service be held responsible. Point two on the group's list enjoined the Forest Service from conducting any programs "to destroy (plant life) without the consent of the people of the entire area involved after all the hazards or outcome of any such project is known beforehand and understood elsewhere." Point three forbade the government from using Globe inhabitants as "guinea pigs." The fourth point reiterated the second, making it even more explicitly: "That it be made a law that no spraying or herbicidal use or destruction of plant life can be done until a complete analysis is made public and approved by the public prior to any action."[26]

McKusick's threat to shoot the next person to spray herbicides on his property, and Shoecraft's orneriness, meant that neither had an extensive network of friends or acquaintances in the broader Globe community. This made their success at the council meeting even more impressive and indicated the degree of unrest in the broader community. Together, they proceeded to pursue several strategies in their efforts to get the spraying halted.

McKusick and Shoecraft educated themselves about chemical herbicides, mobilized the community, contacted multiple public officials, and eventually considered a lawsuit. They obtained Dow Chemical product information on Silvex, a chemical compound closely related to 2,4,5-T; zoology bulletins from the University of Arizona; and USDA user manuals for applying chemical herbicides. They sent samples of plants and soils to a California lab to be tested for chemical contaminants. The lab found no evidence of Silvex but did find 2,4,5-T in samples from Shoecraft's blackberry bushes. The two landowners circulated a petition and convinced two hundred people to sign. They sent the petition, record of spray damage, and their concerns to local elected officials. One of the most effective protests consisted of delivering the dead plants and animals in coffins to the Forest Service offices in nearby Phoenix, with the intent of embarrassing Tonto's chief forester, Robert Courtney, and generating publicity for the cause. The funeral procession for the plants and animals proceeded despite Courtney's appearance in Globe on the day of the action. Shoecraft did not let his appearance disrupt the planned demonstration. The hearses were loaded with coffins, and the "funeral" procession made it to Phoenix in time to be covered by the local press.[27]

Residents' protests brought about an unexpected and welcome result, as spraying on Pinal Mountain was halted. Appearing before a crowd of 150 people at the Globe High School on July 28, Tonto Forest supervisor Courtney announced that there would be no more spraying on Pinal Mountain for the rest of the year, or in 1970. Later that year, in October, Courtney decided to halt herbicide spraying and burning in "brushy forest areas" throughout Tonto.[28] Burning of cut brush, in specific pine tracts, fuel breaks, and control areas continued. This later announcement appeared to expand the area that would not be sprayed. Courtney acknowledged that the "public criticism" had prompted the decision. "Because of this broadened interest, we feel it is time to take a short breather to provide information and the opportunity for observation to those interested conservation organizations and to use their suggestions, critique and assessments."[29] When making the announcement, Courtney took the chance to educate the public about the Forest Service's chaparral program, which had begun in 1961. He listed the hoped-for goods the program was trying to achieve. Residents appeared to have won a small victory, although they now faced months of scrutiny from local and state officials, who often attacked residents' claims of harm.

Over the next several months, Shoecraft and the other sprayed Globe residents received repeated visits from various constituencies, many of which Shoecraft labeled as "task forces." Dow Chemical sent Keith Barrons, the company's director of plant research, who visited the sprayed area with Forest Service officials. According to Barrons, there were no signs of herbicide damage. Although Barrons did not meet Shoecraft, he noted that she had been identified as the person making "a lot of noise."[30] Dow also sent Ross Wurm, a public relations consultant from California, to Globe. Wurm attributed most of the problems to poor public relations and an angry, determined Billee Shoecraft. As he described her: "This woman, Mrs. Shoecraft, who is the formidable opposition, has done a tremendous job of pouring out bad information."[31] Wurm minimized the evidence of herbicide damage and the presence of silvex in a water sample from the Shoecraft homestead, dismissing concerns as misunderstanding and "lies."

Other interested parties tried to evaluate possible harm. After an initial stop by Ranger Moehn and Grazing Officer William Fleishman, Shoecraft received visits from the Arizona Fish and Game Department, range management specialists from the nearby University of Arizona, a group sent by the governor, and a joint party composed of Forest Service and USDA personnel. The findings of a Forest Service scientific task force were typical. In their report released in December 1969, the committee judged the herbicide damage to be "extremely limited."[32] All of the committees investigating residents' claim of harm dismissed any claims of herbicide damage to the Shoecrafts' and other affected residents' plants and animals. The most damning assessment came from a team sent by the USDA and headed by Dr. Fred Tshirley.[33]

The national attention Globe residents had achieved provoked a response by the USDA. For Tshirley, the chairman of USDA Agricultural Research Service, his appointment to the interdepartmental committee meant a return to Arizona, where he had performed tests on the chemical control of range weeds at the Agricultural Research Service station located in Tucson. The Department of Defense had sent Tshirley to South Vietnam when pressed by scientists concerned about the destruction of the country's ecosystem. Committee members represented various areas of expertise, specifically focusing on plant, insect, water and land animals; soil, air, and water pollution; and toxicology and teratology. The task force evaluated plants and soil, effects on wildlife, air pollution, and health complaints.

The task force interviewed nine Globe physicians and considered other public health factors as well. Most of the physicians surveyed indicated they had not seen an increase in miscarriages or other illnesses, such as birth defects, unusual skin rashes, or muscular weakness, among Globe residents. The report included specifics on several individuals and families that had been "the most vocal in their complaints."[34]

The report also noted residents' emotional intensity and volatility, and suggested that evaluations of respiratory illness be conscious of that context. The report's concluding remarks generally dismissed claims of harmful effects on plants, animals, or human beings. It reasserted the safety of the phenoxy herbicides and noted the difficulties of detecting and analyzing dioxin contaminant. The report received widespread attention in the media, although this may have been in part because of a concurrent visit to Globe by Congressman Richard D. McCarthy, a well-known critic of chemical and biological warfare, and of the war in South Vietnam.[35]

The events in Globe appeared tailor-made for McCarthy's investigation. A Democratic congressman from Buffalo, New York, McCarthy had served in the Navy after the end of WWII. He worked as a newsman and publicist before winning his seat. McCarthy attributed his own awareness of chemical and biological weapons (CBW) to an investigative report that aired on NBC's *First Tuesday*, at the time the network's counterpart to CBS's *60 Minutes*. The secrecy that shrouded military tests represented one of McCarthy's greatest concerns regarding the United States' use of CBW. In his perception a significant gap existed between the American public's disdain for CBW and the military's commitment to developing such weapons.

Upon his arrival in Washington, McCarthy received so many inquiries about chemical warfare that he decided to write a book to help inform the public. Putting his journalistic skills to good use, McCarthy wrote and published *The Ultimate Folly: War by Pestilence, Asphyxiation and Defoliation* in 1969. In it McCarthy described the origins of America's postwar policy. He identified the military's various chemical weapons research centers located around the country. These included the Edgewood Arsenal, just northeast of Baltimore; Fort Detrick, a 1,200-acre compound only forty-five miles from the White House; the Pine Bluffs Arsenal in Little Rock, Arkansas; the Dugway Proving Grounds in Dugway, Utah; and Eglin Air Force Base in Florida.

The book examined the host of CBW programs, including Fort Detrick's chemical defoliants and anticrop herbicides. McCarthy referenced the studies produced by American scientists like E. W. Pfeiffer and J. B. Neilands, along with the studies commissioned by the Department of Defense. He noted his own opposition to Operation Ranch Hand. "I consider the use of herbicides and defoliants—the new chemical weapons—on the scale that we have used them in Vietnam without knowing what the long-range consequences will be to be irresponsible."[36] McCarthy warned that using chemical defoliants invited a biological arms race, and that biological agents designed to blight plants and crops could be used against the United States.

McCarthy's visit received regional and national coverage and gave Globe activists a bully pulpit to air their concerns. Both Shoecraft and McKusick spoke to the press in the days before the informal hearings. McKusick, identified as the leader of the Informed Citizens Union Committee, recounted anecdotal stories of increased hemorrhaging in women who had been exposed to the herbicides, along with an increased number of hysterectomies. Citing no source, one news article noted that there had also been an increased number of "unexplained miscarriages" in the area as well. One of the task forces, led by University of Arizona ecologist Dr. Paul S. Martin, indicated that one in four people showed symptoms that could be attributed to herbicide spraying. McKusick detailed other health problems, including respiratory trouble, skin rashes, eye irritations, and chest pains.

Globe activists saw McCarthy's visit as validation of their concerns. While the governor had created and sent at least one of the scientific task forces to study the problem, elected officials had steered clear of the growing controversy, despite being contacted by Shoecraft and others. Other state agencies had been conspicuously absent as well. McKusick described the group's treatment in one newspaper interview as very poor, saying that they had received little aid. "We got nothing from the government except a cold shoulder. We also received no support from the State Health Department."[37] The only bureaucrat who had responded, Congressman Sam Stieger (R-Arizona), had been the one to bring the problems in Globe to McCarthy's attention. McCarthy's hearing gave Globe activists the eyes and ears of the nation.[38]

McCarthy set the tone for the hearing in his opening statement, which noted the National Cancer Institute's study and the hazardous health effects of 2,4,5-T. Six individuals were called upon to testify on the first day of a two-day investigation. Yale biologist Arthur Galston, one of the vocal scientific critics of the phenoxy herbicides, testified both days of the hearing. In his second appearance, Galston confirmed that in his opinion there was "no doubt about damage" done by herbicides to local plants. Galston went on to question whether the Forest Service program had actually increased water runoff, expressing "puzzlement" that it was thought of as a viable program. In contrast to Galston's formal expertise, Shoecraft gave dramatic and emotional testimony that included her own biopsies, which showed 2,4-D present in her tissue. Local veterinarian F. L. Skinner was quickly dismissed when he said he had seen no harmful effects of herbicides on local animals. One news account characterized hearing testimony as critical of the Forest Service and the agency's carrying out of the herbicide spraying program. While touring two of the areas sprayed, McCarthy's group did find an herbicide barrel, which had been cleaned and used for park trash. This use, however, violated policy that required all barrels to be crushed and disposed of properly. One stunning admission came from Forest Service Assistant Regional Commissioner John Pierovich, who learned about the test results on 2,4,5-T from the press instead of the USDA.[39]

Globe's activists received mixed media coverage in which they revealed simplistic scientific thinking. One news article framed McCarthy's visit as evidence that determined conviction had overcome "government apathy." Associated Press coverage described the group more negatively, such as when they described Shoecraft as the "woman who started the whole fuss." A later editorial commentary questioned the scientific validity of residents' concerns and made veiled accusations of planting evidence, suggesting that even McCarthy thought he might have been played. A news brief in *Time* magazine openly mocked the residents, pointing to their suspicions of a visiting environmental group from Washington, DC, as evidence of the group's paranoia. The piece included an unflattering picture of Shoecraft, whose friends urged her to sue the magazine. A misunderstanding had taken place between Environmental Action and the Globe activists, and Shoecraft had confusedly accused the group of being connected to the chemical industry. *Time's* mockery still seemed excessive, especially when McKusick had publicly threatened violence.[40]

McKusick also appeared as an extremist, as when he admitted in a *New York Times* interview that he was prepared to shoot down any aircraft that attempted to spray his land again. The *New York Times* coverage appeared more sympathetic, however, as the article recounted residents' charges of harm to animals and humans. Residents also appeared naive in their understandings of science, although their research had revealed that the barrels used by the Forest Service did not have customary product warnings, or that barrels were being reused instead of properly destroyed. The unkind media coverage, along with the Tshirley report's negative findings, failed to stop two significant actions after McCarthy's visit: congressional hearings on the use of 2,4,5-T and its effects on humans and the environment, and the lawsuit Globe citizens would bring against the government.[41]

Several factors had influenced the decision to hold congressional hearings. These included the February 1969 release of the National Cancer Institute's report on pesticide testing done by the Bionetics Research Laboratories (the study referenced in McCarthy's opening remarks). The Bionetics study had identified 2,4,5-T as causing nongenetic birth defects in multiple species of test animals. The lab's research on 2,4-D suggested it too might cause birth defects and recommended further study. Responding to the growing concerns raised by the Bionetics study, the Richard Nixon administration banned the use of 2,4,5-T on food crops, although it was still sprayed on range pastures and forest lands. Meanwhile, 2,4-D continued to be sprayed on food crops, particularly rice, wheat, and corn. The critiques being made by scientists were appearing in the media. McCarthy also followed up on his own research on CBW, in which he included the phenoxy herbicides used for defoliation with his informal Globe hearing. Sen. Philip Hart (D-Michigan), chairman of the Subcommittee on Energy, Natural Resources and the Environment, admitted, however, that "much of the impetus for the hearing had come from an article by Thomas Whiteside in the Feb. 7 *New Yorker* magazine."[42]

Whiteside's article, "Defoliation," provided an in-depth examination of the US military's use of chemical herbicides in South Vietnam. The article noted that the military did not consider the herbicide spraying missions a form of chemical or biological warfare. Whiteside detailed the history of the phenoxy herbicides, from their extensive domestic use on lawns, fields, and ranges, to their current use in South Vietnam. The essay included the disagreement among scientists like Arthur Galston and Fred Tshirley regarding the potential health hazards the chemicals posed, and informed readers about the labyrinthine route the Bionetics study had taken before being released to the public. Whiteside condemned the continued use of 2,4,5-T and its dioxin contaminant, given the growing body of evidence on dioxin's causation of birth defects and other illnesses. He also questioned what was being done in Vietnam. He asked, if one of the most prosperous countries in the world could not measure the chemicals' potential risk, "How is one ever to measure the harm that might be done to unborn children in rural Vietnam, in the midst of the malnutrition, the disease, the trauma, the poverty, and the general shambles of war?"[43] Whiteside's essay brought the concerns about and possible dangers of chemical herbicides to the reading public, in the process making a complex topic understandable. It also indicted several government agencies for failing to investigate, share information, and protect the public. This led to the congressional hearing being convened.[44]

Shoecraft attended the April 1970 hearing in Washington, listening to public officials, scientists, industry representatives, and citizen advocates debate the safety of the phenoxy herbicides. The hearing began with the testimonies of two public interest lawyers associated with consumer advocate Ralph Nader. Harrison Wellford highlighted the easy availability of weed products containing 2,4,5-T, despite the Nixon administration's ban of the chemical in nonpopulated areas. Many of the products were poorly packaged, leaking from paper bags and with no warning labels. The men deplored the continued use of 2,4,5-T given the Bionetics laboratory results showing extreme toxicity, findings that had been confirmed by the National Institute of Environmental Health and the Food and Drug Administration. They argued that thousands of American families were being exposed to the chemicals, and children were being born with preventable birth defects.[45]

In contrast, the USDA director of science and education, Dr. Ned Bayerly, and his assistant director Dr. T. C. Byerly, argued that it was the dioxin component that caused the problem. This had been addressed, they maintained, through better manufacturing processes that had decreased the amounts of dioxin contaminant below harmful levels. Hearing chairman Hart challenged this stance, pointing out that earlier testimony highlighted the poor testing standards used to measure dioxin content, and uncertainty over the dioxin's persistence in the environment. In a later conversation with Shoecraft, Byerly lectured her with Latin phrases and claimed a position of scientific skepticism. Advocates of 2,4,5-T were equally dismissive of opposing scientists as well. The USDA officials' testimonies were supported by that of Dow Chemical vice president Julius E. Johnson, who asserted that Dow workers manufacturing the chemicals had shown no signs of illness.[46]

Several scientists presented evidence disputing the chemical's safety. Galston's testimony at the Globe hearing was submitted as a part of McCarthy's testimony. His "puzzlement" about the effects of taking water from one place to another provided an ecological perspective. Dr. Jacqueline Verrett, an FDA scientist studying 2,4,5-T's toxicity, was scheduled to present the findings of her experiments with chick embryos. Her findings appeared to be remarkably similar to the conditions reported in Globe animals, including edema and slipped tendons. Dr. Samuel Epstein, a professor at Harvard Medical School, gave technical testimony explaining the ways chemicals, through one of three possible causes, produced teratogenic effects in human beings, or more simply, how chemicals could cause noncongenital birth defects. Epstein defined 2,4,5-T's dioxin contaminant a risk, in part because much remained unknown about the compound. He concluded his testimony: "Finally and critically, available data on the toxicology of the dioxins, and more importantly *on the lack of data* on the toxicology . . . indicate an urgent need for restriction of human exposure to dioxins."[47]

The congressional hearing revealed some of the issues posed by chemical regulation and exposure. There were several dramatic moments, such as when Dow's Johnson acknowledged that the company had known about 2,4,5-T's dioxin contaminant and its potential toxicity since the 1960s. He admitted the company had notified public and private Michigan organizations, including the state Department of Health, yet failed to notify the USDA or FDA. Johnson also chose not to disclose where one set of the company's dioxin studies had been done: using prisoners incarcerated at Pennsylvania's Holmesburg Prison. United States surgeon general Dr. Jesse Steinfeld shocked hearing participants when he started his testimony with the announcement that the USDA was immediately suspending the use of 2,4,5-T on waterways and around homes. The hearing pitted those who feared the toxicity of 2,4,5-T (and 2,4-D) against those who proclaimed the chemicals' safety. Uncertainty figured prominently in both sets of testimonies. McCarthy, Verrett, and Epstein all emphasized how much remained unknown about 2,4,5-T and 2,4-D and the need for caution while further research was done. Chemical advocates also cast doubt. They downplayed the reliability of chemical tests like Verrett's even as they proclaimed the chemicals' apparent record of safety. One of the major problems in chemical knowledge remained the uncertainty of animal studies and inability to test the chemicals on humans—at least publicly. Questions of toxic uncertainty haunted citizens even as they were used by industry to maintain the status quo.[48]

Billee Shoecraft struggled to translate her experience of chemical-caused illness into terms that would be acceptable, or at least understandable, to scientists and policy makers. Shoecraft attempted to bridge this gap in her seemingly eccentric focus on her son's guinea pigs in *Sue the Bastards!* Including stories about her son's guinea pigs let Shoecraft make multiple claims. She offered the animals' physical deformities as proof of harm from chemical exposure, an informal scientific experiment. Here, Shoecraft's actions fit within a broader context of citizen science as lay people increasingly challenged scientific expertise during the 1970s, most often in the form of popular epidemiology. But in Shoecraft's account, the people of Globe, secretly exposed to several spring sprayings since 1965, also existed as guinea pigs, subject to the whims of the United States Forest Service and its spraying program. Harrison Wellford, the public interest lawyer, expressed the standard approach to chemical regulation as "the public should be exposed first and the experiments done afterwards."[49] Shoecraft connects her own physical suffering with that of the guinea pigs. "It's August 8, 1970, and yesterday I brought the guinea pigs up the mountain to see if the sick ones might get well, or if their hair would grow back if I removed them from the area that was sprayed at my home. . . . One of them had three babies yesterday; one albino born dead and one albino with no eyes. He lived ten hours. One baby, although very tiny, is still alive."[50] She then linked the guinea pigs' maladies along with multiple listings of the animals displaying birth defects or that died after the spraying took place.

Commenting on the leaked Bionetics study, Shoecraft connected her guinea pigs with formal scientific studies, even as she declared her description as "unscientific language." Significant portions of the narrative focus on the experimental nature of the Forest Service's herbicide program, the ignorance of various task force officials (there were a total of eight investigations), and the contradiction between directions given on product labeling and the ways and conditions in which the phenoxy herbicides were applied. Shoecraft repeatedly highlighted medical and scientific officials' inexperience with the chemicals. These same individuals professed unfamiliarity regarding the chemicals' effect on watersheds, or that "these herbicides require further testing for the much different climate of the Southwest, but we resent the research being done on us!"[51] Throughout the book, Shoecraft emphasized the continued uncertainty of the scientific knowledge, in direct contrast with officials' proclamations of safety for animals and human beings. Shoecraft unsuccessfully tried to undermine these proclamations, and when it became clear official policy would continue to allow spraying in areas like Globe, she and the other affected citizens turned to legal means to achieve their goals.

Threats of lawsuits had appeared early in the conflict between Globe citizens and the government. They became a reality after the congressional hearings when six Globe families filed two lawsuits totaling $9 million. The first lawsuit was brought against Dow Chemical and three other manufacturers of the herbicides sprayed, along with the company responsible for spraying the chemicals, and a state agency that had helped finance the program. The six Globe families wanted recompense for property damage and personal injuries. The herbicides either were sprayed directly on private property or had drifted onto private lands. The chemical companies failed to fully test or properly label the chemicals sold, "thus endangering the health of the general public."[52] In a separate action, residents filed additional claims against the federal government as the ultimate authority over the Forest Service, which was responsible for the herbicide spraying program. The plaintiffs claimed that the "federal government knew prior to the aerial spraying that began in spring 1968 that the chemicals were potentially dangerous to human life and property."[53] In this second action, the families offered a powerful condemnation of the government and its failure to protect citizens from harm. Some press coverage of the lawsuit included residents' charges and described the various incidences of illness. Interviewed when the lawsuits were filed, Shoecraft was quoted as saying the chemical burden in her body would shorten her life. While the cause would be contested for the next decade, she was right about the time left to her. It would be short.

Shoecraft and the other plaintiffs spent most of the 1970s pursuing their legal offensive, even as Dow Chemical fought to keep selling 2,4,5-T. Out-of-court settlements with the Salt River Project and three of the four chemical companies provided funds to keep the lawsuit against Dow and the federal government alive. At the same time, Dow was challenging the suspensions and cancellations of 2,4,5-T's registration. In 1974 the company forced the EPA to retreat. The agency decided it needed more information before it took action, and it cancelled public hearings on the chemical. Dow continued make and sell 2,4,5-T and silvex relatively uninterrupted.[54]

That same year Billee Shoecraft's condition worsened. Over the next two years, she was diagnosed with breast cancer. She and her daughter-in-law went to Mexico so Shoecraft could receive Laetrile, the controversial cancer treatment. Later that year uterine surgery revealed that she had multiple tumors, so many that surgical removal was not an option. Willard Shoecraft took Billee to San Diego, and over the summer of 1976 he smuggled Laetrile across the border. The couple returned to Globe that fall, but by December they had gone to Oregon to seek more Laetrile treatments. Billee Shoecraft died January 6, 1977. Four years later the original Globe lawsuits would be settled out of court for an undisclosed amount. Just two years later, in 1983, a second, virtually unknown lawsuit was brought against Dow and the Forest Service by four young men who walked through the 1969 spray area and who later fathered children with birth defects. This second $8 million suit was also settled out of court, in 1985. All told, as a codefendant in each case, the Forest Service had paid $275,000 for the two lawsuits.[55]

The *New York Times* article reporting the case settlement described Shoecraft's physical suffering before her death. While observers understood why the remaining Globe residents had settled, some noted that Shoecraft would not have settled. She had wanted more than financial restitution, and her experiences moved people both before and after her death. Shoecraft reached hundreds of people through public talks and her book, *Sue the Bastards!*, published in 1971. Her challenge to the government over the rights of citizens and pesticide policy continued to resonate within the broader populace.

While the lawsuits dragged on, Shoecraft put the time to good use giving public lectures and writing her book. The title for *Sue the Bastards!* came from a comment made by environmental lawyer Victor Yannacone, who had been instrumental in lawsuits against DDT on Long Island and in influencing the state of Wisconsin in banning the same pesticide. Shoecraft spent much of the summer of 1970 writing. Established New York publishers expressed reluctance when Shoecraft made it clear she would not allow the manuscript to be edited. Instead, she approached a small Phoenix press, paying $15,000 for five thousand copies, most of which were bought by others involved in antitoxic campaigns. Shoecraft acknowledged the importance of Rachel Carson and *Silent Spring*. Given the fact that pesticide use had increased in the years after *Silent Spring*'s publication, and that industry and government officials refused to heed Carson's warnings, Shoecraft offered a new battle-cry, one less "genteel." While the book rambled, contained false and inaccurate information, and was often unclear, it still provided a moving portrait of Globe's troubles. It also revealed Shoecraft's global awareness of the phenoxy herbicide problem, as she cited the disparate beings who had experienced herbicide exposures: expectant Vietnamese mothers, German factory workers, Swedish reindeer, Mexican Americans in Globe. Shoecraft identified herself as a "poet, designer, architect, writer, mother and 'friend.'" Equally important, however, she considered herself "an American, and a citizen of the world."[56] This status gave her the right to challenge authority.

Before her death Shoecraft spoke at a variety of different venues, and the book attracted national attention. The University of Arizona hosted a talk after the publication of her book in 1971, sponsored by the campus chapter of the Sierra Club. More than two hundred people attended her talk. Shoecraft accused the government of lying, noting that like people in Vietnam and other parts of the country, audience members had been exposed to dangerous chemicals. She called upon audience members to "do something about it."[57] Shoecraft spoke to science classes at Eastern Arizona, participated in herbicide demonstrations in Portland, Oregon, and Canada. Noted ecologist Frank Egler wrote an introduction for *Sue the Bastards!* In it Egler bemoaned the indiscriminate use of the phenoxy herbicides. He described his conversation with Rachel Carson that the users of the chemicals showed poor judgment, with the thought being "Spray: 'kill' brush; get grass and conifers; now," and "Do not ask about other plants, animals, man, economics or the future. That would upset the simple 'design of the experiment.'" Egler ended with the observation that more Billee Shoecrafts were needed. Beyond writing the introduction, Egler mentioned Shoecraft and her book in his review of other works critical of US herbicide use, validating her work. Reviews of the volume appeared in gardening sections of newspapers as well. One reviewer thought the book deserved a "wider audience" even as she noted that not only could herbicides be used in Illinois and cause problems, they already had. One California woman bought one hundred copies of the book to distribute to friends and libraries. People from all over the country contacted Shoecraft and shared their own herbicide horror stories.

And attitudes started to change. For many Arizonans, Shoecraft and her Globe neighbors were nothing more than cranks who they wished would quit causing a fuss and disappear. Starting in the 1960s and until the mid-1970s, however, several activist groups formed to protect Arizona's natural environment. These groups included advisory organizations like the Southern Arizona Environmental Council (1971) and the Southwest Environmental Service (1974). Along with the previously established Arizonans for Water Without Waste (1966, later the Arizonans for a Quality Environment), these groups advocated for a variety of environmental causes. This meant that members sometimes focused on broader environmental issues than just water supplies for Tucson and Phoenix.[58]

Attitudes within the Forest Service and other public agencies shifted as well. A 1974 editorial in Flagstaff's *Arizona Daily Sun* applauded the concerns raised by the Regional Forester Bill Hurst regarding a plan by the Arizona State Water Commission's and Water Resources Committee's plans to significantly remove vegetation to increase water runoff. The plan required that a third of the Mogollon Rim's ponderosa pines be removed. US Forest Service staff cautioned that "the public would not accept such a drastic clearing, especially by herbicides."[59] Forester Hurst recognized a broader public than the urban residents wanting increased water supplies, supplies they refused to conserve. The proposed cuts would affect "pine stands . . . right in your backyard," warned the writer. The piece ended on a critical note. "If the watershort [sic] valley is so desperate that they want to eradicate the natural resources in our backyard, perhaps it's time to pull the plug on all of the private swimming pools and development lakes down there."[60] Echoing Arthur Galston's concerns raised during the McCarthy hearing, the curator of the Museum of Northern Arizona questioned both the viability and the economics of stripping vegetation to increase water supplies. Speaking at a 1975 water conference, William J. Breed described the negative effects such projects would have on wildlife and the natural environment. By 1983 the Sierra Club's Grand Canyon chapter won a halt to the spraying of at least one aerial herbicide on Arizona lands by the Bureau of Land Management pending an environmental review. From this vantage point, Shoecraft, McKusick, and the other Globe residents had led the charge in challenging the feasibility and right of government to carry out such projects.[61]

Globe's angry citizens successfully halted herbicide spraying in their corner of the world, and they exacted some measure of monetary recompense. Billee Shoecraft had led a valiant fight to protect her land, and to argue for the rights of citizens in influencing federal chemical regulation. Before her death, the story of Globe and its affected residents drew national attention to the problem of domestic herbicide use. In the process, Shoecraft also found a kindred soul who was fighting a similar battle in California. In Los Angeles early in 1971, she and consumer advocate Ida Honorof spoke out against herbicide use. Shoecraft recounted the birth defects seen in Globe animals, including the guinea pigs. Honorof condemned the USDA, claiming: "These are deadly chemicals. There is no way the Department of Agriculture can tell us they're helpful to mankind."[62] Honorof would lead a campaign against the phenoxy herbicides used in California for a variety of reasons and in a multitude of settings—fires, fields, and forests.

CHAPTER FIVE

Fires, Farms, Forests

Ida Honorof and Herbicide Use in California

Writing in her monthly newsletter, *A Report to the Consumer*, activist Ida Honorof expressed strong opinions about the failure of the Environmental Protection Agency (EPA) to properly protect citizens and the environment from harmful pesticides. She singled out industry and government scientists as representing a significant part of the problem. "Reports by scientists have been accepted in the past without challenge because of the common faith in the ethics of science. Bitter experience has now shown that such faith is not enough. Not all scientists adhere to professional ethics."[1] Honorof even called for legislation to hold scientists accountable for misinformation, be it the suppression of data, the misreporting of findings, or publishing of distorted versions. She thought that a scientist who failed "to fully disclose findings of his experiment, or make claims which are counter to his data he ought to be held accountable for his deception, just as others are whose actions lead to losses and damages." Accountability, in Honorof's view, even included jail sentences for those who exposed the public "to great harm."

While Honorof's criticism focused on unethical scientific practices, it also offered a challenge to the broader Cold War political consensus. In the postwar period, science often advanced Cold War political objectives as it remained independent of them. In exposing the failures of the EPA and its convenient relationship with industry, Honorof, like Billee Shoecraft, based her rebellion against that consensus within an environmental setting and in opposition to US chemical regulation policy. Herbicides were used in southern parts of the state to combat wildfires, responsible for what historian Mike Davis called Los Angeles's "ecology of fear."[2] They were also used in Central California's

Figure 7. Ida Honorof continued her environmental activism even after she moved to Eureka, California, where she helped pulp paper mills remove dioxin from their manufacturing processes. Courtesy © 2007 Los Angeles Times. Used with permission.

agricultural industry, a highly specialized system of crop production heavily dependent on chemicals and human labor. And the herbicide spraying was a routine part of Northern California's timber industry operations.

The widespread use of the phenoxy herbicides in California meant that exposures to the chemicals happened all the time, and through a variety of venues that included air, food, and water. They were becoming an increasingly routine aspect of life. Like Billee Shoecraft, consumer activist Honorof serves as the focus for the protests over herbicide use, although many Californians raised questions about the herbicides' safety and challenged public officials, scientists, and industry starting in the 1960s through the 1990s. As state residents questioned the reasons to use the herbicides, they exposed the Cold War military-industrial-academic complex, aided and abetted by powerful agricultural and timber lobbies. The watchful eyes and loud voices of activists like Honorof kept Californians informed and prepared to challenge the uses of these chemicals.

Although Honorof appeared to be little known beyond California circles, her concerns and actions exemplified many of those held by Californians in the 1970s and 1980s. Honorof was born in Chicago and had early experience picketing when she accompanied her mother at a bread protest in 1915. She worked for a Chicago slaughterhouse and fought for workers' rights in the 1930s. She married and moved to California in 1961. In 1968 she divorced her husband and shortly thereafter became a volunteer at the progressive radio station Pacifica KPFK 90.7 FM, writing and producing a *Report to the Consumer*, which aired weekly. Although the station had a relatively small audience, it gave Honorof an intellectual and physical space to speak out about issues of concern to her, most of which dealt with health, particularly the pervasive presence of chemicals in the daily lives of her listeners. Honorof built up a reputation for her "hard-hitting" show and was featured and recognized in venues beyond the radio station. She also appears to have used her radio broadcast research as the basis for a monthly newsletter, *A Report to the Consumer*, mailed to subscribers across the country. Honorof's involvement with Pacifica and her newsletter made her part of the emerging alternative press of the 1960s.[3]

Her passionate views sometimes led her to associate with organizations that had less-than-respectable reputations, such as the National Health Federation, a group dedicated to alternative medicine and often labeled as offering quack medical advice. Yet she was also acknowledged as a local environmentalist and gave talks throughout the region on food, nutrition, pollution, and chemical pesticides. She spoke out early against the contamination of beef with diethystilbesterol (DES), the hormone added to cattle feed to promote increased meat production. She worked hard to inform the public and took pride in the work she did. A relative latecomer to the California environmental scene, Honorof still recognized the many challenges confronting the state in its use of chemicals to control fires, increase agricultural productions, and prevent forest fires. Honorof seemed more willing to accept the realities of the California climate rather than seek to control it. In this perspective she differed from the many decades of previous settlers in the Golden State of the West.[4]

The climate of southern California puzzled nineteenth-century Anglo-American settlers. Los Angeles, and much of southern California, lies within a Mediterranean ecotone. This means part of California experiences the same rare environmental system as only four other regions of the world: the Mediterranean rim, South African Cape region, central Chile, and southern and western Australia. The same mix of vegetation appears in all these regions, which includes "chaparral, foothill woodland, coastal scrub, and montane evergreen forest."[5] Three regional variations modify California's climate: the coast moderates temperatures from west to east, rainfall decreases moving north to south, and both these axes are affected by elevation. The higher the elevation, the lower the temperature and the more the precipitation until above eight thousand feet, where Californian mountaintops become "cool deserts."[6] Important to our story, key environmental elements—lightning without rain, low humidity, dry brush, and strong winds—make fire an essential component of the Mediterranean climate. Before the 1920s wildfires were an ordinary natural event in California, and certain areas burned every ten to fifty years. Excepting wetlands, deserts, and higher elevations, fire was a common occurrence. California vegetation adapted to these climate conditions, and many plants not only survived fire but needed it to propagate. Indigenous peoples understood this cycle and often set fires to preserve understory vegetative growth.[7]

Even the proper geological language necessary to describe this Mediterranean climate eluded Anglo-American settlers. As Mike Davis notes, "By no stretch of the imagination, for example, is an arroyo merely a 'glen' or 'hollow' —they are the result of radically different hydrological processes."[8] Geology also meant that Northern California fires occurred in greater numbers, while fires in Southern California were greater in size and intensity. The California landscape reflects its geographic complexity in its plant and animal life-forms. But perhaps no one could adequately understand the wildfires created by these climatic conditions, and only an artistic mind fully captures the experience.[9]

In his 1978 novella, *Angels Burning: Native Notes from the Land of Earthquake and Fire*, California author Thomas Sanchez tells the story of two famous Santa Barbara fires, the Great Coyote fire of 1964, and its 1978 descendant, which became known as the Sycamore Fire for a geographic area where one hundred homes burned down. The Sycamore Fire flared up when a kite's string, tossed about by strong sundowner winds, became entangled with power lines charged with up to 1,600 volts, creating a shower of sparks that started the inferno. *Angels Burning* captures not only the environmental devastation caused by the wildfire but the emotional one as well. Sanchez's dinner at a local café is interrupted when he hears a cacophony of sirens in the evening air and immediately seeks to return to his home before roads are barricaded. "Sundowner winds . . . were blowing fifty miles an hour, bending one-hundred-foot eucalyptus trees along the road nearly double. It was clear, with sundowner winds there was no chance for what was burning to stay in one place."[10] Sanchez notes the main road to his home, Casa Coyote, was already burning, "a mile-long red tongue lashing down the cañada."[11]

Sanchez describes the landscape just an hour later: "Thick darkening clouds piled a thousand feet overhead. It seemed only we were left against a growing wall of flame feeding off one house after another like the devouring breath of a giant dragon."[12] He continues his account. "The fire was totally out of control. There was no way to stop it before morning, no telling which way it might turn its fury. All inventions of man were powerless to halt such a wanton force of nature."[13] Sanchez, like others before him, wrestled with the complete loss of his home and all his possessions. Initially he thought, "Let it burn."

Sanchez began a dialogue with the fire, telling it to take everything, denying it joy in its destruction. Sanchez taunts the fire, telling it he refuses to rebuild. And then he realizes what a liar he is. "Of course, I would rebuild, plant the scorched earth. It's a funny thing, when you are burned out of a place your instinct is to return, not to run."[14] He included a story from the Great Coyote fire of 1964, where a man had made ten trips, evacuating his home. On the next trip, police stopped him because it was too dangerous. He screamed at them. "Come on up and help save my house, you bastards!" He drove through the barricade, only stopping when a sheriff shot his dashboard. He was led off to jail, in a dazed state. California's wildfires evoked passionate responses, both defiant and despairing, to the environmental and emotional loss of people's homes. But policy and greed meant that the entire Southern California ecotone would be in constant jeopardy of fiery devastation.

During World War II, urban planning designers and municipal officials recognized the very real challenges facing California in the upcoming years. Important agricultural areas in the San Fernando and San Gabriel Valleys represented attractive land ripe for real estate development. The approach undertaken by Los Angeles city planners used "a zoning strategy that opened the [San Fernando] Valley to hundreds of thousands . . . but concentrated new development at medium-density levels around 16 existing suburban nodes permanently separated by 83 square miles of citrus and farm greenbelts."[15] Deliberate efforts by other government entities undercut the attempt to preserve "Los Angeles agricultural periphery," such as when the county tax assessor classified farmland as prime real estate, with subsequently inflated taxes. By the mid-1950s, citrus groves in San Gabriel Valley disappeared; agricultural activities declined by 52 percent between 1940 and 1960. The construction of highways and homes consumed land. Despite attempts to preserve remaining greenbelts, environmental activists were thwarted by the corruption of the Los Angeles County Regional Planning Commission. Other events in the 1950s gave greater impetus to Forest Service fire suppression policy, events that made controlling California fires a Cold War priority.

By mid-century, fires were being recognized as potential Cold War threats. While incendiary weapons were developed in World War I, few were used. Fire as a wartime weapon reappeared in World War II, however, with the extensive use of napalm and incendiary bombs. The fire bombings of Dresden and Tokyo in 1945 proved that the use of traditional means proved to have devastating effects second only to atomic weapons. Napalm, a form of jellied gasoline, proved significantly cheaper than the atomic bomb and required only an air force. Once mass fire was recognized as a potential weapon, it demanded the attention of first Civil Defense and later the Forest Service. This awareness meant that research on how to defend, or prevent, such fires appeared as a part of the postwar agenda. Historian Stephen Pyne noted this new consciousness and differing responses of officials and ordinary citizens to the threat of fire and other kinds of chemical fallout, which included "pesticides, herbicides, and industrial pollutants."[16] California, with its annual outbreaks of rural-suburban fires, presented the exemplary case study. Given the number and severity of fires the state experienced over the decades of the twentieth century, it made sense that the Forest Service would be seriously engaged with responding to this destructive natural phenomenon. A 1956 fire started by an Inaja Indian teen, coupled with two 1957 Malibu fires just weeks apart, attracted national attention and concern. California congressman Clair Engle criticized the Forest Service's fire suppression policies, provoking intense scrutiny of fire prevention practices.[17]

The blazes also marked an important transition in fire policy from a focus on rural wildfires to those occurring in urban and suburban spaces. In an environmental system designed for periodic fires, California public officials and economic leaders had filled the valleys and hills with homes, homes that would be at risk for destruction with little recourse in prevention or control. By the 1950s Southern Californian homeowners had created a deadly mixture of suburban homes and natural chaparral, in a region that (re)grew fuel as quickly as one season. The typical suburban layout presented a landscape mostly resistant to fire, as its buildings are spread out, with pavement covering much of the ground surfaces, and home lawns and trees that did not burn well. In California, however, homeowners left brush in place that would normally be removed. They considered it a usual part of the natural scenery, one that created boundaries between the wealthy and less-than-wealthy developments. Building materials also promoted fire, most particularly wood-shingle roofs. These conditions made the use of herbicides to remove suburban brush and prevent such fires almost a no-brainer. It also represented some of the state's first herbicide wars.

The stretch of coast just outside of Los Angeles known as Malibu has burned again and again and again over the course of the twentieth century. Perhaps no place better represents the challenges of urban-suburban-rural fires in California's Mediterranean ecotone. Major firestorms, those greater than 10,000 acres, have devastated Malibu thirteen times over six decades from 1930 through 1996. Approximately more than five thousand homes have burned in the same period. Like those who insist on living in floodplains, wealthy Malibu homeowners have received subsidies for land use, insurance, and disaster relief that help them rebuild, despite the fires' reoccurrence in equal or greater intensity as long as the suburban sprawl exists.[18]

Responding to Congressman Engle's criticism of the United States Forest Service (USFS) and its commitment to fire suppression, the San Bernardino *Sun* ran a twelve-part series in 1957 examining various opinions on the benefits of controlled burns to remove brush, a practice that had some measure of success in Northern California. Along with Engle, some of those interviewed advocated controlled burns along the lines of those done by Native American and early Mexican and Anglo settlers. Many of the experts interviewed also noted the challenges associated with controlled burns, most particularly weather conditions, such as humidity. In Southern California, chaparral represented the most significant obstacle. S. A. Nash-Boulden, a veteran Forest Service official interviewed in part one of the series, emphasized that brush had to be eliminated and structural changes made to better contain fires once they started. He noted there were some aids to improve the response and outcomes to wildfires, including the "application of chemicals and other known methods."[19] It was where and how such chemicals were applied that concerned activists like Honorof.

Writing more than a decade later in a 1971 issue of *A Report to the Consumer*, Honorof cited a California Department of Agriculture requirement that areas sprayed with pesticides be posted with a warning notice that included the date of application and prohibited people from entering the area for at least two weeks. She then noted that no such warning signs were posted in places like the San Bernardino Forest after they had been sprayed. When questioned about this decision, the local United States Department of Agriculture (USDA) Forest Resource Officer in San Bernardino, Hatch Graham, stated that the information the USDA had received showed that 2,4-D and 2,4,5-T did not remain as soil contaminants.[20]

The USDA planned an expansion of the spraying program in Maloney Canyon, Fredalba, Penrod Canyon, Meyers, and near the Bear Burn Fuelbreaks. In the same newsletter, Honorof included updates on several scientific and governmental reports, including information on the toxicity of the phenoxy herbicides and other herbicides and pesticides used on California public lands. She decried the continued use of the phenoxy herbicides domestically when spraying those same chemicals had been "banned" in South Vietnam by the Richard M. Nixon administration. A telegram, signed by several thousand Californians and people from other states, had been sent to President Nixon on January 28, 1971. It asked that 2,4-D and related herbicides that had been banned in South Vietnam also be proscribed in California and other parts of the United States. The plea fell upon deaf ears, and Honorof and other activists continued to fight herbicide use throughout the state.

Southern California officials charged with controlling fire found their jobs unpopular at best and often the subject of intense public criticism. While not about the phenoxy herbicides, one example of a public hearing on herbicide use highlights such conflicts. In January 1972, seventy thousand Los Angeles County homeowners were notified that they must remove weeds, brush, and rubbish from their properties or be subjected to herbicide sprayings conducted by the county fire department. Chief Nino Polito was quoted in the piece informing the public that the herbicides would decompose after six months and become "nutrients for the soil" after decomposing. Honorof contacted Polito and discovered that Simazine, a photosynthesis inhibitor manufactured by Ciba-Geigy, was the selected herbicide. Honorof then wrote to Ciba-Geigy about the claim that the herbicide decomposed to become a soil nutrient, a claim promptly denied by the company. Honorof listed the product's warning label, which advised that the herbicide should not be applied to food or feed and be kept away from water supplies such as lakes, streams, or ponds. By March county supervisors had declared a moratorium on herbicide use pending a public hearing to investigate the safety of the herbicides being used throughout the county. The comprehensive moratorium lasted only five days, when the same board of supervisors revoked it for all uses except those by the fire department's weed abatement program. Honorof and other activists spoke at the subsequent public hearing, making the case that the proposed chemicals were too dangerous for use.[21]

The public hearing featured the testimonies of the chemical industry, state agricultural officials, university researchers, and environmental and health activists. The newly constituted Herbicide Hearing Committee was composed of county administrators from the public health department, the farm advisor, the agricultural commissioner, and the veterinarian. The committee heard testimony on April 4 and 5 about the use of Simazine and other herbicides, including 2,4-D and 2,4,5-T, and produced a report that they submitted to the Los Angeles County Board of Supervisors several months later. Among those testifying were locals such as Malibu resident Dr. Charlotte Taylor, a UCLA biochemist; private citizens like Billee Shoecraft; and various representatives of health and environmental groups like "Cancer Victims and Friends" and the Angeles Chapter of the Sierra Club. Speakers also included officials from agricultural, industry, and county agencies.

Honorof also spoke, with her listed affiliation as radio station KPFK-FM where her consumer reports were broadcast. According to Honorof's account of the hearing, although only Simazine had been singled out for scrutiny, the meeting rooms were filled with representatives from chemical companies like Amchem, Ansul, DuPont, Chevron, and the like. These individuals had traveled to the hearing from outside the state, a routine part of their corporate duties. DuPont had originally been scheduled to speak along with a Ciba-Geigy rep but chose not to, according to Honorof, because they recognized that attention was focused on Simazine, and they wished to avoid scrutiny of their own products. Honorof expressed an even more scathing opinion of the representatives from agribusiness in universities and colleges, questioning whether these individuals did not have a conflict of interest. She detailed the testimonies of those opposing the use of herbicides.

Honorof recounted some of the critiques offered by the citizen activists in her newsletter. Dr. Ruth Harner, the author of *Unfit for Human Consumption*, opened the hearing. Government agencies, she charged, had allowed themselves to become "pesticide pushers" for the chemical companies. Two other witnesses asked what was known about simazine—how did it work, what were its effects, what plans were in place for protecting children, a group especially sensitive to the chemical? This question appeared especially important considering potentially seventy thousand backyards would be sprayed. Dr. Granville Knight (Billee Shoecraft's California physician) introduced the subject of 2,4-D and its harmful effects. When it came time for Honorof to testify, she too indicted the phenoxy herbicides—which had been banned in Vietnam—and the chemical companies. She noted "that the USDA and the chemical agribusiness [had] but one objective in life and that is to sell chemicals." She contrasted this stance with those testifying against CBW who had no such self-interest.[22]

Among the groups Honorof claimed had been affected by pesticide spraying were Swedish railroad workers who had died after being exposed to 2,4-D and 2,4,5-T. Honorof also included a scornful, if brief, summary of the chemical company representatives' testimony. Dr. Frank Lyman, testifying for Geigy, provided a stupefying answer when asked why spray warnings were posted for livestock but not for places where children might be exposed. He explained, "Because we are not cannibals."[23] Honorof condemned state officials she pejoratively labeled "Agribusiness" who supported chemical companies' assertions of the various pesticides' safety. As she pointed out, these public officials were supposed to protect the public, not the chemical industry. She ended the newsletter with the hope that the testimonies provided had given the Herbicide Commission pause, and that Los Angeles County would stop the "chemical biological war being waged on the people of the United States by Agribusiness."

Overall, the Herbicide Hearing Committee issued a seemingly balanced report. While it still endorsed the use of Simazine and the 2,4-D herbicides, the report acknowledged concerns over 2,4,5-T and its dioxin contaminant. The committee recognized the need to establish baseline levels of the chemicals, the need for trained personnel to properly apply chemical herbicides, that proper warning signs be posted for animal and human exposure, and a ban against aerial application. The committee recommended the creation of a permanent body to oversee the issues raised. It affirmed the right of private citizens to refuse the county's application of herbicides on their private property, although the homeowner would be required to pay for any alternative method of weed and brush removal. The committee's decision obviously privileged individuals with more wealth. The antiherbicide activists displayed an equally troubling class bias when they urged that welfare recipients and prisoners be used to provide manual labor to remove weeds and brush. The committee estimated a switch from chemical control to manual labor would increase costs from $170,000 to $2 million. Although the report sanctioned the status quo, it acknowledged the concerns raised by those opposed to herbicide use and the need to monitor both the chemicals being used and those applying those chemicals.[24]

When Honorof labeled the group of chemical industry, elected officials, and university scientists as "Agribusiness" she spoke both literally and in condemnation. With crops valued at $42.6 billion in 2012, agricultural represented one of, if not the, preeminent businesses of California, and the agricultural lobby carried tremendous influence and power. Dating back to the late nineteenth century, Californians had taken several actions that by the postwar period created a farming landscape composed of a patchwork of monoculture fruit and nut crops. Steven Stoll's study of California growers dominated the production of fruit crops that included apricots, prunes, plums, lemons, raisins, and nut crops of almonds and walnuts, providing between 60 and 100 percent of the national total. While this kind of intense crop cultivation appeared all over the state, it was especially prevalent in California's Central Valley, with its interior plains fed by two major rivers and their tributaries— the Sacramento River flowing southward in the northern region and the San Joaquin River, which flows northward in the southern region.[25]

As Americans moved westward, constantly looking for land, the Central Valley promised good farming land. Early Anglo settlers planted wheat crops but realized that the crop depleted soil fertility and hindered settlement. The realization that the San Joaquin River could be used for irrigation versus transporting crops to market helped end wheat's reign. Engineering canals to distribute river water supported breaking up the large wheat empires into smaller farms of approximately 220 acres, and irrigated lands rose from 1.2 million acres in 1900 to 4.2 million in 1920. The creation of smaller farms produced rural settlement, especially once land developers realized the opportunity to profit from a change from large-scale agriculture to one of intense cultivation of monocrops. Central Valley's population increased significantly, from 54,000 in 1890 to more than 800,000 in 1930.[26]

The settlers attracted to this area viewed it as an investment, and many improved their lands and sold them again for a profit. By the 1920s much of the farmland was owned by single-owner growers and corporations. Unlike with farm ownership in the rest of the country, California flipped the statistics, with only 30 percent of owners working the land, and 70 percent of the labor provided by fieldworkers or tenant farming, a phenomenon especially pronounced in the post-1945 period. The Imperial Valley, a landlocked desert 110 miles east of San Diego, followed this settlement and farming pattern once the region was irrigated using the Colorado River in 1901. By 1934 journalist Carey McWilliams described the settlers of California's Imperial and Central Valleys as something different. "The old-fashioned farmer has been supplanted by a type to which the term can no longer be applied with accuracy. The new farmer is a grower. He is only semi-rural. Often he regards his farm as a business and has it incorporated."[27] McWilliams continued with his description, emphasizing that this new agriculturist resembled a bookkeeper or banker more than a working man, a businessman adept in red-baiting, manipulating public perceptions, and exerting influence in state politics.[28]

Although the phenoxy herbicides were a postwar addition to California agriculture's chemical arsenal, their use followed the early pattern that emerged between California growers and state universities, another crucial player in the emergence of "agribusiness." Specialized crops meant that while California agriculture avoided complete monoculture, there was still a significant decrease in natural flora and fauna. This "simplified ecology" made fields, orchards, and vineyards vulnerable to numerous non-native insect and plant life. The initial distrust growers displayed toward experts slowly dissipated. The University of California started the Agriculture Experiment Station, which became a major ally in the growers' battles with insects and weeds, along with scientists, universities, state agencies, and industry. "The university, largely by winning the allegiance of orchardists to chemical cures, rendered itself the central institution in the matrix of industrial farming."[29] Charles W. Woodworth's arrival at the university shaped the role university scientists would play in the California agricultural landscape. The university increasingly helped regulate industry sprays, but when orchard spraying became commercialized, the university no longer could guarantee the quality of products. Woodworth saw a vital role for private industry, as it could reach so many more growers and disseminate the information and research done at the experiment stations to hundreds of growers, significantly greater numbers than the university itself could reach.[30]

Research scientists did investigate the use of chemicals in making more land available for crops and grazing. As with the situation in Arizona, increasing water yield through brush control represented one of the US Forest Service's major charges. Lack of water affected Southern California most acutely. Studies conducted in the 1960s suggested that grass cover worked better than other types of brush or naked soil in increasing water yield. In promoting grasslands and removing brush, the Forest Service had followed a specific procedure in California. The first step involved clearing fire break and "mashing the brush with a bulldozer."[31] Next a controlled burn would take place, and then the area would be planted with grass. Finally, herbicides would be sprayed to minimize brush regrowth.

One case study performed in the early 1960s experimented with skipping the mechanical removal or controlled burn steps and using herbicides to remove brush. In places where the use of fire was not recommended, or with rough and hilly terrain that made mechanical removal difficult, the approach made sense. The test case showed good results, and it marked an increased attention to converting brush to grass. As one study published in 1969 noted: "Approximately ten million acres of California land supports a vegetation cover of chaparral. These lands are increasingly important as watersheds, grazing lands, recreation areas, homesites, and for their esthetic value."[32] The 2,4-D was sprayed on plots of land located in western Los Angeles County to see if soil moisture, necessary for grass cover, could be improved. Results suggested that sprayed lands saw better grass growth, an argument for the use of 2,4-D.[33]

Chemical herbicides and pesticides became a part of these new "farm factories," and their use went almost unnoticed in the postwar period. As historian Linda Nash observed in her study of environment and disease in California's Central Valley, complete records of pesticide use in California exist only from the 1990s, and the piecemeal earlier records do not indicate the amount of chemicals used. Growers had applied the insecticides available since the early twentieth century because, as many noted, "It paid to spray."[34] But as early as 1948, individual farmers protested the unintended casualties of bees, alfalfa crops, citrus groves, and vineyards from 2,4-D use. Complaints from San Joaquin "vineyardists" regarding financial losses in the thousands, possibly millions, from 2,4-D spraying that destroyed their crops, along with losses suffered by cotton growers, prompted the state to toughen regulations applying to individuals who used the chemicals. Still, herbicides like 2,4-D were touted as "promising" by some farmers for crops like alfalfa.

Unlike the negative environmental effects on crops, the health concerns posed by the phenoxy herbicides got lost in these early postwar decades, because so many more obviously lethal chemicals were being used. Parathion, an inexpensive, persistent, and toxic organophosphate, represented one of the most significant problems, identified as the causative agent in several cases of group illnesses after exposure to the chemical. But as the example of Los Angeles County Herbicide Committee suggests, by the 1970s California public officials were starting to pay attention to issues of groundwater contamination, possible illnesses, and the need to regulate aerial spraying of herbicides.[35]

By 1970 the dangers of at least one of the phenoxy herbicides was brought to the attention of Californians, and Americans more broadly. Congressman Richard McCarthy's visit to Globe, Arizona, in February 1970 (discussed in the last chapter) prompted national coverage. Coverage of an announcement by White House science adviser Lee DuBridge on the Bionetics report and its findings showing that 2,4,5-T with regard to birth defects explained the growing concerns about the herbicide's use. State agricultural weed control expert Murray Pryor noted that 2,4,5-T was primarily used by ranchers in foothills to improve grazing. The chemical was expressly forbidden in areas containing grapes, cotton, and other vulnerable crops. According to Pryor, the California State Division of Highways had sprayed 2,4,5-T but was now reconsidering its use. He reassured Californians, stating that any spraying of large amounts of the chemical required a permit from the county agricultural commissioner, and that the spraying needed to meet state regulations.[36]

Yet just a little over a year later, Honorof's newsletter suggested far more widespread use of 2,4-D and 2,4,5-T on California farmlands, challenging Pryor's reassuringly vague information. Honorof claimed that the USDA paid farmers $9 million to spray 2,4-D and 2,4,5-T on 2.5 million acres of land. "A California State wide reporting system reveals that for Jan. and Feb. 1970, 400,000 acres were treated with one and/or the other or the combination of the two (teratogenic) chemicals."[37] Ranchers received $3.7 million for participating in a brush and weed removal program sponsored by the federal government's Conservation Program. Honorof too quoted from the Bionetics report regarding the hazards of both 2,4-D and 2,4,5-T. Despite the report's findings, the USDA approved the phenoxy herbicides for numerous food crops, including fruits like apples and pears; vegetables like tomatoes, cabbage, and beets; grains; livestock feed; and timber. It was probably no coincidence that new, stricter spraying rules were announced just a week after McCarthy's visit to Globe. Multiple sources reassured Californians that while the herbicides were safe to use, reasonable precautions were taken to minimize potential harm. University scientists represented one of the largest groups voicing the safety of herbicide use in California.[38]

Two events within ten days of each other exemplified the state of affairs in California in the early 1970s. In mid-January 1971, a three-day meeting in Sacramento of "weed men" showed the heavy involvement of the academic community in agricultural discussions of herbicide usage. The meeting offered a half-day "plant science school," which included talks by University of California at Davis (UC Davis) faculty. Among those presenting were UC Davis botanists, an extension weed expert, a UC Davis agricultural pesticide safety specialist, and an agricultural engineer. Also featured on the conference program were representatives from Dow Chemical and Hercules, Inc., along with employees from state agencies, primarily agriculture. Yet just ten days later, Billee Shoecraft and Honorof were quoted in an Associated Press story warning about the dangers of chemical herbicides. The same "dangerous herbicides used in Vietnam as defoliants" were sprayed on "food crops, forests, and watersheds in California," according to the women, identified as consumer advocates in the article. Their charges prompted a response by both the US Forest Service and the California Department of Agriculture, whose spokesmen denied any "deformities or deaths attributable to the herbicides [2,4,5-T or 2,4-D]."[39]

Honorof continued her food consumer activism beyond her radio broadcasts throughout 1971, giving lectures in settings like the Hollywood Hilltoppers Republican Women's Club, or emceeing the showing of *Action for Survival*, a documentary featuring Ralph Nader, Congressman James J. Delaney (D-NY), Eddie Albert, and Adelle Davis. The film examined water, soil, and food pollution. Honorof's exposé of another chemical episode would win her reporting honors.[40]

In 1973 Honorof led the investigation of contaminated lettuce from California's Imperial Valley that had been shipped to cities across the county. Although the lettuce was not contaminated with either of the phenoxy herbicides, the case demonstrates Honorof's degree of engagement and legitimacy as an investigative reporter. Monitor 4 (M4), the pesticide used, was an organophosphate insecticide, an especially toxic group of chemicals. The Food and Drug Administration (FDA) had set limits of one part per million (ppm) as the maximum residue level. It made national headlines when FDA testing showed contamination ranging from 3.4 ppm in Oxnard (California), to four ppm in St. Louis (Missouri), to a high of six ppm in Rochester (New York). The announcement led to a February press conference held by farmworker activist César Chávez, president of the United Farm Workers union (UFW). Chávez and UFW members had raised questions over Monitor 4 in 1972, when several fieldworkers discovered lettuce that had been burned after the pesticide's application. The workers themselves then experienced a range of symptoms, including nausea, dizziness, and burned skin—evidence of toxic poisoning. The California Agriculture and Food Department tested the lettuce but found no evidence of M4 contamination. Later investigations by the Environmental Protection Agency discovered that this negative result was accomplished by removing the outer leaves of the lettuce. The agency subsequently banned M4, but the ban was removed in March 1973. California congressional representatives appeared with Chávez at the press conference, along with Honorof, who charged that over forty thousand acres of Imperial Valley lettuce had been sprayed with M4. Honorof put her bully pulpit to good use.[41]

Honorof's investigation into the lettuce contamination was based on leaked documents and interviews with various officials, including the UFW and EPA. She charged the Agriculture Department with deliberate negligence, claiming that officials had known about M4 residue on lettuce since December 1972. Even worse, the department received communications from Chemagro, the company that manufactured M4, urging an immediate cessation of M4 spraying on lettuce until further notice, in late January of 1973. Excerpts were reprinted in *A Report to the Consumer*, and Honorof further explained how a rig operative had noticed burned lettuce in the fields, conditions that were reported to the chemical companies. Hundreds of acres had already been harvested and were in warehouses across the state. Honorof summarized her discussions with agriculture officials, who tried to justify their shoddy testing for chemical contamination.

Honorof also reported on a study done on migrant workers in Texas and Florida that documented the "deplorable state of health and welfare among migrant farm workers."[42] Honorof noted that big business had overtaken family farming and that this phenomenon had consequences. One of the most negative results appeared to be the millions of taxpayer dollars that went to agricultural research, conducted at state universities by "mad captive scientists" who worked for chemical companies like Dow, Monsanto, Thompson-Hayward, Shell, and Dupont. "They care very little about the planet, or about the health of the people." Instead, this group cared only for "*yield per acre. Make as much money as possible, worry about effects later.*"[43] Honorof ended the newsletter with an idea. If entomologist Dr. Blair Bailey really cared about the health of farmworkers, he would go work in the fields, harvesting crops. To make it a valid experiment, his entire family—wife, children, grandchildren—should be in the fields with him, living in inadequate housing and eating a subsistence diet. Honorof was convinced he would change his mind and practices. Her investigative reporting was recognized with an Associated Press award for journalism.[44]

The Monitor 4 episode marked the intersection of Honorof's critique of big agriculture's use of chemical pesticides—which in parts of the state heavily depended on the phenoxy herbicides—and agricultural workers' demands and activism for safer workplace environments. The best-known agricultural worker protests over pesticides would be the series of strikes and boycotts led by Chávez and the UFW in three periods: (1) the initial organizing done around grape crops in 1963; (2) the successful UFW strike and boycott of grapes and Gallo wine from 1965 to 1971; and (3) the protracted boycott that started in 1984 and finally ended in 2000. These protests present one significant problem for this study with respect to several factors. The phenoxy herbicides *harm* grape vines, cotton, and many fruit and nut crops that were the focus of UFW action. California grape growers routinely complained about phenoxy herbicide drift that killed their vines. The acute toxicities agricultural workers dealt with demanded attention be given first to other chemical entities like the organophosphates. Work done by Laura Pulido, Linda Nash, and

Adam Tompkins explores the role of the UFW's labor and environmental activism and is discussed in greater detail in chapter 9.[45]

Honorof did more than simply mobilize the public over various chemical contaminations, including the phenoxy herbicides. She acted as well. At a March 1973 legislative hearing on the safety of Monitor 4, Honorof led a rowdy group protesting chemical company claims of M4 safety. The previous day farmworkers and physicians had testified that they had experienced a host of ailments, including dizziness, nausea, and vomiting. The Friday session featured witnesses from the Chevron Corporation, the manufacturer of Monitor 4. Denied the right to testify first, although she had exposed the story, Honorof still made her presence felt. "Several times during the hearing Mrs. Honorof interrupted with shouts such as: 'One drop on your hand will kill you.' The spectators, many of them young, backed her up with cheers and clapping."[46] While her broadcasts on KPFK 90.7 FM and her newsletter continued to disseminate her message, Honorof also joined other Californians in their protests against chemicals. Given her early and loud condemnation of the phenoxy herbicides, her voice was especially prominent in the last area of contested use of 2,4-D and 2,4,5-T in California, in its national parks and forests.

Forestry experts debated the merits of controlled burns, also known as broadcast fires, as seen in a 1969 article in *Science*. In California, University of California Berkeley forestry professor Harold Biswell advocated for controlled burns as a means of reducing underbrush, dangerous fuel for fires. But the US Forest Service officials in California offered reasons to use controlled burns judiciously, among them fear that local landowners might set broadcast fires, the fear of runaway fires, the cost of the conducting controlled burns, or the risk of undercutting national programs like Smokey the Bear and Keep America Green that worked to decrease accidentally set fires in recreational areas. As one Forest Service official put it, "The public . . . can't see why we allow a fire we set in November to burn when we'll jump in with a thousand men to put out a wildfire that starts in June."[47] Moreover, the Forest Service was exploring a host of alternatives to human-set fires, including "cheap herbicides to remove undergrowth." At the very least, many considered herbicides to play a role in preparing an area for a controlled burn. But by 1970, when the disturbing reports out of South Vietnam about the possible toxicity of the phenoxy herbicides, particularly 2,4,5-T, began to appear, university and Forest Service scientists exercised caution when conducting controlled burns.

Given the Cold War concerns over incendiary weapons, California's powerful agriculture industry and its need for water and grazing lands, and the need to remove brush to reduce wildfires' intensity and frequency, the involvement of the Forest Service in testing and using the phenoxy herbicides should not be surprising. Experimental Forest and Range Stations offered spaces to test the use of the phenoxy herbicides, like the one in southern California's San Dimas Forest. Here, research foresters worked on replacing brush with grass to increase water yield. The Forest Service faced challenges in the demand for more trained foresters; the uncertainty about when and what kind of vegetation should be preserved or removed; and the recognition that more research was needed.[48]

One of these early experiments with the phenoxy herbicides was undertaken by the Forest Service at Lassen National Forest from 1951 to 1960. The use of the herbicides was considered "one of the fastest, easiest, and cheapest ways of increasing the grazing capacity rangelands."[49] The authors noted that the phenoxy herbicides worked well in removing several species of unwanted brush, although the full extent of the chemicals' range needed to be further studied. The results, however, suggested that the chemicals appeared promising for brush control for stockmen and wildlife management. Later studies supported this position. Research on brush control continued throughout the 1960s.[50]

Spraying 2,4-D represented a routine component of the controlled fires and brush control efforts undertaken in California forests and rangelands. The California Division of Forestry offered its services to local landowners regarding the best herbicides to use in brush eradication. Along with the Forest Service, vegetation researchers and technicians worked to increase water yield, decrease chaparral brush, and increase grazing rangelands. The chemical herbicide was used after fire and mechanical means had cleared brush and was intended to prevent the growth of weeds in the newly cleared land. One treatise on the chemical control of chaparral calculated that an investment cost of $30,000 ($10 × 3,000 acres) would be repaid in about nine years. Although California FS rangers lagged in their support of controlled burns, also known as broadcast fires, by 1970 they had started trials at the Fire Laboratory, located in the San Jacinto Mountains. The Forest Service was joined in its brush removal efforts by researchers from the University of California Agricultural Extension Office and the Division of Forestry; the agencies held a "brushland conversion" workshop in June 1972. A representative from Holm Timber Industries presented at the workshop, telling attendees about his company's costs in aerially spraying herbicides to begin the process of brush conversion. The decision to use herbicides for forest and range brush control put the Forest Service and its allies into direct conflict with Californians, who were increasingly concerned about chemical sprays by the 1970s.[51]

While Honorof appeared as one of the most visible critics, ordinary citizens echoed her concerns. A January 1971 headline in the San Bernardino County Sun proclaimed, "Dangerous Herbicides Being Used in State." Both Billee Shoecraft and Ida Honorof were quoted in the article, with Shoecraft connecting the herbicides to those being used as defoliants in Vietnam. She recounted the deformities she had found in pet guinea pigs, goats, and pigeons. US Forest Service officials denied that any deformities or deaths had been caused by the chemicals. They acknowledged that thirty thousand acres of the state's seventeen federal parks had been sprayed. Honorof criticized the Department of Agriculture for its proclamations of safety, although state agricultural official Si Nathenson noted that neither 2,4-D or 2,4,5-T were used on food crops. The only deformities, according to Nathenson, happened when lab animals were injected with too much of the chemicals. Two letters to the editor published by the Sun later that year reiterated Shoecraft's and Honorof's points. The first writer quoted from noted environmental and occupational health physician Dr. Samuel Epstein and noted that 2,4,5-T had been linked with birth defects in animal studies. The extensive and increasing use of both 2,4-D and 2,4,5-T justified more research. The letter charged that the "continued use of these herbicides in the environment constitutes a large-scale human experiment in teratogenicity" and ended with the question "Do you want your unborn baby to be a possible victim in this experiment?"[52]

The second letter emphasized a different concern, the potential environmental harm the chemicals presented. It connected the chemicals to the situation in Vietnam, where it claimed forest regrowth had been delayed for as much as ten years. The chemicals represented "a renewed threat to the balance of nature and to human life."[53] The situation went from words to actions when protesters demonstrated outside Cleveland National Forest supervisor Kenton Clark's offices just a month later. While Clark acknowledged that the picketers were most likely sincere people, he thought the Forest Service had made the right decision. Environmentalists continued to challenge that decision, with mixed results.[54]

Conservation groups brought a lawsuit to ban the use of 2,4-D and 2,4,5-T that were supposed to be sprayed on three thousand acres of the Cleveland National Forest. Led by the Sierra Club, the groups made the phenoxy herbicides' effects in Vietnam central to their case, voicing concerns about birth defects and miscarriages as well. The conservationists even went so far as to bring Billee Shoecraft to testify. Reversing his previous decision, Clark acknowledged that "because of the extreme emotionalism" spraying plans had produced, the Forest Service had decided not to use 2,4,5-T. They would, however, still use 2,4-D and 2,4-DP, a chemically related herbicide. Clark noted that while the concentration of 2,4,5-T would have been diluted compared to that used in South Vietnam, and he justified the change in part because less than 10 percent of the area would have been sprayed with 2,4,5-T. Clark also took pains to emphasize that the revised plan applied only to the specific project. The news coverage of the hearing and outcome followed up on the previous demonstration outside Clark's offices. The group included Mrs. Virginia Taylor, a San Diego mayoral candidate, who decried what she called the Forest Service's "public-be-damned" attitude.[55]

Protesters kept the pressure on, fighting against another planned spraying of 2,4-D and 2,4-DP in the Cleveland National Forest in 1973. In this second case, Clark now argued that only the Environmental Protection Agency could ban chemicals, remonstrating that those opposed to the herbicides should report problems to that agency. The Sierra Club had remained engaged in the herbicide challenges, and the second spraying had mobilized a group of San Diego county women to circulate petitions making the same claims of reproductive harm. Just weeks later an environmental group filed a lawsuit seeking to halt the spraying of the phenoxy herbicides.[56]

People for Environmental Progress (PEP) represented the youth of America. An incorporated nonprofit, the six-hundred-person group was led by sixteen-year-old Eric Ellenbogen, a Beverly Hills High School student. The group had singled out the Forest Service, because it used the most 2,4-D and 2,4,5-T in Southern California. The chemicals were applied using airplanes, which guaranteed chemical drift. While the chemicals were intended to protect waterways, they were not supposed to be used near water supplies. Although the average member might be in the seventh grade, the group attracted local allies like the local city attorney. The group's concerns aligned with national, better-known advocates like Ralph Nader, who had brought a lawsuit against the EPA. Another group provided an important source of support—biological scientists. Dr. Richard J. Vogl, a professor of plant ecology and editor of the journal *Ecology*, noted the dangers of the forest spraying: "I wouldn't take my family anywhere near a forest sprayed with these substances."[57]

Honorof disseminated Vogl's assessment via her newsletter, which cited a news article highlighting his concerns regarding the herbicides' dioxin contaminant and which also included National Cancer Institute study confirming his assessment. Vogl offered a long list of problems associated with herbicide use. One dealt with the ineffectiveness of using the chemicals to prevent fires. As detailed in the lawsuit, the chemicals preserved flammable brush and intensified the volatility of brush with their own flammable nature. Worse yet, the chemicals made subsequent fires to clear brush more toxic. Effective fire control needed to use unadulterated fire to control it. Other "noted environmental scientists [had] filed depositions or [would] appear in behalf of the cause."[58] Two experts submitted testimony in support of the PEP's claims of harm.[59]

Samuel Epstein's work on the chemical causes of cancer made him a well-recognized voice in health and environmental groups. The other expert who submitted supporting testimony, however, was less well known, but as a Food and Drug Administration (FDA) division of toxicology official, her opinion carried weight. Dr. Jacqueline Verrett's affidavit reported on findings from her laboratory that showed 2,4-D and 2,4,5-T showed teratogenic effects. Verrett noted, "From our testing we must conclude that dioxin, an ever-present contaminant of 2,4,5-T is some 100,000 to 1,000,000 times more potent in its capacity to cause birth defects in the species tested than thalidomide."[60] Epstein made four major points in his submitted testimony, emphasizing the toxicity of the chemicals, the dioxin contaminant, and the potential harm done by grazing cattle on lands sprayed with the herbicides and the potential for increased amounts of toxic dioxins from incomplete burning of brush sprayed with the herbicides.

PEP had not only scientific allies but political ones as well.[61] Political discussions of the phenoxy herbicides' safety reflected the same divisions as scientific ones. At the same time as the legal hearing, state senator Alan Robbins had introduced a resolution banning herbicide spraying in California national forests. Robbins expressed special concern about spraying in the Angeles National Forest, both because of its extensive waterway network and because of its proximity to the heavily populated San Fernando Valley. In his introduction of the legislation, Robbins invoked the familiar charges against the chemicals offered by activists and used them to support his proposed legislative solution. In contrast to Robbins's opposition to the use of the phenoxy herbicides, former state assemblyman Charles Conrad defended the use of the chemicals. Conrad charged that scientists' criticism of the chemicals was politically motivated, although he failed to explain further. He also

challenged the idea that the chemicals had been used only in Vietnam, pointing out that the herbicides had been routinely used in agriculture before the
war. The most significant critique Conrad offered of the scientific studies in-
dicating teratogenic effects focused on the concentrations used. The amounts
administered in a laboratory setting bore no relation to the much lesser concentrations used by farmers and the Forest Service.[62]

Conrad and Honorof exchanged words in letters to the editor, with Honorof citing a National Institute of Environmental Health Sciences study showing the phenoxy herbicides' teratogenic effects. Conrad responded by comparing Honorof's claims to those he heard uttered by Jane Fonda upon her
return from North Vietnam. He applauded the judge's decision to not halt
chemical spraying. And he argued that he would be glad to be sprayed with
and drink water sprayed with the chemicals. Along with her personal appearances and letters to the editor, Honorof also covered the Forest Service
spraying project in her bi-monthly newsletter, marshalling the extant scientific findings on the hazards of the phenoxy herbicides and calling for them
to be banned. She, like other Californians, continued to protest chemical use
despite the failure of PEP's lawsuit.[63]

In 1973 Citizens Concerned with Pesticide Poisoning (CCPP) formed in
response to paraquat spraying in Los Angeles parks, signifying the ongoing
concerns chemical sprayings provoked among citizens. The group shared the
same concerns as PEP as well as some of the same people. Honorof, identified
in newspaper coverage as a consumer activist focused on herbicide use, appeared at the first demonstration, while Senator Robbins was slated to speak
at the second protest. The second rally was expected to attract several hundred people, and California senator Alan Cranston had pledged to send a representative. Cranston also released a statement urging mandatory agency review of pesticide use. The rallies were intended to support City of Los Angeles
Parks and Recreation employees like Harry Fales, who had been hospitalized
after spraying paraquat. The city had denied his workman's compensation
claim based on the lack of proof that the chemicals had caused his illness.[64]

Shortly after the public rallies, the Department of Parks and Recreation
was ordered to stop spraying paraquat because a product-label warning explicitly advised against the chemical's use in areas where children and animals
play. The ban came after a prominent agency official had publicly defended
the safety of the chemical. Just a few months after the CCPP protests, city
councilman Joel Wachs called for an investigation into the use of chemicals
by all city agencies. He argued that the city needed to know what chemicals

were being used, to better safeguard the health of both city employees and the public. His actions, along with Robbins's, suggest one way grassroots activists positively affected local policies, even as their attempts within the legal arena failed. While the chemicals might not be banned outright, protesters drew the attention of public officials and placed the question of chemical safety on the agendas of city, state, and national officials.[65]

Ida Honorof's activism in California continued the work of Rachel Carson in its focus on chemicals and their use in the natural and built environments, and in her efforts to use radio and print media to arouse citizens. Honorof became part of a network of local and regional activists protesting chemicals in general and the phenoxy herbicides specifically, as her appearances with Billee Shoecraft provide evidence of their joint efforts to alert the public about these hazardous substances. Like Shoecraft, Honorof challenged postwar chemical policy and regulation. An October 1976 newsletter reported on one of the victories in the fight against wanton pesticide use. By that time the war in Vietnam had officially ended, and it had been more than five years since defoliation missions were halted. The phenoxy herbicides had drawn more negative attention, specifically 2,4,5-T and its dioxin contaminant. Where and how should these chemicals be disposed of? Air Force officials found it difficult to dispose of the remaining military supplies given concerns over 2,4,5-T's dioxin contaminant. One "hop-head," as Honorof called him, in Corvallis convinced the Air Force to ship surplus defoliants to him for experimental use. Honorof had tried to bring criminal charges against the professor, a "consultant for the forests of Western Oregon," but the man remained free, publicly proclaiming that "Agent Orange is harmless!"[66]

Honorof reviewed the current state of affairs, noting studies that showed contamination of Vietnamese shrimp with dioxin. Based on information provided by Shoecraft, Honorof speculated that recently sprayed forests had probably released significant amounts of dioxin after burning. She also praised the efforts of a grassroots group, Citizens Against Toxic Sprays (CATS), located in Oregon and fighting to stop the spraying of the phenoxy herbicides in the Siuslaw National Forest. The group, organized and led by a young wife, mother, and aspiring writer, challenged the "bitter fog" of the phenoxy herbicides as embodied in the policies and practices of the USDA and US Forest Service.[67]

CHAPTER SIX

Timber and Rights-of-Way

Carol Van Strum and Herbicide Spraying in Oregon

Carol Van Strum began her 1983 book, *A Bitter Fog*, describing a typical childhood activity. "On a spring morning in 1975, thirteen years after the publication of *Silent Spring*, four children went fishing in the river bordering their home in the wilderness of the central Oregon coast."[1] She went on, though, to describe the way this idyllic day turned into a nightmare that would consume years of her life. "On this particular morning, however, a heavy tank truck . . . inadvertently [sprayed] the four children fishing down below."[2] The spray contained 2,4,5-T, which was being applied to remove roadside brush. Along with 2,4-D, the chemicals were routinely used to remove weeds along roadsides and railroad rights-of-way. The children spent the night with headaches, vomiting, nausea, and cramping. Van Strum herself became ill as well, as she had been exposed when she went through the brush to bring the children home.

Over the next few weeks, the Van Strums watched while plants died, farm animals were born with deformities, and the family dog suffered sores and paralysis, all from unknown causes. Van Strum's experience with growth-regulating herbicides proved to be a typical one, although her decision to confront public officials over the routine use of such herbicides was not. Her fight against the phenoxy herbicides challenged scientists, industry, and public officials over the safety of potentially toxic chemicals and helped mobilize the broader Oregon populace. In the process, and influenced by Rachel Carson's *Silent Spring*, Van Strum emphasized the ways democratic decision making had been subsumed by technocratic bureaucracy. Like Shoecraft and Honorof, Van Strum and other Oregon activists fought over chemical use and regulation. Like

Figure 8. During her fifty-plus years of activism, Carol Van Strum amassed more than twenty thousand documents dealing with chemical use and regulation, including documents from her protests over the use of Agent Orange herbicides. Courtesy of Risa Scott Studios.

Shoecraft, Van Strum and others decried the intimate relationship US Forest Service personnel had with the timber and chemical industries.

Unlike Shoecraft and Honorof, however, the Oregon activists achieved some visible successes that directly affected chemical policy regulation. In publicizing her personal fight against herbicide sprayings, Van Strum's book chronicles what did not change after *Silent Spring*'s critique of the overuse of chemical pesticides and the chemical industry that promoted such practices. Like Carson, Van Strum revealed an understanding of democracy that required an informed citizenry, and like Shoecraft and Honorof, she continued Carson's challenge to the military-industrial-academic complex and the Cold War political consensus. The banning of 2,4,5-T in 1985 represented the culmination of activists' demands to stop using the phenoxy herbicides. The ban came after an especially vigorous defense of 2,4,5-T by various industries. The story of these successes begins with the mobilization of citizens who challenged state officials and industry representatives, challenges that led to significant changes in chemical regulation.

Prior to the creation of the EPA in 1970, the USDA was charged with the regulation of agricultural chemicals and pesticides. Historically, different divisions within the USDA focused on solving different threats to farming and the agricultural industry, which included both insect and plant life. Most of these divisions were consolidated in 1953 in the Agriculture Research Service (ARS); the ARS depended heavily on its entomology section and drew personnel from land-grant universities in the Midwest and West. At the time, ARS scientists experienced congenial if not downright cozy relationships with chemical industry researchers. These good relations came about in part because the USDA saw itself as responsible for helping farmers improve productivity, and chemical pesticide control was considered an essential part of this.

As mentioned previously, the USDA and the Forest Service it oversaw conducted massive regional spraying campaigns against fire ants, gypsy moths, and Japanese beetles throughout the 1950s. When *Silent Spring* raised questions about the safety of the chemicals used in these spraying campaigns, the USDA found itself acting less as a regulatory agency and more as an industry public relations firm as it generally dismissed the public's concerns. The response of top ARS officials in attacking Carson and *Silent Spring* supports Carson's inference that the chemical industry, rather than the public, constituted the USDA's major constituency. While the Forest Service used the phenoxy herbicides to increase water supplies in Arizona, and to control fires in California, in Oregon the chemicals were primarily used to cultivate the state's most valuable natural resource, its timber.[3]

The Forest Service represented a major player in determining forest use and timber policy. The Forest Service's presence parallels the ascent of the timber industry in Oregon from the turn of the century onward. Both government service and corporate business reflected the broader societal attitudes that valued capitalist enterprise and cooperation between the federal government, which provided expertise, and industry. In its partnership with industry, the Forest Service collected and provided data and tried to implement best practices in properly conserving the nation's diminishing forest resources. Such services might include, as multiple Oregon papers announced in 1905, the Forest Service hiring "forest guards" to identify and prevent destructive forest fires.[4] This also meant that early foresters shared the same ideological outlook as their contemporaries in industry and academic settings. When Forest Service inspectors clashed with land speculators in the early 1900s through the 1920s in Oregon, the agency was balancing numerous ecological and business interests. Charges of land fraud plagued both the timber industrialists and public officials, and many Oregon settlers suspected that good farmland was, in effect, being stolen for use as forest reserves. Foresters also had conflicting charges. They were supposed to manage and conserve the forests, maximize the potential of valuable resources, and preserve forest reserves as a means of protecting western water supplies for agricultural irrigation. In Oregon, the US Forest Service worked to conserve forests and make them productive.[5]

Forests and fields represented the two most significant economic sectors in the state of Oregon. Harvesting timber like Douglas firs, what historian Emily Brock has called "money trees," represented the major economic activity in Oregon. The timber industry dominated the entire Pacific Northwest region and had for the majority of the twentieth century. Industry had tried unsuccessfully for several decades to impose greater cooperation on its members, but it was not until WWII that the industry achieved its goals of merging, consolidating, and "increasing the dominance of large-scale operations."[6] The state's 1941 Forest Conservation Act effectively decreased the amount of federal regulation, and the timber industry acted on the decreased oversight. By the 1950s the timber industry was centered in southwestern Oregon, with more and more lumber being harvested from public lands. Agriculture, particularly specialty-crop farming, represented one of the state's other significant economic engines, even though Oregon could not compete with California's agribusiness juggernaut. Still, crops remained important to many Oregonian livelihoods. Both industries were plagued by problems with insects, weeds, and brush, and both turned to chemical and mechanical means to address these problems.[7]

The decreased regulation during the postwar period also saw greater intervention on the part of foresters as they sought to improve forest conditions even while attitudes within the Forest Service focused on greater timber production. Historian Nancy Langston provides one example of the way foresters sought to manage forests with often unexpected consequences. These scientists viewed a healthy forest as a productive one, and they worked to increase the productivity of species determined to be valuable, like the Douglas fir. In their understanding, diversity meant trees at different stages of growth, conditions mostly unseen in old growth forests. It meant that foresters depended on fast-growing fir trees to restore forest lands to a productive state. This also meant that foresters needed to decrease insect pests, competing brush, and fires to protect monocultural timber stands. In the process of promoting these goals, however, foresters misunderstood the interlocked ecological systems that promoted species diversity, a system of checks and balances (e.g., prey/ predator cycles), and forest growth and decay.[8]

The experiences of foresters working in the 1960s captured these tensions.

Seeking to enforce erosion controls in Northern California's Humboldt County, US Forest Service forester Ronald McCormick stopped timber harvests that led to a local sawmill's shutdown. He was called to a meeting with the assistant regional forester and the lumber purchaser, where he was told that shutting down the mill meant that there was no work for local community members. McCormick was transferred without promotion two months later. He described the lesson learned as the concerns of the timber business often overrode those of forest management. "I began to think of this as the 'timber imperative.' This hard reality was to be reinforced in the years ahead and assignments yet to come."[9] During the same period and just one county to the south, Union Lumber forester Dennis Tavares described the joy of surveying an old growth forest that had been properly managed, spurring regrowth of about 1 percent per year, which represented a substantial amount of valuable timber, worth more than an equivalent new growth stand because of the quality of the harvested lumber. Government and industry foresters were committed to managing the forest to the best of their ability, although government foresters were charged with juggling multiple public interests.[10]

Concurrent developments in timber management had seen an increased use of mechanical and chemical means to control fires and promote the growth of valuable timber stands of Douglas fir, spruce, and redwood. In 1976 the Bureau of Land Management (BLM) calculated increased production numbers based in part upon herbicide spraying, which justified harvesting more timber.[11] One problem that plagued loggers was the lack of access roads, which made it difficult to reach valuable old-growth timber. Roads were built, and access maintained, through the application of herbicides. The Forest Service also began spraying timber stands with herbicides to achieve better brush control and to lessen the risk of fire. The use of 2,4-D and 2,4,5-T raised some of the same early concerns as those voiced in California in the 1950s, primarily over drift and the destruction of other agricultural crops. The sheer amount of land sprayed with herbicides in Oregon, as much as a million acres in 1957 and even more in 1958, revealed the industry's dependence on the chemicals.[12]

Both groups of foresters erred in attempting to reduce the complexity and redundancy of old growth forests to easily manageable processes. Brush control to improve timber stocks and to prevent fires—both admirable goals in a worldview that valued producing timber—increasingly relied upon chemical agents like the phenoxy herbicides. As a 1965 piece in the Eureka Humboldt *Standard* explained, herbicide spraying was part of a "long range program to release Douglas-fir stands which are now over-topped by tanoak and other brush."[13] At the end of the process, the valuable timber received more "sunlight and nutrients." The opportunity for conflicts increased as timber interests along with the US Forest Service increasingly used these chemicals to manage forests, becoming one of the most likely sources of disagreement between agencies and industry, and again between government and industry experts and the general public.[14]

Economic decline and changes in scientific thought complicated forest management in the 1970s and 1980s. Beginning in the late 1960s, timber interests had begun to decline as measured by many more sawmills closing than opening. These realities resulted in the Forest Service and timber industry responding to an emerging crisis. The development of new, non-wood building materials in the 1970s, industry consolidation, and the high interest rates of the 1980s further stressed the timber industry. Industry responded to these changes and newly enacted legislation designed to protect endangered species such as the spotted owl by seeking cheaper production costs. Such measures included the practice of clear-cutting, which produced unproductive, unsightly, and unusable landscapes. The Forest Service also approved these changes in logging practice. Clear-cutting required increased herbicide use to remove unwanted brush in harvested areas.[15]

The Van Strums along with many of their neighbors sought to make a living either by subsistence farming, by raising specialized produce and livestock, or by a combination of both, or by supplementing their farming efforts through the support of another nonfarming income. Prior to moving to the Five Rivers Valley region of Oregon, the Van Strums had protested the Vietnam War during the 1960s, owned a bookstore in Berkeley, California, and farmed in Northern California.[16] They moved in 1974 hoping to create a retreat in the natural beauty of Oregon's central coastal region. In this respect the family was part of an ongoing "back to the land movement" that started in Oregon at the end of WWII, continuing the state's long history of utopian experimentation. During the early 1980s, Van Strum wrote children's-book reviews for the *New York Times* and served as a department editor for the *Co-Evolution Quarterly*, an intermediate incarnation of Stewart Brand's influential *Whole Earth Review*. The countercultural magazine melded faith in new technologies with environmental awareness. The Van Strums hoped to raise their children in the peaceful setting they had found in Five Rivers. The family bought a farm in what Van Strum described as "real wilderness country." Their nearest neighbor lived half a mile away. Even with the advent of paved roads and electric lines in the 1970s, the Five Rivers Valley region remained isolated and underdeveloped.[17]

Once that peace was broken by the June spraying, the Van Strums began researching the phenoxy herbicides. In the beginning of their inquiries, the Van Strums contacted several experts, including local, regional, and federal authorities, for information on the chemicals. They were reassured by the USDA that the herbicides used were safe for humans and animals. The couple requested a copy of the 1976 Environmental Impact Statement (EIS), which led them to focus on the phenoxy herbicides. The couple found a significant amount of scientific evidence that questioned the safety of 2,4-D and 2,4,5-T. One unexpected supporter was Thomas Whiteside, the reporter for the *New Yorker* who had written exposés of the phenoxy herbicides during the war and, later, a 1971 book. Whiteside suggested the names of more scientists who were researching the herbicides. The chemicals had been banned in South Vietnam since 1970, and a whole host of local, state, and federal public hearings had been held regarding the potential hazards of the chemicals. Despite this, the official response by the USDA remained one of reassurance regarding the chemicals' safety.[18]

Concerns regarding the chemicals' safety intensified when the Van Strums responded to an article about herbicides and the forest industry that prominently featured Dr. Michael Newton, a "forest weed ecologist" and Oregon State University professor. Newton had already gained notoriety for trying to use discarded surplus Agent Orange herbicides on local timber land. According to a news release, Newton had declared that the contested herbicides posed no threat to public health or wildlife. He based this opinion on his longtime experience with weed control herbicides, which he had been studying since 1958. Newton had engaged in the wartime herbicide debates, publicly responding to calls for caution in using the defoliating herbicides. He published a response to one such article, written by Yale botanist and herbicide critic Arthur Galston, that appeared in *Science* in 1970. Newton's coauthor, a US Forest Service administrator from Oregon, Logan A. Norris, exemplifies the strong ties between academic scientists and the forest service industry. Newton had also been a member of a 1974 National Academy of Science research team sent to assess the possible harm done to the South Vietnamese people and countryside by Agent Orange spraying operations.[19]

Carol Van Strum characterized the couple's reaction to Newton's article, noting: "Those statements by an 'expert' presented an incomplete, misleading picture of herbicides to a trusting public."[20] Newton's proclamations of safety angered the Van Strums, and the couple's letter to the editor roused the community. In it the Van Strums' identified 2,4,5-T and its contaminant dioxin as a serious problem with past and planned spraying programs. They concluded by noting: "We who live and work in these forests and suffer directly from the effects of spraying receive no financial support to fight these practices. Indeed, our tax money supports the very persons, such as Professor Newton and the Forest Service, who are using us as guinea pigs in an experiment designed for their own profit."[21] Responses revealed a community deeply concerned about harmful effects that members attributed to the spraying of the phenoxy herbicides.

The Van Strums had lived in the Five Rivers Valley for only two years and did not know many of their neighbors. Their letter changed that. Two local women, unhindered by employment in the timber industry, arranged a public meeting to discuss the letter. At the meeting, the Van Strums shared their research and the Forest Service's EIS. Steve Van Strum claimed residents' experiences gave them an equal status with federal and state employees. "The people of the Forest Service and road department and the people at E.P.A. who regulated the chemicals, do not live in the forest. . . . But we do. We work here, fish and hunt here, grow our food, hike with our kids."[22] Longtime residents shared stories that included bees dying after spraying, heifers lost during the spring spraying, miscarriages, deformed domestic animals and wildlife, and human illnesses. Anxious about the harm they now feared came from herbicide sprayings, the group struggled with what to do next. In the end they documented their stories and sent them to the Forest Service, in the hope that these public officials would at the very least investigate further.[23]

These concerned citizens were dismayed to find officials nonchalant and unwilling to intercede to stop the planned spring spraying. One local farmer-logger expressed the sentiments of many: "It was a shock, don't you see, to do this thing that was such a big step for us and then they didn't even care and wouldn't listen."[24] In what was a familiar story, bureaucratic indifference led to community mobilization and then political action. Town members from Five Rivers and Deadwood formalized their protest activities with the creation of Citizens Against Toxic Sprays (CATS). Like other grassroots environmental groups before them, CATS turned to legal remedies and began looking in nearby Eugene for a lawyer who could represent them. They found Bruce Anderson and proceeded to negotiate with the Forest Service to try and stop herbicide spraying. After a breakdown in those discussions, CATS brought a lawsuit against the Forest Service. Eugene residents provided invaluable support,

joining CATS and raising money through donations, bake sales, and benefit concerts.[25] The Oregon Environmental Council and the Hoedads, a collective labor group that ran a tree-planting cooperative, also joined the lawsuit.

The trial revealed interesting alignments and the contours of the legal battle. The groups charged that the Forest Service EISes from 1975 to 1977 had
been inadequate. The Industrial Forestry Association joined the Forest Service as a defendant, using a law firm that had previously represented Dow
Chemical Company, the manufacturer of the 2,4,5-T and 2,4-D herbicides.
CATS and its allies filed a new motion that sought to ban 2,4-D; 2,4,5-T; and
silvex, a related compound.[26] In the courtroom, CATS's thorough preparation
allowed it to discredit the various government, academic, and industry witnesses. Dr. Patrick O'Keefe, a Harvard chemist and member of the EPA Dioxin Monitoring Group, testified dioxin had been found in local samples, including breast milk. The sample had been tested as part of an EPA contract
awarded to O'Keefe and Matthew Meselson, the biologist who had headed
the American Association for the Advancement of Science's Herbicide Assessment Committee, which evaluated the US defoliation program in South
Vietnam. These positive test results were omitted in the Forest Service's EIS
on the basis that only negative results were deemed reliable.

Van Strum made an interesting observation about O'Keefe's testimony that
demonstrated activists' view of what role scientists should play in the debate.
"He told what he knew and with equal candor told what he didn't know. He
drew no conclusions, took no sides. None were needed."[27] O'Keefe's testimony
documented the fact that EPA tests on wildlife from the Siuslaw National Forest had shown dioxin present in samples collected in 1974. In stark contrast,
Dr. Ralph T. Ross, formerly director of the Dioxin Monitoring Program, re-
nounced a 1975 study on the presence of dioxin in beef fat taken from Texas
cattle. He now claimed the memo was based upon inconclusive evidence.
Many in the courtroom were surprised, although as one CATS member noted,
the only thing that appeared to have changed was Dr. Ross's employer, as he
now worked for the Department of Agriculture. The Forest Service offered
up the testimony of two witnesses, Dr. Newton, a familiar figure in the conflict, and Dr. Fred Tshirley, as noted in chapter 4, a longtime researcher connected to Agent Orange.[28]

The trial was covered extensively in the regional press, and its influence
could be seen in numerous ways. CATS began networking with other environmental groups, like the Environmental Protection Center in Fort Bragg,
located in Northern California's Mendocino County timber region. There the

center's Betty Lou Whaley spoke about growing community concerns over herbicide use, in January 1977. Just a month later, the Group for Organic Alternatives to Toxic Sprays (GOATS) met for the first time in Arcata, California, to discuss their concerns regarding herbicide use "around their homes and on public lands," with the group committed to seeking safe and economic alternatives and educating the public. A representative from CATS, Steven Van Strum, attended the meeting and spoke at a subsequent press conference. The GOATS and CATS speakers handed out a fact sheet, and Van Strum spoke about the pending lawsuit against the Forest Service. Both Van Strum and the GOATS representative, Ilene Mandelbaum, raised concerns about 2,4,5-T's dioxin contaminant. Mandelbaum argued that physical removal offered a safer and more economical (only a ten-dollar increase over the $50 cost of herbicide sprays) way to remove timber brush, especially in an area with a high unemployment rate.[29]

The trial also publicized the Harvard dioxin testing conducted by O'Keefe and Meselson, with news accounts beginning to cover the story early in 1977. One San Antonio paper reported on the findings of dioxin in breast milk, noting that the samples had been collected by the La Leche League, a national group that promoted breast feeding. Along with previous tests done at Harvard and Dow Chemical that showed dioxin contamination in beef, the breast milk samples demonstrated the ways 2,4,5-T might be contaminating the nation's food chain. The CATS lawsuit served to educate and mobilize others even as it sought to stop the Siuslaw Forest spraying program.[30]

CATS won its case, winning a one-year stay, but conflicts over herbicide spraying continued. In March 1977 Judge Otto Skopil issued a temporary injunction banning the use of 2,4,5-T and Silvex pending Forest Service explanation of the effects of dioxin on human and animal health. The following year Skopil was forced to approve the revised EIS, removing the injunction against herbicide spraying. Skopil administered a very public and blistering lecture to the Forest Service. While the agency had complied with the letter of the law, Skopil chastised them for their secrecy and failure to deal transparently with the public. The judge confessed: "If I were the person responsible for making the decision of whether herbicides containing this substance [dioxin] should be broadcast sprayed from helicopters over our national forests, I would be extremely reluctant to allow it."[31] Attempts by the Oregon legislature to ban 2,4,5-T remained stymied, as a May 1977 hearing demonstrated.

Oregon lawmakers were considering House Bill 3230, a proposed ban on all herbicides containing dioxin. In a state heavily dependent on the

timber industry, the legislation represented a major policy change. Once
again Dr. Michael Newton appeared as an expert witness, testifying before
the committee considering the bill. Newton asserted that 2,4,5-T and its dioxin contaminant had been "rigorously evaluated by the scientific community including toxicologists, biochemists, ecologists and many others."[32] Newton presented the findings of a few studies to support his claim, including
personal conversations with researchers assuring him of 2,4,5-T's safety. He
concluded by noting that while dioxin had been found in beef samples the
EPA had announced a "'no effect' level of TCDD [dioxin] in dietary fat."[33]
Newton reassured the committee that the EPA could be trusted to ban substances on the potential for harm to human health. "The continued registration by EPA of 2,4,5-T and silvex as general use herbicides reflects the lack
of documented hazard."[34] By the fall that very same agency had restarted its
administrative review of 2,4,5-T and caused consternation in the hearts of
industry executives.

Spraying 2,4,5-T continued to be controversial in the state. While the CATS
legal victory curtailed the extent of the Forest Service's planned spraying,
three-fifths of the original 150,000 acres would still be sprayed with 2,4,5-T.
Even though the Forest Service quickly prepared to carry out the program,
one later observer noted that the "spray program never regained the footing
it had lost."[35] Forest Service enthusiasm may have also been dampened by the
threat of violence. In September that year, three men armed with shotguns
barricaded a road in southern Oregon with their cars. They were waiting for
a local forester whose wind speed and humidity measurements would determine whether 590 acres of forest would be sprayed with 2,4,5-T. After dis-
charging a weapon in the air, the men warned the ranger that there would
be people hunting rabbits in the spraying area all day. The ranger returned
to his office. The Forest Service called off spraying the tract that day, and for
the rest of the year. The contracted helicopter service expressed concern about
being shot out of the air.[36] Citizen protests of herbicide spraying, including
threats of violence, had erupted in several hotspots.

Environmental activists in British Columbia, Minnesota, and Alaska all
reported acts of civil disobedience, threats of violent deeds, and attempts to
regulate chemical herbicides. Residents living next to Lake Okanagan in Canada protested the use of 2,4-D for eradicating water weeds, especially as the
lake provided drinking water. In Minnesota an organic farmer and his allies
staged a sit-in at a national forest spraying site. The group succeeded in stopping the planned herbicide spraying for only two days, as helicopters began

spraying again. The farmer shot blanks at the aircraft and was charged with aggravated assault. Later tests showed that streams had been contaminated, and the farmer was declared "not guilty" at his trial. Public hearings were held in Juneau, Alaska, regarding a proposed bill to stop the spraying of 2,4-D and 2,4,5-T. The bill died in committee, and many credited the testimony from an organization, the Council for Agricultural Science and Technology (CAST), as carrying great weight. The various pro-herbicide industries continued their campaign throughout the West.[37]

✿✿✿

The CATS lawsuit represented just one of several events taking place over the course of the 1970s that culminated in the emergency banning of 2,4,5-T in 1979 and eventually a permanent ban in 1985. The nexus of these events would be an EPA hearing scheduled to be held early in 1978. The agency required chemical manufacturers to submit new safety data to renew 2,4,5-T's registration. The phenoxy herbicide's dioxin contaminant became the focus of industry, activists, and the regulatory agency.

Dow's dioxin dilemma dated back to the use of the herbicides during the war, and the company spent decades defending 2,4,5-T. During the 1960s the company had worried that Agent Orange would be targeted in the same way student protests over napalm and defoliation had broken out on campuses across the country. Concerns about the chemical's effects on human beings increased in the late 1960s after the Bionetics report raised questions about 2,4,5-T's role in birth defects and reproductive problems. The situation worsened in 1970 when Dow Chemical's vice president Dr. Julius Johnson testified at a congressional hearing that Dow had known about 2,4,5-T's dioxin contamination well before the company supplied the chemical for the defoliating missions flown in South Vietnam.[38]

In 1977 Dow faced a new problem when the company found dioxin contamination in the Tittabawassee River, the major waterway in which Dow discharged its Midland, Michigan, plant's manufacturing wastes. In the summer of 1978 the Michigan Public Health Department issued a warning to Midland residents, home to Dow's headquarters and a manufacturing plant, that dioxin had been found in local fish. While company officials asserted that the small amounts detected were not a health hazard, they confessed that they did not know how the river had been contaminated, or how much of the waterway was affected. Dow officials eventually announced they had found the cause, which they proclaimed was the "entire environment." Dow went on to offer a creative explanation of the dioxin contamination. According to Dow, it was combustion, or fire.[39]

Dow identified a whole host of potential sources for the chemical contamination. The sources that tested positive for trace presence of the chlorinated dioxins included the to-be-expected culprits, things like refuse incinerators, chemical tar burners, fossil-fueled power plants. More surprisingly, the list included items such as charbroiled steaks. The theory provided an explanation that exonerated Dow. It also allowed the company to "naturalize" the chemical, since dioxins had, in the words of one Dow chemist, "been with us since the advent of fire."[40] Dow scientists offered up the theory as one reason for the widespread environmental presence of dioxin contamination. The discovery proved quite convenient for the company. As John Davidson, a Dow chemist, observed: "We learned so much about dioxins in order to defend our pesticides."[41] At the same time the Forest Service was amending its EIS in Oregon, the Dow Chemical Company was arguing that dioxin's "presence is due to the existence of a natural phenomenon, trace chemistries of fire."[42] Even as late as 2002, Dow's official position was that the EPA had failed to achieve a balanced scientific judgment on acceptable dioxin levels, claiming that veiled advocacy was "still evident in the interpretation of the data."[43]

With the timber industry as one of its major economic mainstays, Oregon citizens had a lot to lose if the industry's productivity decreased. Claims that timber production might decline by as much as 50 percent if a ban on 2,4,5-T was enforced represented a powerful argument to continue current forestry practices that relied on the phenoxies. By the fall of 1977, the EPA was responding to the Harvard study that had detected trace levels of dioxin in breast milk from Oregon and Texas, evidence that if validated supported claims that herbicide spraying exposed individuals to the dioxin contaminant. Press coverage of the issue framed it as a contest between safety and economics. The timber industry in Oregon had experienced a 22 percent decline in harvests and the proposed ban both intensified and accelerated this problem. The downturn in lumber production had closed six mills in Oregon, and increased shortages affected company profits and sustainability, workers and layoffs, county finances, and vital forest timber receipts that paid for Oregon's roads and schools. The required EIS, the result of the CATS lawsuit, meant that summer herbicide spraying had not yet been conducted.

CATS had based its case primarily on the *environmental* harm the phenoxy herbicides represented, but within a year of their lawsuit, claims of human harm surfaced. In the spring of 1978, Bonnie Hill and eight other women living in Alsea, Oregon, just a few miles west of Five Rivers, contacted the EPA concerning thirteen miscarriages that the women had experienced over the last five years. Hill had miscarried in the spring of 1975. She first began questioning the potentially harmful effects of herbicides in 1977 after reading a fact sheet compiled and distributed by CATS she had found at a local craft fair. Hill continued her research, finding studies done by University of Wisconsin scientist Dr. James Allen that suggested that the phenoxy herbicides affected reproduction in rhesus monkeys.[44]

As a local teacher, Hill knew of several former students who had miscarried. She met with these women to verify their miscarriages and obtain their medical histories. She also collected information from the Forest Service, Bureau of Land Management, and private timber companies about spraying patterns. Including only miscarriages that had been confirmed by a physician, Hill then analyzed the information and discovered a pattern. As in other cases of popular epidemiology, Hill created a chart that correlated thirteen miscarriages experienced by the group of nine women, all of which occurred shortly after the spring spraying season, and thus with exposure to 2,4,5-T and 2,4-D. Hill mailed a letter and the chart to the EPA, where it might have been lost except that it came to the attention of Friends of the Earth research associate Erik Jansson.[45]

Hill's study arrived at the EPA at an opportune moment and initiated a major health study. The EPA had announced that it would finally be holding a hearing to renew 2,4,5-T's registration with manufacturers required to supply new safety data. The hearing had been delayed since 1974, even after an industrial accident in 1976 released a dioxin cloud that exposed the Italian city of Seveso and nearby towns to 2.2 pounds of dioxin, a substance considered hazardous at concentrations in the parts per million. The accident presented the largest and most significant case of immediate exposure to dioxin. Thousands of animals died within days, numbers of children were hospitalized, over eight hundred people were evacuated, and pregnant women were offered abortion services (no small thing in the nominally Catholic country). The event drew international coverage and put 2,4,5-T and its dioxin contaminant once again on regulatory agencies' agendas.[46]

The letter's deliberative reasoning and epidemiological support helped Jansson's efforts to publicize its findings. Whether it was the miscarriage information, the attention generated by Jansson, or both, the EPA decided to conduct a formal health study. The EPA health team administered a thorough health questionnaire. Coupled with county health records, the information was sent to EPA experts for analysis. This group refused to assign causality to the chemicals but did think the evidence presented merited further investigation. The EPA's study began with a more extensive compilation and examination of data. It analyzed greater numbers of soil, water, animal meat, and human milk samples to determine the presence of dioxin. This larger and more intensive study would be known as Alsea II. The EPA issued an emergency suspension of 2,4,5-T and Silvex except for use in rice crops and rangeland while it conducted its study. The March 1979 action drew national media attention to herbicide spraying in Oregon, and the ire of the Dow Chemical Company.[47]

As the reason for the emergency suspension, the first Alsea study had attracted attention. The mainstream news media ran investigative pieces such as a syndicated column by Jack Anderson and a story on the ABC news series *20/20*. CAST, the chemical industry ally, weighed in on the debate. CAST criticized the *20/20* segment and the scientific validity of the EPA's Alsea studies, based on analysis offered by Dow scientists and others. These critics deemed both studies inadequate, especially the methodologies used to identify and record miscarriages. Several factors made collecting accurate numbers difficult, among them the timing of the pregnancy loss, as many miscarriages occur early and spontaneously, and the nature of the loss. Women hesitated to speak about the miscarriages, even when offered an explanation that absolved them from individual responsibility for the terminated pregnancy.[48]

Dow immediately went on the offensive, claiming that both Alsea studies were "unscientific" and "seriously flawed." Dow chairman Earle B. Barnes betrayed a more revealing attitude in a written response to a clergyman. In it he characterized the Alsea area as one comparable to "northern California and other northwestern states in growing marijuana in open spaces and in forests." These illegal marijuana crops, worth $900 million in California alone, were susceptible to the 2,4,5-T herbicides used by the Forest Service to kill underbrush. The opposition to herbicide spraying came because of monetary, not health, concerns. Opponents to the grassroots health and environmental groups had already raised this charge, and in the process tainted activists as radicals and/or hippies. These efforts had been successful enough that even Bonnie Hill thought of CATS members as "a bunch of radicals and freaks who were living out there in Five Rivers."[49] Barnes implied the reports, which had shown "no valid relationship between the spraying and miscarriages," as one of financial interests rather than health concerns.[50] Dow went on to propose that nature itself had caused the environmental contamination and its ill effects, not humankind.[51]

Questions regarding the scientific neutrality of phenoxy herbicide supporters arose. Dow flew Oregon State forestry expert Michael Newton and his wife to Midland, Michigan, to strategize.[52] CAST also attracted negative attention of its own when the full membership of the organization was publicized. While the organization included twenty-five independent societies and scholars, it also included more than two hundred agricultural businesses and industrial trade organizations. These members provided 57 percent of its operating budget. The group's corporate membership and its share of funding raised questions of credibility and the group's claim to scientific neutrality.[53]

Industry also had begun gathering evidence for the EPA 2,4,5-T registration review. A June 15, 1978, letter went out to trade associations like the Western Timber Association asking for help. Dow, as the "leading producer of 2,4,5-T herbicides . . . assumed responsibility for rebutting alleged health hazards."[54] Industry allies could assist in identifying the "economic and other benefits" of the weed killer. For this information, the company looked to the "agricultural and allied sectors, including extension personnel, growers, foresters, right-of-way managers, grower organizations, and the USDA National Assessment Team." The letter urged these individuals and groups to send their comments directly to the EPA, focusing on three major areas: (1) the kinds of crops, number of acres, and effectiveness for which the 2,4,5-T herbicides were used; (2) what alternatives existed, including availability and costs; and (3) the effect on agricultural and timber products and what the effect on consumer prices would be if 2,4,5-T were removed from the market.[55]

The letter showed Dow's continued confidence in making the case for the herbicide's safety, and expectation that industry allies would help show the costs of replacing 2,4,5-T. There were both government and industry foresters who challenged the idea that effective alternatives for 2,4,5-T existed. Others acknowledged, however, that not all costs could be calculated. Dr. Frank Dost, a professor of agricultural chemistry at Oregon State University, noted that while replacement costs were relatively simple to calculate, quantifying risk remained difficult to compute. This made it virtually impossible to compare the totals. Some scientists even went so far as to say that no level of dioxin should be acceptable. The report of miscarriages in Alsea meant that evidence now existed that weakened the industrial-academic complex claim of safety.[56]

But official scrutiny of the Oregon case transformed 2,4,5-T's review. Even as the EPA was evaluating further investigation into the situation in Oregon, Friends of the Earth's Jansson was urging the EPA to consider a spraying ban for the 1979 season. His promotion of Hill's letter and the resulting attention to Hill's popular epidemiologic study interrupted the expected routine approval of 2,4,5-T. The press coverage of the EPA's decision to gather more information about the Alsea case increased scrutiny of 2,4,5-T and studies showing it caused birth defects and stillbirths in laboratory animals such as mice and birds. On August 15 a *Washington Post* article prompted attention from newspapers across the country, with news outlets either reiterating (or reprinting) the story or reporting on local cases of herbicide spraying. Bonnie Hill even appeared on the NBC News national broadcast on August 21, 1978, in a two-minute interview. (One measure of the changed understandings of chemical contamination can be seen by comparing this national coverage to *Time*'s news brief mocking Billee Shoecraft eight years earlier.) Some officials acted, such as establishing new spraying criteria that specified that 2,4,5-T would not be sprayed within a half mile of waterways or within a mile of "permanent habitation."[57]

Assistant Secretary of Agriculture M. Rupert Cutler announced the change on August 7, 1978, which included other criteria, such as increased monitoring before, during, and after spraying; greater supervision during spraying; better delineation of buffer areas; and more information about the costs associated with different weed removal programs. Internal correspondence from industry associations reveals the concerns generated by the increased attention from the press and regulatory agencies. Industry representatives expressed their unhappiness over the kind of news coverage the issue received. They also noted that a CAST press conference had been scheduled to address the science of the issue. As one letter noted, the initial network coverage did not mean that the issue would attract national attention.[58]

Over the next several years, Dow and its industry allies launched an all-out campaign to defend 2,4,5-T, a campaign complicated by concerns raised by a new group potentially affected by the chemical contaminant dioxin. The initial Dow efforts, already discussed, focused on discrediting the EPA's Alsea studies. Press coverage continued to highlight local stories, almost always connecting the domestic spraying of the phenoxies with the herbicides' use in South Vietnam. "Agent Orange still used in US" proclaimed an October 1978 front-page headline. The Texas newspaper article highlighted information from the Veterans Administration that five hundred Vietnam veterans sought disability benefits for assorted conditions they all linked with exposure to Agent Orange. The health problems listed ranged from "nervous disorders to deformed children." This new group, American Vietnam Veterans, protesting 2,4,5-T increased national attention and official scrutiny of the chemical.

The AP wire service used by papers across the country noted that while herbicide spraying had been halted in Vietnam, the domestic use of 2,4,5-T had continued. "At home, foresters and farmers resisted the environmentalists' push for a ban. They said they needed 2,4,5-T to clear brush, weeds and hardwoods. . . . Used properly, they said, 2,4,5-T was safe."[59] In contrast, a multitude of citizens joined veterans in blaming 2,4,5-T for illnesses. Among those quoted was Bonnie Hill, who stated the case of the Alsea women showed "an incredibly close correlation" between spraying and the miscarriages experienced by local women. The newly identified concerns expressed by veterans gave added gravitas to the charges identified by the Friends of the Earth and the Alsea case. A brief look at the criticism leveled at the ABC news magazine *20/20* report on 2,4,5-T, proves illuminating.[60]

Explicitly addressing the *20/20* episode on 2,4,5-T, Accuracy in Media, Inc. (AIM), the conservative media watchdog founded by economist and vehemently anticommunist Reed Irvine, addressed what it claimed were unfounded charges dating back to the Vietnam conflict. Citing the 1974 National Academy of Sciences report on herbicides and South Vietnam (itself controversial), AIM argued that the study had been misreported, and that criticism of herbicide use in Vietnam had been politicized, was discredited, and was unwarranted. This meant ABC's report represented a "propaganda campaign against the peaceful use of a chemical that has added very significantly to the productivity of our forests and which performs an essential function at low cost for farmers, highway departments and utility companies."[61]

Of the many cases presented in the ABC report where 2,4,5-T and/or dioxin had caused problems, AIM noted that several had been debunked by CAST. The newsletter's rebuttal of the story's claims that 2,4,5-T and dioxin caused health problems included speaking to the ABC researcher who had worked on it as well as the program producer. Both men were excoriated as being biased and antibusiness. AIM ended by urging its readers to contact ABC's chairman to ask that *20/20* "ceases to be the vehicle for anti-business, anti-science programs that it has been so far."[62] Even more important, ABC news personnel had admitted that they were trying to influence the decision of a government agency (the EPA RPAR review of 2,4,5-T). AIM ended with a plea that the "other side of the story deserves a fair shake." This meant that "qualified scientists [should be given] an opportunity to reply to '20/20's' hatchet job on the herbicide, 2,4,5-T."[63] The defense of 2,4,5-T continued.

Groups that supported the continued use of 2,4,5-T used a variety of strategies in their efforts to vindicate the chemical over the next several years.

These included airing questions about EPA studies, public defenses in industry media, legal maneuvering, and personal attacks. The EPA's emergency suspension of 2,4,5-T alarmed industry. A March 17, 1979, edition of *Science News* reported on the 2,4,5-T controversy with the headline "Dow Attacks Study Used to Ban 2,4,5-T." The news brief noted the previous controversy the herbicide had attracted, and that the dioxin contaminant had been shown to cause "birth defects, miscarriages and tumors in laboratory animals."[64] But the piece also publicized Dow's complaints about the study, which included the accusation that the EPA was "gerrymandering data." Dow made sure its challenges to the scientific veracity of the EPA studies appeared in the press. Other publications went on the offense, proclaiming the safety and benefits of 2,4,5-T.

American Forester ran a two-part series on 2,4,5-T. Part 1 of the series had focused on the chemical's risks. In "'T' on trial, Part 2," the article made the case for the importance of 2,4,5-T to commercial timber production and government forestry management. The EPA was charged to review chemicals when animal studies indicated a substance's potential harm to humans. Such findings required the agency to begin the RPAR process. But questions remained about the applicability of study results to actual exposure and measurable harm. RPAR review also included an assessment of benefits, which in the case of 2,4,5-T meant evaluating economic costs. According to industry claims, "the financial loss for the first 50 years after the removal of 2,4,5-T would be more than $1 billion."[65] The Forest Service estimated costs of another $5.6 million annually. While only 15 percent of forests were treated with the herbicide, an industry spokesman claimed that these were the most productive lands, the ones that needed herbicide treatment the most. One forest-industry spokesman likened the process to a grand jury, although as the Friends of the Earth's Jansson noted: "Chemicals don't have rights."[66] As such, they could not be innocent until proven guilty.

Another industry newsletter published a four-part series on the phenoxy herbicides, written by Newton. Ralph Hodges, of the National Forest Products Association, wrote about the controversial coverage of 2,4,5-T, much of it from "irresponsible news reports."[67] By the summer of 1979, the National Forest Product Association's Forest Industry Chemicals Committee had circulated a "'T' Memo" with updates of official actions regarding the use of 2,4,5-T and even a round-up of articles that charged the government with causing cancer scares through the release of "inaccurate, misleading reports."[68] Informational pieces like these ensured that industry rank-and-file members understood the terms of the debate and the consequences of losing. The battle for 2,4,5-T also occurred in more traditional legal settings.

In April 1979 Dow Chemical forced the EPA's hand with a motion to initiate cancellation hearings while withdrawing from the emergency suspension hearing with which it had been engaged. This raised the stakes, shifting the case from adjudication by a three-person panel chosen by the EPA to a bench trial overseen by an administrative law judge (ALJ). Dow's actions followed the company's failed petition to a US district court in Michigan to lift the suspension earlier that month. Dow had participated in only two days of the suspension hearing before deciding that it would be better to proceed immediately to a cancellation hearing. This decision was made in part, according to Dow's attorney, because it considered the suspension hearing a waste of time, as the EPA administrator had already determined the fate of the herbicide, and agency employees might face undue pressure to overturn his decision. Dow challenged the EPA's "scientific objectivity" and openly expressed the expectation that the ALJ should resolve the case before the 1980 spring spraying season.

The company recognized that such maneuvers helped their efforts to save 2,4,5-T. In withdrawing from the suspension hearings, "Dow also [withdrew] its offer to make its confidential and trade secret data available for use in these suspension hearings."[69] The support of fifty-four industries and trade groups like the National Forest Products Association suggests that Dow spoke for industry in its fight to preserve 2,4,5-T's registration. Company officials also grasped that the legal process overseen by an ALJ favored the side with money—which paid for the very best, and most favorable, studies and lawyers. Dow would also use its legal might in pursuing another strategy in its defense of 2,4,5-T.[70]

Chemical manufacturers became especially adept at harassing scientists as a means of discrediting their research in the early postwar period. Rachel Carson represents one of the best-known examples of this, as she endured numerous, and gendered, attacks on her scientific credentials, character, and temperament as a woman. This was not the same thing as funding competing science or distorting science by funding it, but rather making the personal political (which is not the same thing as the feminist insight that the personal was political). In the case of the phenoxy herbicides, Dow set its sights on one of the most compelling research agendas supporting the link between exposure to the phenoxy herbicides and their hazardous reproductive effects.[71]

In their suspension of 2,4,5-T, the EPA had relied upon the research of

Allen, the University of Wisconsin researcher investigating the effects of dioxin on rats and rhesus monkeys. In 1978 Allen had been found guilty of filing false travel requests that reimbursed him for $892, which he had used for two ski trips. The grant monies funded work unrelated to his dioxin research, and Allen pled guilty to four counts of fraud. He repaid the government $892 and a $1,000 fine. Dow later used this transgression as a means of questioning Allen's status at the University of Wisconsin as it sought to gain access to all his research records and discredit him personally and professionally. In comments filed as a part of Dow's suit to gain access to Allen's research records, EPA counsel offered one motivation for the company's actions: "The recent pleadings indicate that Dow is dissatisfied with the information that Dr. Allen has produced to date, that Dow is curious, perhaps even anxious, about the results of ongoing studies."[72]

In February 1980 a USDA official shared the 1978 court documents dealing with Allen's plea agreement with Dr. Alvin Young, the Air Force major long associated with Agent Orange and its use in South Vietnam. The accompanying note mentioned that Young might address the Allen issue in his own statements at the hearing.[73] Industry and government had closed ranks and succeeded in making Allen's life hell, not to mention exerting enough pressure that his academic position might be revoked. It also potentially cast a shadow over his research.[74] Despite all these efforts, the protests and negative coverage of the phenoxy herbicides continued to escalate.

Industry records offer one way to measure the multitude of media and groups critical of 2,4,5-T and its dioxin contaminant. The Forest Chemicals Communications Task Group sent the Chemicals Committee, the industry group responsible for the "'T' Memo," a summary of press coverage on 2,4,5-T in a January 8, 1980 letter. These stories appeared in local, regional, and national outlets. The *Willamette Week*, a Portland, Oregon, paper, ran a three-part series, "The Spraying of Oregon," while the *Los Angeles Times* examined herbicides as a part of their "The Poisoning of America" investigation. A *McCall's* magazine feature, "What Happened to My Baby?," brought the story of Bonnie Hill and the Alsea women to the women's reading public. Late in 1979 PBS's *Nova* documentary series broadcast *A Plague upon Our Children*. Part 1 focused on dioxin herbicides and included interviews from Oregon residents and Vietnam veterans, all of whom expressed concern about their exposure to 2,4,5-T. The documentary won the DuPont Columbia Journalism Award, considered to be the Pulitzer Prize of broadcast journalism. A group of women in the small town of Condon, Montana, near Missoula, filed a class action lawsuit against Dow Chemical in 1980, claiming that exposure to 2,4,5-T and 2,4-D had caused their miscarriages. The continued controversy regarding phenoxy herbicide use prompted a recalculation of the chemical's cost-benefit ratio.[75]

By 1982 critics started questioning whether the problems associated with 2,4,5-T and 2,4-D made them wise economic choices for timber management. The Council on Economic Priorities (CEP) was founded in 1969 as a public service research group. It initially focused on the defense industry but later expanded to include social, corporate, and environmental responsibility. In 1982 the CEP funded a study examining the use of the phenoxy herbicides in western Oregon. Addressing both public health concerns and the chemicals' effectiveness, *Forests, Herbicides and People* assessed what alternatives existed to maintain effective timber management if the phenoxies were permanently banned. As the report noted, the safety of the phenoxy herbicides could not be guaranteed or disproven and the uncertainties about them pitted anxious parents against forest managers in fundamental disagreement about whether the chemicals should be used.[76]

In ten chapters the authors reviewed and analyzed the existing literature on the chemicals' health effects and effectiveness in timber management. They considered the outcomes of a regulatory complete ban on both phenoxy herbicides, an aerial ban on the spraying of 2,4,5-T and 2,4-D, both with and without an associated ban on ground applications of 2,4,5-T. It concluded that while a total ban of the chemical herbicides resulted in no exposure to residents to those chemicals, it would mean an increased exposure to other chemicals. The CEP report represented the best attempt to recognize the uncertainties that plagued herbicide use and safety, recognition of incommensurability of understandings of herbicide safety, and a fact-based assessment of what different chemical bans entailed. But its tone of measured reason was lost as revelations about 2,4,5-T and dioxin studies became public.[77]

A host of information about 2,4,5-T and its dioxin contaminant came to light in the spring and summer months of 1983 that undermined industry's defense of the herbicide. In March it was revealed that an EPA official shared a study of Michigan dioxin contamination with Dow scientists that resulted in the claim of Dow's culpability for the contamination being removed from the final report. Later coverage of the Michigan dioxin contamination noted that Dow had challenged the EPA study, although lawsuits had made public internal company documents that showed deep concern over dioxin toxicity expressed by Dow scientists dating back to the 1960s. In June evidence emerged that both the EPA and Dow had kept silent about unpublished British research on dioxin's effect on the body's immune system. Although the research was more suggestive than definitive, it showed dioxin, and Dow, in a negative light.[78]

Press coverage of the internal Dow documents on their own dioxin research intensified, with news accounts emphasizing the company's expressed awareness of dioxin's toxicity and the high concentrations in the herbicides supplied to the US government. Facing a court battle with thousands of sympathetic Vietnam veterans, and after years of defending 2,4,5-T, Dow announced it would stop production of 2,4,5-T, and its voluntary cancellation of all registration for products containing 2,4,5-T. The ongoing battle had cost Dow $10 million and hurt its public profile. With the company's concession on 2,4,5-T, the EPA announced it would ban all sales of the herbicide by other manufacturers.[79]

Like Billee Shoecraft and Ida Honorof, Carol Van Strum was influenced by Rachel Carson's work and life. Like these women, she sought to raise public awareness, to rouse the consciousness of Oregonians, and, later, Americans. At the end of her book Van Strum argued for a constitutional amendment that would give citizens the right to be informed about possible exposures to chemicals that might harm them, a kind of informed consent. All three women, and many others, saw themselves as citizens engaged in the democratic process, and concerned about their potential exposure to hazardous chemicals, to health and environmental risks they had not chosen. From the late 1960s through the mid-1980s these activists had challenged industry, state agencies, and national chemical regulation. Their challenges sent ripples throughout several powerful industries, companies already concerned about a changed legislative environment that they considered antibusiness. Here, business concerns merged with political concerns, as both sectors identified the counterculture as an enemy to be attacked. Once again, the phenoxy herbicides would be used both internationally and domestically as a part of a new initiative, the war on drugs.

PART THREE

The Phenoxy Herbicides' Toxic Legacies, 1970–95

This last part returns to the theme of war and conflict. An exodus north took place as the counterculture of the 1960s ended, with numerous "hippies" settling throughout Northern California. These new settlers brought new attitudes about families, work, and the environment with them, attitudes that put them into direct conflict with longtime farming and ranching families. The "herbicide wars" were fought there in California and south in Mexico as a part of Richard M. Nixon's war on drugs. American Vietnam veterans found themselves fighting the deadly fog as suspicions grew that their exposure to the phenoxy herbicides in the form of Agent Orange was causing their ill health. Tragically, these chemicals appeared to have affected the next generation of Vietnamese, farmworkers', and veterans' children, as these groups struggled with the unexpected casualties of chemical exposure.

CHAPTER SEVEN

The War on Drugs

The Phenoxy Herbicides in Counterinsurgency and the Counterculture

Published more than a decade after the events in Oregon, a short story written by Carol Van Strum captured the distress local community members experienced when told their opposition to the herbicides was motivated by protecting recreational drugs versus responding to illness. In "Melyce," Van Strum described a young mother overwhelmed by the county commissioner's public claim that smoking too much marijuana was the cause of community illnesses, not herbicide exposure. The climax of the story highlights the tensions between public officials and community members when Melyce marches into the commissioner's office carrying the bodies of dead birds, along with her infant son. She demanded he explain how the animals died from smoking too much pot. Melyce then thrust her son's dirty diaper at the official. "You tell me this child has bloody shits day after day from smoking too much marijuana. Tell me to my face, Mr. Commissioner!"[1] Even when not facing overt official aggression, activists often experienced community disapproval, either of their opposition or their lifestyles, or both. The Alsea women told of their fears in angering the local Forest Service and powerful timber industry: "Anything against the timber industry was a negative thing of course . . . we were the enemy."[2] Yet they also rejected any association with groups they considered hippies, showing that government and industry's attacks on the counterculture had at least partially succeeded.

Van Strum's short story revealed a broader cultural conflict that underlay the fights over the use of potentially toxic chemicals, one that took place within a regional setting. In timber and farming communities throughout Northern California and southern Oregon, longtime residents clashed with the influx of newcomers displaying different values and practicing sometimes radically different lifestyles. Some of the early complaints against the phenoxy herbicides in Oregon came from organic farmers, and the Van Strums themselves represented the incoming wave of new residents who were returning to the land, disillusioned after the military war in Vietnam and culture wars of the late 1960s and early 1970s. Widening the focus of the anti-phenoxy protests uncovers a broader political, economic, and social conflict occurring in the postwar period. This domestic battle pitted capitalist industry and its government allies against the group(s) Dow's public relations director Ned Brandt labeled "the Flower Children."[3] And it would become an integral part of President Richard M. Nixon's "war on drugs."

✿✿✿

By the mid-1970s some of the focus within the arena of international Cold War had shifted from the jungles of South Vietnam to the poppy fields of Southeast Asia and Latin America. Phenoxy herbicides continued to be used internationally as an auxiliary means of counterinsurgent action in these new locations of concern. The chemicals had also become a crucial weapon in attacking the newly perceived domestic threats that had emerged in postwar American society during the Vietnam War. Nixon publicly identified drug addiction as a major problem to be addressed, which could be achieved in part through the eradication of international and domestic drug crops. At the same time, the chemical, agricultural, and timber industries identified the remnants of the counterculture as the problem. Even though the environmental battles over the spraying of the phenoxy herbicides 2,4-D and 2,4,5-T had intensified after the 1979 EPA study showed a link between miscarriages and herbicide spraying, the agricultural, timber, and chemical industries fought for their continued use. The chemicals would also be an important part of a new government initiative. Responding to claims that returning Vietnam veterans were addicted to drugs, Nixon launched a campaign in 1971 to address the problem. He called his initiative the "war on drugs."

Nixon was not the first, nor the last, political leader to advance his agenda by condemning what society perceived as moral decay. Nixon may have very well borrowed an approach used by Ronald Reagan in his 1966 and 1970 gubernatorial campaigns when Reagan attacked student unrest on California's college campuses. Reagan began focusing on the issue of campus turmoil after he observed that wherever he was campaigning in state the first question of the night was, "What are you going to do about Berkeley?"[4] Both the city of Berkeley and University of California at Berkeley had harbored New Left activists throughout the 1960s, and the university had been the site of the student Free Speech Movement in 1964. Although Reagan never successfully pacified students, his attacks embodied "the populist themes of [his] campaign: morality, law and order, strong leadership, traditional values, and anti-intellectualism."[5] They also allowed him to besmear liberals, distract attention from his cuts to education funding, and highlight his leadership. Nixon "improved" on Reagan's tactics in his antidrug agenda.

Given the revelations exposed in an interview with top Nixon aide John Ehrlichman (publicized again in a 2016 *Harper's Magazine* assessment of American drug policy), it appears very possible Nixon used his war on drugs to launch devastating attacks on the New Left and the African American community. Ehrlichman laid out the strategy in the 1994 interview. "The Nixon campaign in 1968, and the Nixon White House after that, had two enemies: the antiwar left and black people. You understand what I'm saying? We knew we couldn't make it illegal to be either against the war or black, but by getting the public to associate the hippies with marijuana and blacks with heroin, and then criminalizing both heavily, we could disrupt those communities. We could arrest their leaders, raid their homes, break up their meetings, and vilify them night after night on the evening news. Did we know we were lying about the drugs? Of course we did."[6] Although former Nixon aides have questioned this story, the disclosure probably did not surprise African Americans, Latinos, political dissidents, or scholars of illicit drug regulation. It only strengthens the critiques like the one offered by law professor Michelle Alexander, who has likened the war on drugs to the "new Jim Crow" in exposing the intent to harass was present from the beginning. At roughly the same time, corporate America became concerned about the perceived threat to capitalism posed by the successes of the progressive Left. These concerns, those of the state and industry, merged in the arena of drug policy and attacks on drug production and consumption.[7]

This is the story of the ways the war on drugs perpetuated Cold War counterinsurgency programs abroad and solidified attacks on the New Left counterculture domestically, and the ways this "war" was supported by the chemical, agricultural, and timber industries. The long decade of the 1970s saw three key events crucial to these attempts: (1) hippies returned as part of "back-to-the-land" movements in Northern California and southern Oregon;

(2) Nixon launched his war on drugs, supposedly in response to addicted veterans, which conveniently attacked the antiwar and minority communities; and (3) the US pursued an aggressive program of drug destruction as a means of Cold War counterinsurgency, particularly in Mexico. While historians have examined US drug policy as a counterinsurgent tactic, and others have studied the domestic battles over drug regulation, criminalization, and normalization, the war on drugs combined these campaigns. Once again, the phenoxy herbicides united international and domestic chemical use and policy in the continuation of counterinsurgency programs abroad and in addressing the challenges to the Cold War political consensus posed by the New Left counterculture. The control of alien others—Asians, African Americans, Mexicans, and hippies—who endangered American morals, societal order, and political stability required increasingly militarized responses to contain the threat.

✳✳✳

Conflict between conservative landowners and counterculture newcomers, later known as the "herbicide wars," broke out in Northern California and the coastal Pacific Northwest in the mid-1970s. Local residents protested both the spraying of chemicals by the timber industry for access to forests, and to promote better timber stock, and law enforcement removing illegal marijuana crops. These battles built upon the protests that had already taken place across the West, as exemplified by the stories of Billee Shoecraft, Ida Honorof, and Carol Van Strum, and significantly intensified after the 1979 EPA emergency suspension of 2,4,5-T. In the process of examining this environmental conflict, the influences of the New Left and generational change can be seen in the attitudes of residents who questioned the values of their parents, expressed their sexuality differently, regarded the environment in a new way, and used different recreational drugs. Provided is an examination of industries' responses to the fear that the countercultural and environmental movements had attained political power, and what this meant for capitalism and regulation.[8]

From the beginning, the fears underlying marijuana interdiction were connected to race. When Americans began regulating potentially harmful or addictive foods and chemicals with the passage of the 1906 Pure Food and Drug Act, what to do about marijuana use emerged as a major problem. Reformers wanted to include cannabis in the list of substances regulated by the 1914 Harrison Act, but the pharmaceutical industry opposed its inclusion. The federal government, however, did receive complaints. "Many Californians, particularly in San Francisco, were frightened by the 'large influx of Hindoos . . . demanding cannabis indica' who were initiating 'the whites into their habit.'"[9] Historian Isaac Campos has described the ways that Mexicans understood cannabis as causing madness and violence, and that this understanding shaped Americans' experiences (both regulators and consumers) with the drug. The loudest and most intense protests arguing for controlling marijuana use came from those parts of the country with the largest numbers of Mexican immigrants, who often used the drug for leisure entertainment and relaxation. By the mid-1920s Mexican criminality and deviancy was being directly connected with marijuana use. Various municipalities around the country criminalized marijuana as it became associated with certain racial minorities—Middle Easterners, Mexicans, African Americans.[10]

The fifties marked a shift in the ways illicit drug use and addiction were understood, with a medical model replacing the previous criminal one; federal officials continued to use cultural discourses to identify the perpetrators of the drug wars. The Federal Bureau of Narcotics used popular magazines, such as *Reader's Digest*, and a new genre of nonfiction to warn the public. One of the best examples of this can be seen in the work of crime reporter Frederic Sondern Jr., who spent the decade helping create the narrative of the "drug war." Sondern's 1959 book *The Brotherhood of Evil: The Mafia* exemplified this new genre, which became known as true-crime stories. The decade's preoccupation with the "dope menace" began with concerns about organized crime and narcotics but transformed into anxieties over marijuana and youth culture. Relocating the origins of the drug war to the 1950s reveals its connection to the Cold War, and the hegemonic thinking required of all Americans in the fight against communism and radicalism. Interdiction activities began in the cities.[11]

As white, middle-class Americans moved to the suburbs, urban spaces became increasingly occupied by nonwhite minorities. Urban police departments began cracking down on drugs, often aggressively through physical force. Even a leftist magazine like *Ramparts* located the epicenter of the heroin crisis in the "destructive funkiness of the black ghetto."[12] The use of heroin in suburbia drew the attention of national leaders as it spurred a significant increase in crime, as addicts stole an annual amount worth $2.5 billion, and as young people between the ages of eighteen and thirty-five died from drug addiction. Both conservatives and liberals reached agreement on the issue of illicit drugs. Conservatives supported aggressive state intervention focused on interdiction and carceral punishment. Liberals condoned strong state measures as a means of bolstering federal powers. All this happened at the same moment that white youth celebrated drugs, primarily psychotropic compounds like peyote, marijuana, and lysergic acid diethylamide, better known as LSD.[13]

Ьжжьь

The New Left challenged the postwar political consensus in several ways. Intellectuals charged that American military actions constituted imperialism, which was antidemocratic and which tainted the legitimacy of the war in Vietnam. Protests in the form of civil disobedience occurred when young men burned draft cards. And these radicals viewed culture, too, as a means of transforming American society. Countercultural defiance of the Cold War political consensus via social and cultural means focused on several specific arenas and represent what historian Doug Rossinow calls a "strategy for achieving social change."[14] One change entailed new kinds of living arrangements, such as shifting from the structure of the nuclear family to one of communal living, or the less extreme measure of rejecting traditional marriage for couples. These transformations were aided by new musical expressions, and new ways to "experience" the world.

One of the pathways taken by the counterculture dealt with altering consciousness through drug use. Military experiments using LSD, a synthetic psychedelic, introduced the chemical to a generation prepared to experiment and harness its psychotropic effects to achieve enlightenment. The drug typically caused distinct physical, psychological, and sensory effects, including hallucinations, an altered sense of time, experiencing radiant colors, and both good and bad "trips." Harvard psychiatrist Dr. Timothy Leary, who exhorted thousands to "turn on, tune in, drop out," and others promoted the drug as a means of achieving personal liberation.

The peak of the counterculture came in 1967, known as the "Summer of Love," when over 100,000 youths descended upon the Haight-Ashbury neighborhood of San Francisco. The drugs of choice were LSD and cannabis, and the fact that marijuana came from natural, organic sources made it an ideal choice for getting high. Even as the counterculture combusted under its own internal pressures and the external scrutiny of the state, the proliferation of "head shops"—places where marijuana paraphernalia like pipes, papers, posters, candles, and incense—helped mainstream cannabis culture, catering to a growing number of users.[15]

As the decade waned, believers found alternative spaces as mainstream

America and the disillusioned Left saw the counterculture discredited. Film and fiction reflected the contradictions of rebellion, drug use, and the failed idealism of the 1960s youth movement in movies like *Easy Rider* (1969), and fictionalized life like Hunter S. Thompson's *Fear and Loathing in Las Vegas* (1972). There was the real-life tragedy of the 1969 Altamont Speedway Free Festival, where the concert ended in violence and death as the Hells Angels, the infamous motorcycle gang, protected the stage, supposedly for $500 worth of beer, while the Rolling Stones played. In San Francisco itself, the epicenter of the 1967 Summer of Love, drugs had led to increases in crime, addiction, and death. In Northern California, seven people were attacked—five died—from December 1968 to October 1969 by a serial murderer known as the Zodiac Killer; the last confirmed victim was a twenty-nine-year-old graduate student in San Francisco. These visible and dramatic depictions and episodes suggested the counterculture had failed and disappeared. The reality, however, proved less definitive. The 1970s became a decade when hundreds of Americans, from both the political left and right, retreated to remote places to carve out communities that reflected their values. Pot played an important role in each.

Many of those attracted to the burgeoning counterculture in California simply drifted northward as the scene in San Francisco turned bad. These "Back-to-Landers" moved into northern California counties like Humboldt, Mendocino, and Trinity. Memoirs and fiction illuminate the domestic scene of illicit pot production. Known as a part of the "Emerald Triangle" of marijuana cultivation, Mendocino County, California, became a refuge for those fleeing the increasing chaos of the 1960s. Novelist Kevin Stewart based his fictionalized account of pot growers living in Mendocino County on extensive interviews with local community members. In his novel, protagonist Duncan Easley slowly migrates northward of San Francisco. A self-identified member of the counterculture, Easley sees his involvement in growing marijuana as an act of rebellion. He is not growing the illegal substance for money, but rather as a continuation of his enlightenment. But his love of farming and the money that could be made from growing marijuana motivate his new career as a pot grower. Stewart's novel provides in-depth descriptions of how to grow, harvest, and sell marijuana, and what it was like to live with the increasing realities of armed violence. While Stewart offers a seedy window on marijuana growers, other accounts told the story of the flower children's retreat to Northern California and their lives in these remote communities.[16]

Like hundreds of other Haight-Ashbury refugees, art student Mare Abidon left San Francisco in 1970 to go live on a commune in Gopherville, located in Humboldt County, California. Here she and about twenty other adults and a few children lived at a closed logging camp, raising vegetables and marijuana and supplementing the commune's finances with the welfare money some of the women received. Mare found a kiln and began making pottery. She considered this period to be one of the happiest times in her life and mourned it when religious extremists forced nonconverts out of the commune. She and her common-law husband drifted apart, and she camped in the summers and sold pottery at an annual Bay Area Renaissance fair. By the mid-1970s, however, she had found a new means of supporting herself—growing and selling marijuana. Local pot growers had figured out, supposedly with help from drifters, how to grow new and more potent strains of pot. Weed represented a valuable crop, one whose value only increased with the launch of a new initiative designed to decrease drug use in American society. Destroying Mexican marijuana and opium crops represented a major component of this war on drugs. But continuing the counterinsurgent techniques learned in Vietnam to destroy drugs in Mexican fields had unintended consequences.[17]

Seemingly prompted by the "epidemic" of addicted veterans returning from Vietnam, Nixon increased his efforts to address the problem of drug addiction. On June 17, 1971, Nixon delivered a special address to Congress concerning Drug Abuse Prevention and Control. In his speech, Nixon noted the increasing number of deaths due to narcotics, and while noting the overhaul of drug enforcement already accomplished by the administration, he maintained that it was now clear that more needed to be done. He also asked Congress for $155 million to "tighten the noose around drug peddlers and thereby loosen the noose around the necks of drug users."[18] The reason prompting this heightened concern appeared to be Vietnam veterans returning to the United States addicted to heroin. Historian Jeremy Kuzmarov has argued that a "myth of the addicted army" was created by both the political Right and the Left to advance their agendas. Americans opposed to the war uncovered connections between the Central Intelligence Agency (CIA) and illicit drug smuggling by high-ranking South Vietnamese officials, such as Ngo Dinh Diem's sister-in-law Madame Nhu, rumored to have profited by as much as $250 million to $500 million. The *Ramparts* article that had identified heroin as previously a problem of the ghetto made these very accusations, detailing the Southeast Asia "opium war" that the United States CIA played a central role in overseeing. Evidence that the CIA helped Hmong farmers grow and transport opium has been documented along with further activities undertaken by the CIA. But radical critics erred when they claimed the US military used and became addicted to the smuggled drugs. This misrepresentation paralleled similar efforts ongoing among urban law enforcement, which had begun to raise concerns about the threat posed by ghetto drugs like heroin and its use by suburban, white America.[19]

Following the lead of antiwar leftists, the myth gained credence and spread within American society. Newspaper commentary addressed the problem of "too many people both in uniform and out [who] have been hooked by heroin."[20] It was recognized that the problem had both international and domestic dimensions, and Turkey and France were identified as major players in the production and processing of hard drugs. Domestically, there were calls for increasing the penalties for narcotic sales, and for stronger sentencing. "The courts in this nation must get tough with pushers. Such criminals cannot be wet-nursed by bleeding-heart judges."[21] The severity of the problem demanded that a united citizenry must take action, reporting on trafficking, supporting medical intervention and research, and working in tandem with the federal government. A James Reston column charged that the military had a "narcotics crisis" and that the best way to "duck combat" was to "get hooked on drugs."[22] Reston noted that soldiers serving in Vietnam faced greater challenges, since the country produced large quantities of heroin. While Nixon's administration had been working to address the international drug trade, it had failed in Saigon, according to Reston. He too echoed the charge that soldiers were "ruining their lives by drugs" to avoid combat. They returned to communities ill prepared to handle the problem. Nixon's intent to increase rehabilitative services for addicts seemed to be lost in the frenzy to stamp out the drug trade—pushers, drugs, and all.[23]

Distinguishing between kinds of illicit drugs appeared to be one of the problems with the proposed war on drugs. Interviewed about the plan, then-congressman Gerald R. Ford identified four major points of Nixon's plan. These included renewed efforts to "dry up" overseas sources from Turkey, France, Thailand, and Mexico; intensifying law enforcement efforts; increasing rehabilitative services, especially for GIs; and an antidrug informational campaign among young people. One article in the *Salt Lake Tribune* expressed concerns about the ways heroin and marijuana use were treated as the same, despite significant differences in demographics. The administration included marijuana in its drug war based on the assumption that addicts start with stepping-stone drugs before they graduate to more hardcore substances. Nixon acknowledged this and a concern for America's young people at his

June 1 press conference. "I can see no social or moral justification whatever for legalizing marijuana. I think it would be exactly the wrong step. It would simply encourage more and more of our young people to start down that long dismal road that lead to hard drugs."[24]

Marijuana, though, according to the article, had become a part of youth culture, like long hair and blue jeans. Statistics from California undercut the argument that use of marijuana led to heroin addiction. Between 1961 and 1968, marijuana arrests had increased by 103 percent, while heroin arrests decreased by 7 percent. But, as the article noted, even the spaces where the drugs were consumed differed dramatically. Heroin addicts came from urban spaces and were predominantly Black and Spanish-speaking Americans. Given this fact, the piece charged the president with seeking to increase his popularity by attacking unpopular groups. Marijuana, on the other hand, was consumed in the relatively benign setting of schools and colleges, unlike heroin users' underworld in urban ghettoes. Nixon's crackdown on stopping drug production and increasing criminal penalties on consumption relied in large part on the use of the phenoxy herbicides abroad and at home.[25]

The Nixon administration identified the heroin in Vietnam as one of the major causes of the addiction epidemic, but it failed to acknowledge the United States' role in supporting the region's drug trafficking. This reality resulted from diplomatic and security decisions made in the 1950s that focused on anticommunism efforts and mostly turned a blind eye to the narcotics trade. These actions reflected a certain measure of irony, as the United States had positioned itself as the moral leader of international drug-control efforts and even went so far as to dictate policy to other nations. What US policy makers and enforcers often missed or ignored were other nations' social, economic, and political conditions. Along with applying pressure on individual producer nations, the United States pursued its crusade at the United Nations. Efforts to affect international policy had begun before World War II, but it was only until after the war that the United States had enough influence to advocate for prohibition in drug control programs.

As the French connection—narcotics grown in Turkey and processed in France—was disrupted, marijuana crops and heroin from Mexican poppies drew the Nixon administration's attention. In September 1969 Nixon inaugurated Operation Intercept, a unilateral campaign by the US designed to confiscate drugs before they could enter the country. The program focused mostly on marijuana but included heroin and cocaine smuggling as well. Ostensibly designed to crack down on marijuana supplies, the program focused on better monitoring of small aircraft flying from Mexico and inspection of automobiles at border crossings. Deputy Attorney General Richard Kleindienst characterized Intercept as designed to stop "the source of the problem, . . . not to make thousands of arrests."[26] He later admitted the real goal of the action was to achieve "meaningful cooperation with the Mexican government" and stop production of drug crops there. An editorial in the Fremont (CA) newspaper noted that while increased border enforcement might effect a shortage of marijuana, the real solution lay in giving the Mexican government cause to cooperate with the US to "stem this tide of narcotics across our common border."[27]

The operation virtually shut down the border, angered Mexicans, and lasted for less than a month. One account of the unintended six-hour border shutdown at the Tijuana crossing described radiators that had boiled over, failed batteries, and drivers pushing their cars into the United States. Another California paper charged that the program hurt innocents instead of catching the guilty. In its first official week, customs agents searched 2.3 million people, with only twenty-six persons arrested for smuggling. Although brief, Operation Intercept did inspire a 1970 song, "Mexico," by the Jefferson Airplane, which lamented the interruption of marijuana from south of the border. Along with its unpopularity with the Mexican government and tourist industry, American businesses complained about it as well.[28]

As one Californian newspaper year-in-review put it: "Drug crackdown snarls border, hurts local economy." A return trip from Mexico that usually took ten minutes became six hours under Operation Intercept. Some local businesses in Chula Vista, California, experienced losses up to 50 percent, with motel business down 25 percent, and businesses overall down 6 to 15 percent. Business still lagged even after Operation Intercept ended. Historian Kathleen Frydl and others have charged that there was no further need for Intercept, as it had achieved its goal, as "the real target of Operation Intercept was the Mexican government, and its success could best be measured in exacting concessions from that government at a time when Mexico sought to disentangle itself from American power and influence."[29] In this respect, it succeeded. The antidrug task force headed by Kleindienst had already identified one important element in any joint program with Mexico. Given the difficult terrain, task force members were encouraged at the potential to destroy marijuana crops through the aerial spraying of chemicals.[30]

Despite American allegations of Mexican laxness with respect to drug production, Mexico had been destroying poppy and marijuana crops as early as

the 1930s and by the 1950s had significantly improved its own crop destruction program. Part of this improvement was the result of the involvement
of the Mexican military with the federal police and by extending the time of
the campaign known as La Gran Campaña. It was estimated that there were
between twenty thousand to forty thousand plots of opium poppies, located
mostly in the states of Durango, Sonora, Chihuahua, and Nayarit, with Sonora as the epicenter.

No real estimates existed for marijuana crops, except that they exceeded
poppy fields. The *campaña* required five thousand Mexican soldiers and air-
men to be on permanent duty and involved in the eradication of poppy and
marijuana crops. They were employed either as troops airdropped into remote
areas charged with destroying crops and/or arresting traffickers, or as a part
of units performing patrols dedicated to finding especially well-hidden plots.
Once in the fields, soldiers used sticks, machetes, and hoes to manually destroy plants. Along with the rugged terrain, soldiers often found themselves
outgunned by drug traffickers or the victims of political and legal corruption.
Another, more difficult problem appeared to be the troops' poor morale. Despite Mexico's considerable success in destroying crops in rugged terrain and
under trying circumstances, the country remained the number one supplier
of heroin to the United States in 1974. This fact added pressure to escalating
interdiction and especially eradication efforts. Further pressure came from
the Mexican government's growing concern regarding their own domestic
drug addiction problem.[31]

To be successful, future eradication campaigns required aerial access, an
effective measure of eradication beyond manual removal, and a response for
significant and armed resistance on the part of growers. Concessions forced
on Mexico by the disruptions caused by Operation Intercept allowed that
program to be succeeded by Operation Cooperation. Operation Cooperation,
which unlike Intercept had the support of both the Mexican and American
governments, intensified US pressure on Mexico to "stop the yield in the
field." Along with interdiction, the program focused on destroying poppy and
marijuana plants. Between 1970 and 1975, the US sent approximately thirty
helicopters and fixed-wing aircraft to the Mexican Attorney General's Office,
along with telecommunications equipment and pilot and mechanic training.
It provided $15,000 to conduct remote sensing experiments designed to iden-
tify hidden plantations, and another $35,000 for equipment to destroy poppy
crops. Equipped with this advanced technology, support personnel, and funds,
in a scenario eerily reminiscent to the one that played out in South Vietnam,

the United States supported Mexican eradication programs with technology, personnel, and aerial spraying campaigns, but in this case the crops destroyed were poppies (heroin) and marijuana instead of rice.[32]

Using technology and practices developed as a part of its global counter-insurgency programs, the United States offered Mexico aid in detecting, ac-cessing, and eradicating drug crops. The early efforts focused on providing air support in the form of helicopters and the creation of staging areas that put troops within twenty minutes of their targets. Then specialized technol-ogy like "remote multispectral aerial photography" allowed poppy fields to be discovered. Refinements in the technology eventually let Mexican authorities detect poppy fields and plants that previously had been invisible before they had begun to bloom. As historian Daniel Weimer described it, the new tech-nology "made the unknown known and illustrated the links between tech-nology, drug control, and state power."[33] The stage was set to initiate the next step, an aerial herbicide spraying program. Mexican president Luis Echeverría held out for several years, hesitant to begin defoliation activities. There were two major reasons underlying the delay. One was denial, as Mexicans strug-gled with acknowledging "the massive extent of drug cultivation and produc-tion taking place in their country."[34] The second reason dealt with concerns over the chemicals' safety and their negative connection with their use in Viet-nam. This foot-dragging about herbicide spraying figured among several criti-cisms made against the eradication program. But using chemicals got results.

Aerial herbicide spraying finally began in the fall of 1976. Just a little over six months after its launch, the spraying operation had wiped out 5,745 acres of poppies, and herbicide defoliation "destroyed 21,405 opium plots from Sep-tember 1, 1975 to August 31, 1976, as compared to 13,580 plots during the preceding twelve months."[35] Incoming Mexican president José López Portillo intensified the drug eradication program in 1977. The new program, known as Operation Condor, was committed to the complete eradication of illicit drug crops in Mexico. The United States matched every $4 spent by Mexico with $1; there was, however, no subsidy for herbicide costs, a fact that would cause trouble later. The US support of international drug eradication drew the attention of domestic critics.[36]

Ida Honorof began her June 1976 newsletter reporting on a "secret war" in Mexico, one that she claimed already had seen casualties of over two hundred people. The Mexican government had received aid from the United States to carry out a war on drugs, specifically marijuana and heroin. Honorof described the spraying campaign in Cuernavaca, Mexico, where 2,4-D and paraquat (a nonselective herbicide) were used to destroy the crops. "Overruling protests, they [the combined US and Mexican forces] contaminated farms, produce, people animals, birds (air, water, etc.), but their insane project continued!"[37] Honorof noted that this "unholy secret war" was funded with $40 million, plus the gift of new helicopters, fixed-wing light aircraft, an "executive jet," and all the fuel needed for this fleet. The US not only provided monetary and technological support but sent troops as well. She decried the Mexican government's complicity, asking, "Are Mexican lives that cheap!!"[38] In the same newsletter, Honorof alerted her reading public to the EPA's recommendation that the surplus Agent Orange supplies be reprocessed and used domestically. Once again, the international and domestic uses of Agent Orange herbicides were intertwined.

Honorof had written previously about US eradication activities in Mexico. A March 1976 newsletter had quoted extensively from a Washington *News-works* article that explained the ways 20 percent of all arable land in the Mexican states of Chihuahua, Durango, and Sinaloa in the north, and Guerrero in the south, were drenched with 2,4-D. "There has been popular resistance to the use of herbicides in the mountain states, but the mountain farmers who have to live with the sterilization of their land, and the possible birth deformities in their children, for generations to come, aren't being consulted!"[39] Mexican peasants had joined other members of the international community in challenging industry and government assurances of the phenoxy herbicides' safety. Honorof made inquiries about the two hundred deaths in Cuernavaca to the Consul General, whom she had met in Los Angeles. The embassy claimed that no such incidents had occurred. "I wish to inform you that there have been no official reports of any deaths of people in Mexico of the kind you mention in your above letter."[40] She continued her efforts to try and speak with the consul about agricultural pesticides, and her determination that "the truth must out!" Despite the response Honorof received, Mexican authorities expressed their concerns over the use of the phenoxy herbicides through their silent inaction rather than by vigorous denunciation of the chemicals.

Two major problems appeared in the herbicide destruction of drug crops. The drug eradication program initially used 2,4-D in destroying both poppy fields and marijuana crops. The destruction of poppy fields was well achieved with 2,4-D, and it was considered ideal for this plant type. Operation Condor directly attacked the heart of heroin production, the states of Sinaloa, Durango, and Chihuahua. But as one contemporary expert on the drug wars

noted, "Enthusiasm, commitment, and cooperation notwithstanding, the total elimination of Mexican poppy cultivation was, and is, a virtual impossibil-
ity."[41] Marijuana crop destruction presented a different problem. While 2,4-D
was also considered ideal to defoliate marijuana crops, it was not used on marijuana. The Mexican government, responsible for herbicide costs, chose to
replace the 2,4-D with the cheaper but more dangerous herbicide paraquat.

Although subsequent research has questioned the herbicide's harmful effects, contemporary studies showed paraquat to be highly toxic to humans,
primarily damaging to the lungs. Paraquat was discovered in 1882, but it was
not until 1962 that it was manufactured and sold for weed control. The compound made an excellent herbicide, as it was fast-acting, worked on numerous grasses and broad-leaf plants, withstood rain, and was partially deacti-
vated in soil. In the case of the marijuana sprayed in Mexico in 1975 through
1978, as much as 21 percent of the confiscated marijuana sampled indicated
paraquat contamination. Inhaling paraquat increased its toxicity, making it
a particularly dangerous herbicide for crops that would be smoked. A March
1979 *Morbidity and Mortality Weekly Report* addressed the issue of paraquat-
contaminated Mexican marijuana. The update reported on the results of a
Center for Disease Control testing program that showed contaminated marijuana in the border states but no contamination in marijuana samples taken
from the Pacific Northwest or the Eastern Seaboard. This meant that pot
grown in Northern California had become an even more valuable commod-
ity, one that dramatically changed the lives of growers.[42]

❧

The development of new potent strains of pot, eradication of Mexican crops,
and risk of paraquat poisoning all significantly increased the value of domestically produced marijuana. This meant that growing pot in what became
known as the Emerald Triangle would create a new major regional economy. "By 1979, the year Congress suspended the paraquat-spraying program,
an estimated 35 percent of the marijuana smoked in California was home-
grown."[43] Marijuana began selling for between $500 and $1,000 a pound,
an increase of five to ten times. The birth of this new agricultural economy,
aided by the decline of regional timber and fishing industries, came about because of blood—the next generation's mixed-culture children—and money.
Hippies and longtimers joined together to produce this lucrative cash crop,
gaining a new moniker that symbolized the ongoing social shift: hipnecks.
And the new economy was acknowledged openly when Mendocino County

agricultural commissioner Ted Ericksen listed the $90 million marijuana harvest in his 1979 crop report under the heading of timber production. The decision drew the ire of the county board of supervisors, who disagreed that marijuana constituted a crop, and the State of California, and others who objected on "moral grounds." Ericksen was pressured not to run for reelection and ended his service in 1981.[44]

After agriculture, forestry represented one of California's major industries. Just a few years prior, in 1974, Honorof had been a part of a group, People for Environmental Progress, that had sued the US Forest Service to stop the spraying of herbicides in the state's national forests. The group reached an agreement with the Forest Service that the phenoxy herbicide 2,4,5-T would no longer be used in the Angeles Forest, but if it were used, the city of Los Angeles would be notified. By 1976 the herbicide issue had heated up in California. The decision by Trinity County officials to ban herbicide spraying happened because of label warnings on 2,4,5-T supplies. The product caution noted "meat animals" should not be allowed to graze within two weeks of application. "Although a number of questions have been raised about potential hazards of herbicide application, the immediate focus of attention is the question of whether deer killed in areas treated with 2,4,5-T may be safely eaten."[45]

The battle to ban herbicides in one California county, Mendocino, illustrates citizens' concerns and the opposition they faced from agricultural and timber interests. Led by a Mendocino County wife and mother, Betty Lou Whaley, residents began asking questions about herbicides as early as 1973. In a 1979 interview, Whaley described herself as "apolitical until she picked blackberries and learned after eating them that they had been sprayed with herbicides."[46] Whaley tried to research the herbicides at the Fort Bragg library, but it had almost no resources. She resorted to ordering information by mail. She helped form Citizens Against Aerial Application of Phenoxy Herbicides (CAAAPH).

Whaley and CAAAPH spearheaded a campaign to stop the spraying of herbicides on county roadsides, clashing with agricultural commissioner Ted Eriksen over the practice and eventually challenging the county board of supervisors. Board member Ted Galetti appeared vulnerable, as he was up for reelection. Early skirmishes between antiherbicide citizens and its supporters played out in the letters to the editor sections of the newspapers. Responding to a woman opposed to spraying the herbicides, a letter writer defended business. He charged she represented "a very small but opinionated minority who try to sell you on the idea that industry, the very fiber and backbone on which this county's economy is based, is out to destroy your environment, pollute your water and cause you all kinds of illness and distress."[47] Just a month later Whaley called for the establishment of a herbicide study group. Whaley urged the county board of supervisors to create a bipartisan committee that would investigate the safety and feasibility of using the herbicides.

Whaley's leadership put her in a media spotlight, which she used to highlight the perceived dangers of the chemical herbicides. These included miscarriages, birth defects, cancer, and nervous disorders. From her perspective, political leaders traded "off economic benefits, supposedly, against public health hazards, without consulting the public, without letting the people vote on it."[48] One newscaster highlighted the significance of the herbicide ban initiative, describing the grassroots campaign as a "classic citizen's revolt underway. . . . For the first time in the United States, voters will decide whether to ban aerial spraying of certain chemicals."[49]

The chemical industry made its opposing position quite clear. A March 1979 Dow Chemical publication, *The Bottom Line*, characterized its critics as "obstructionists" rather than "environmentalists," individuals opposed to technology. Dow suggested that these people were deeply pessimistic, unsure of what they wanted, and hostile to the "nasty chemical companies and horrid pesticide-using farmers."[50] As far as industry was concerned, the "anti-technology folks" wanted the chemical companies and their products gone. The piece ended with a subtle threat, suggesting it was time to find out more about the groups that opposed them.

The proposed herbicide aerial-spraying ban of 2,4-D, and 2,4,5-T, and any other chemicals containing dioxin in Northern California's Mendocino County positioned longtime residents against individuals they considered political nuisances. These second- and third-generation ranching and farming families railed against a minority population they characterized as "long-haired, pot-growing political activists" who had migrated to the county from places like Berkeley and Los Angeles and as far away as New York City. The arrival of these former members of the 1960s political movement, the counterculture, had increased over the course of the 1970s, but their presence had mostly gone unnoticed. Health concerns, and a desire to keep their non-code housing, had ignited political action. Marsha Johnson, a leader of the anti-ban forces, described the problem as an expensive mistake being inflicted on Mendocino County residents against their will. "It's being pushed by a small,

Figure 9. A 1967 glimpse of a Mendocino County commune showing newcomers' non-code housing, the source of much conflict with longtimers. Courtesy of Corbis Historical Collection, Getty Images.

arrogant minority which has recently arrived in our county. I call them the 'hill people.' . . . In no way do they represent a majority of the citizens in our county and they are going to find that out June 5."[51]

The new residents, united by the aerial chemical sprayings and possible building code condemnations, came together and put their political skills to good use. They were able to oust established local supervisor Galetti and replace him with a younger and more liberal alternative. The countercultural forces charged that conservative residents opposed to the ban had received funds from Dow Chemical and other chemical companies, something anti-ban residents denied. When interviewed, Georgia-Pacific, operator of the "world's largest redwood mill at [local] Fort Bragg" admitted that the company refused to state the chemicals were safe. That information needed to come from federal officials. But the company opposed the ban for the same reason the chemical companies did. "If they can call a special election and win on this issue, they can do it on others. What if they develop a proposition which would ban clear-cutting or some other restrictions which would greatly inhibit our operations?"[52] The moment marked the height of what became known as the "herbicide wars."

Rejected Mendocino county supervisor Galetti expressed the conflict as one with people who had no connection to the land. "There's been a change in lifestyles in this area. We have had a lot of people come in with different lifestyles who are not really tied to agriculture or food products." Galetti and other longtime residents noted they had not experienced any apparent ill effects from multiple years of herbicide spraying. Many of those opposed to herbicide use lived in an artist colony established in Mendocino in the 1960s. (Whaley made pottery and stained glass.) One news article describing the divisions among county residents identified the conflict as "the recently developed back-wood counter-culture against Mendocino County's traditional power structure which included longtime farm families, conservative ranchers and giant timber companies."[53] The county's highly profitable timber industries appeared to be lying low. The most intense opposition came from established farm and ranching families, who saw the antiherbicide group members as "recently-arrived, long-haired, pot-growing political activists."[54] Longtime residents appeared dismissive of the CAAAPH's organizing, labeling members with the derogatory "hill people" and suggesting they lived in squalid self-built housing. The "counter-cultural forces" had already flexed their political might by successfully voting out the conservative county supervisor (Galetti) and electing a liberal.

Environmental activists, themselves counterculture refugees or influenced by the movement, joined with those concerned about the safety of the phenoxy chemicals and helped pass the herbicide aerial spray ban in June of 1979. In 1977 a group of individuals concerned about logging practices formed the Environmental Protection Information Center (EPIC). One founder, Robert Sutherland, recounted in an interview that he became concerned about herbicides after encountering a timber crew cleaning a helicopter while hiking on the northern California North Coast. The helicopter had sprayed 2,4-D and 2,5,5-T into the river; the herbicides were commonly sprayed in clearcut forests to prevent invasive species from growing. Sutherland tried to negotiate with the companies to no avail. When the herbicide issue resurfaced, EPIC amassed information on herbicides to help inform residents. The group brought a lawsuit in California that stopped the spraying of 2,4,5-T.[55]

Other activists followed suit. Betty Whaley and CAAAPH began circulating petitions to place an aerial ban referendum on the ballot in September 1979, after county commissioners refused to ban spraying, upon the advice of their attorney. Residents had the help of another grassroots organization, Oregon's Citizens Against Toxic Sprays (CATS), in filing the petition to ban aerial spraying. CAAAPH had the required 4,000-plus signatures just four months later but continued to gather support. By February of the next year, the county clerk certified the petition, ensuring that the referendum would be put to a county vote. Mendocino County resident Kathy Bailey, a former member of Students for a Democratic Society and a concerned mother, represented a typical supporter. One of the initiative organizers, Bailey walked to her neighbor's house to use their telephone. Bailey later joined efforts to regulate sustainable forestry practices, showing the interconnection between herbicide protests and broader concerns over timber industry practices.[56]

Although battered, countercultural activists achieved some environmental victories. Mendocino County's aerial spraying ban represented one of two counties that succeeded in limiting herbicide exposure, while similar measures failed in two other California counties. The ban survived for several years, despite questions about its legality, and was still under review in 1982. Perhaps more importantly, environmentalists on the national level had achieved some significant victories during the Nixon administration. Here, California's herbicide wars composed part of a larger environmental movement, a movement whose efforts to increase ecological awareness and protections scared corporate America. "The environmentalists have had the biggest victories: Ranking jobs . . . have gone to men and women who have . . . lobbied . . . for conservation, protection of wildlife and clear air and water."[57] This observation appeared in *The Coercive Utopians: Their Hidden Agenda*, a 1978 book written by social critic Dr. H. Peter Metzger. In it Metzger described the kind of people now populating Washington, DC, and various government agencies under the Carter administration. He went on to distinguish such individuals as different from usual political appointees, the beneficiaries of the routine spoils system. These members of the counterculture, he warned, wanted to do more than simply enjoy political patronage; they wanted to dismantle the basic mechanisms for generating wealth in the country. These "coercive utopians" sought to achieve their goals covertly, through legislation, which included a reduction in per capita energy consumption, a shift from fossil fuels to solar energy, and an undetermined form of economy that bore no relation to capitalism or private ownership.

Metzger's warnings appeared in a 1979 Dow Chemical Company brochure. They revealed the growing concern the chemical industry in general, and Dow especially, held toward the new regulatory climate that had emerged in Washington during the 1970s. For Dow and corporations like it, new legislation like the 1969 National Environmental Policy Act (NEPA), newly created agencies like the Environmental Protection Agency (EPA), and new advisory bodies like the president's Council on Environmental Quality (CEQ), all heralded a new regime of public interest that would "strangle" society through regulation, review, and repeal of existing and potential chemical licenses and legislation. During the decade of the 1970s, from the banning of DDT in 1972 through the controversy over the herbicide 2,4,5-T and its associated dioxin contaminant at the end of the decade, Dow would fight the "coercive utopians" and their vision of a government more responsive to the health and ecological concerns of ordinary Americans.

Dow joined other industries as they fought Americans' changed environmental consciousness and legislative landscape. In a 1981 issue of *The Bottom Line*, Dow's free newsletter, the lead story examined the "script" written by the Alsea, Oregon, case and the "pesticide road show being restaged in Missoula, Montana; Ashford, Washington and Peevy's Crossing, Oklahoma."[58] The National Forest Products Association's Forest Industry Chemicals Committee "'T' Memo" for its member organizations, for instance, demonstrated the ways industry allies kept in communication. The report updated various industry players on the status of regulation. A June 11, 1979, issue gave a postmortem of the Mendocino County ban, giving CAAAPH credit for the passage of the ban, and calling out San Francisco's KRON-TV for re-airing its documentary on the phenoxy herbicides, *The Politics of Poison* (which itself had been the topic of a previous "'T' Memo").[59]

The documentary aired the night before the aerial ban referendum, and the station claimed that ten thousand contacted the station demanding an EPA investigation into herbicide regulation. In August an American Forest Institute internal memo noted that a documentary film on cancer-causing substances, focusing heavily on herbicides, had received grant funding from the EPA. The memo's author, John Benneth, later wrote a letter to Oregon senator Bob Packard complaining about the use of VISTA monies to hire workers to manually remove weeds. Benneth identified himself as a citizen, not an industry official, and decried the actions of groups opposed to herbicides. Even as industry monitored and protested private and state scrutiny, they appeared to adopt new tactics as well.[60]

Chemical, agricultural, and timber industries spent decades fighting the environmental protection agenda and defending the capitalist system against countercultural values. In the process, companies played a major role in contesting scientific knowledge of herbicide safety, the right of the state to set regulatory policy, and the concerns of ordinary citizens over the safety of everyday chemicals that were sprayed on their fields and trees. Despite significant monetary and political resources, industry appeared to concede the battle in environmental battles and began working with its own grassroots movements.

Speaking to industry groups, former reforester Betty Dennison told about the origins of her group, Women in Timber, at a Pacific Logging Congress. While a minority of industry members had been hostile, Dennison called out the illusion of power, pointing out that the kinds of legislation coming out of Washington were things like the 1964 Wilderness Act. Dennison was inspired to become active when a General Electric executive bemoaned the lack of an active, committed constituency of supporters. Dennison and the women in the timber movement worked to educate the public and lobby legislatures. Dow's 1981 *Bottom Line* applauded the formation of a group in Montana, Citizens for Food and Fiber, dedicated to "prompt collective action."[61] Another citizens group had formed in the state—the Washington State Pest Management Alliance, committed to "scientific reason and rationale" in putting out the anti-pesticide fires. An article editorialized, "It is time for more of this kind of citizen action, where grass roots elements representing farmers, foresters, applicators, agri-women and other proponents of free enterprise defend the agricultural chemical tools that are important to everyone's standard of living." Opinions like those expressed by longtime Mendocino County residents or the Women in Timber group suggest the ways that many of the users of chemical herbicides did not recognize potential harms and actively pushed back against regulatory efforts. These individuals and groups would not join forces with environmental and health activists in their efforts to control herbicide use.[62]

While industry had found its own grassroots activists, the assault on the counterculture continued and changed. State and federal police programs expanded the scope of their efforts to halt the production, sale, and use of illegal drugs into the 1990s. The need to contain the counterculture faded as the "flower people" embraced the consumerism and corporatism of the 1980s. It would be a different group of citizens protesting past practices that emerged in the late 1970s and the 1980s to challenge chemical policy and the toxic legacies of the war. The contested terrain of the war on drugs shifted to a fight against Vietnam's deadly fog of hazardous herbicides.

CHAPTER EIGHT

Fighting the Deadly Fog

Vietnam Veterans Protest Agent Orange Herbicides

Early in 1978, Vietnam veteran Paul Reutershan appeared on NBC's *Today Show* and declared that he had died during his service in Vietnam, even though he "didn't even know it."[1] Reutershan flew helicopter missions in Vietnam in 1968 and 1969. In the process of crewing his helicopter, he was repeatedly exposed to Agent Orange spray in the air. After the war, as a health-conscious twenty-eight-year-old, Reutershan had read the mounting media coverage of the Agent Orange issue and become convinced his own exposure to the chemical had caused his terminal cancer. Reutershan's very public declaration was followed by a documentary from CBS reporter Bill Kurtis. Called *Agent Orange: Vietnam's Deadly Fog*, the documentary included interviews with veterans who believed the various illnesses they experienced were caused by exposure to Agent Orange. By the end of 1978, Reutershan had filed a $10 million lawsuit against two of the major manufacturers and suppliers of Agent Orange to the US government during the war. Although Reutershan died before the end of the year, his lawsuit became a major class action suit, seeking billions of dollars in compensation.[2] The issue reinvigorated a veterans' group, Vietnam Veterans Against the War (VVAW), who took on the issue of Agent Orange and its effects on vets. The lawsuit, which was resolved in a controversial decision by the presiding judge, resulted in a $180 million settlement. America's Vietnam veterans had been alerted to the possibility of ill health resulting from their country's use of hazardous chemicals.

Reutershan's story embodies many of the elements of veterans' protests over Agent Orange use. A young man serving in an unpopular war came home with an unexpected, and to him, inexplicable illness. Veterans returned from the war angry and disillusioned, with some protesting against the conflict before it had even ended. Soldiers' alleged addiction to heroin had prompted President Richard Nixon's war on drugs, and the traumatic nature of the war had led to a new psychiatric diagnosis, Post-Traumatic Stress Disorder (PTSD). The threat of amorphous illnesses that veterans increasingly attributed to herbicide exposure seemingly came out of the blue, increasing the burden veterans encountered as they tried to reintegrate into society. The phenoxy herbicides had been used to environmentally contain civilians and enemy troops during the war, maintain the military-industrial complex as a means of furthering Cold War objectives after the war, and as a means of attacking and constraining countercultural protest. With veterans, the terrain of the battle altered as soldiers questioned what further risks they were being asked to bear and challenged the state's narrative of safety.[3]

After the Vietnam War ended, the US government was confronted with thousands of angry, anxious, and increasingly engaged veterans convinced that the chemicals used in war had harmed them. From the beginning, the US military, the Veterans Administration (VA), government officials and their industry allies ignored, denied, and minimized the possibility that the phenoxy herbicides and the dioxin contaminant that tainted Agent Orange were causing veterans' illnesses. Various state agencies demanded veterans prove exposure using elaborately plotted mission runs. Veterans' participation in Agent Orange protests has perhaps received the most attention within the popular and scholarly realms, in part because exposure has re-emerged as an issue in ongoing wars. Examining the lesser-known actors and cultural discourses of soldiers' protests over the herbicides sprayed in Vietnam puts these protests within the broader context of Agent Orange activism. Despite the controversy then, and now, over Agent Orange's health effects, in this work veterans' claims of harm are accepted. Other scholars, such as Michael Gough and Edwin Martini, take more skeptical stances.[4] Veterans' challenges took several different forms, overlapping in time and actors. The story of the veterans' fight begins with one of the best-known individuals involved with veteran protesters, Maude DeVictor, a VA civil servant working in Chicago.[5]

An encounter with a dying veteran exposed to Agent Orange uncovered the possible link between the phenoxy herbicides and the host of unexplained illnesses being experienced by Vietnam veterans. DeVictor had served in the US Navy, survived breast cancer, and attained a college degree by the time she became a benefits counselor for the Chicago VA office. Her role in connecting

Figure 10. Maude DeVictor participates in the Vietnam Veterans Against the War Winter Soldier Investigation of Agent Orange in Chicago, 1979. DeVictor was considered the mother of the veterans' protests over government neglect regarding Agent Orange exposure and its possible hazards. Courtesy of Vietnam Veterans Against the War.

Vietnam veterans' illnesses to their exposure to Agent Orange in Vietnam led her to being called "the Mother of Agent Orange."[6] Here, according to DeVictor in a later interview, "mother" was used in both the "ghettoese" sense of the word and as capturing the ways she nurtured afflicted men and their families and the larger protest movement her investigations launched. DeVictor's involvement with Agent Orange began in June 1977, when a woman contacted her and said that her husband had just received a diagnosis of terminal cancer. The veteran, Charlie Owens, had served in Vietnam as a finance clerk, and had told his wife if he died from cancer, it was from the chemicals used in Vietnam. Owens died during the processing of his medical claim, which meant the claim needed to be refiled for a surviving spouse. After three months the VA denied the widow's claim on the basis that the government did not recognize Agent Orange–exposure-related illnesses as a valid condition for coverage. Using her right of appeal, Owens's widow asked De-Victor to represent her in challenging the denial.[7]

In the process of determining the facts of the case and any possible grounds for appeal, DeVictor spoke to Air Force captain Alvin Young, who held a PhD in plant physiology and was one of the officers closely connected with defoliation programs. In DeVictor's recollection, Young proceeded to explain the toxic effects connected with dioxin, the contaminant present in Agent Orange. Encouraged by her supervisor, DeVictor documented the information, and it went into the public record. She also continued a series of informal health survey interviews with the veterans and their families whom she interacted with during her routine job duties. She asked questions such as "Has your wife or lady been pregnant and then something happened between the second and third month?" or "Do you have numbness in your hands and feet?"[8] As her inquiries progressed, DeVictor began receiving negative feedback from her supervisors and eventually was told to stop gathering information on Agent Orange cases. It was these further inquiries, however, that brought DeVictor and the issue of Agent Orange to the attention of local television reporter Bill Kurtis, already well known for his investigative reporting. DeVictor had become the mother of a movement.[9]

The investigative documentary helped mobilize veterans. In reporting and producing *Agent Orange: Vietnam's Deadly Fog*, Kurtis nationally publicized DeVictor's work and veterans' fears that they had been exposed to a life-threatening toxin.[10] The documentary and alarm it sounded reverberated throughout the media with important consequences. Checking wire news sources, Agent Orange appeared in story headlines or lead paragraphs only once between January 1, 1974, and March 22, 1978, in a story on the disposal of excess herbicide supplies. From March 23, 1978, to December 31, 1980, Agent Orange appeared as the main topic of over a 150 stories, most dealing with Agent Orange and veterans' claims of harm.[11] The *New York Times* did a major three-part series, "Agent Orange: A Legacy of Suspicion," that documented ordinary citizens as well as veterans' fears of damaged health, in the spring of 1979.[12] As early as October 11, 1978, news sources reported that approximately five hundred veterans had filed Agent Orange exposure disability suits with the Veterans Administration. A little short of two years after *The Deadly Fog* aired, more than five thousand veterans had submitted claims. By this time, VA offices had received briefings on the safety of the phenoxy herbicides, which were used domestically with few problems. Government officials also pointed to the apparent absence of illness in Ranch Hand pilots, a group that would be expected to show symptoms from their exposure. Another element in the awakening of veterans gained national attention with a dramatic proclamation.[13]

In the spring of 1978, Paul Reutershan made his dramatic statement on *The Today Show*. Reutershan, a health-conscious twenty-eight-year-old, had read the media coverage of the Agent Orange issue and become convinced his own exposure to the chemical had caused his terminal cancer. Working with a friend, he began publicizing the problem using public access television. As recounted in Peter Sills's history of Agent Orange, Kurtis knew about Paul Reutershan's campaign against Agent Orange and had mostly accepted dismissals of Reutershan as a "kook."[14] Kurtis changed his mind regarding Reutershan's claims while interviewing veterans and researching the phenoxy herbicides in preparation for the documentary. DeVictor, Kurtis, *The Deadly Fog*, and Reutershan's lawsuit all served as successful catalysts focusing national attention on what would become a major toxic tort case and what some considered a national scandal.[15]

The documentary and subsequent lawsuit prompted differing responses from various groups. The VA and official government position seemed to be one of denial. When the local Chicago VA office was inundated with veterans' calls after the airing of *The Deadly Fog*, national VA official Dr. Paul Haber emphasized that scientific literature thus far supported the relative mildness of any toxic reactions.[16] Just a month after the documentary aired, however, the VA had formulated an official policy for Agent Orange claims. The regulations recognized only chloracne—a skin disease resembling severe acne and characterized by cysts, blackheads, and pustules. This signature symptom of dioxin poisoning would be the only condition covered as a result from Agent Orange exposure. Responding in the fall of 1978 to the increasing number of veterans' disability claims, Major General Garth Dettinger, the deputy surgeon general of the Air Force, admitted that while extensive amounts of the herbicide had been used in Vietnam, "Our best evidence right now is that we do not have a problem."[17]

VA officials dismissed the multiplying number of veterans' Agent Orange claims, pointing out that by the fall of 1978 only five hundred veterans had filed claims out of a total of the 700,000 veterans who had visited VA hospitals. By the spring of 1979, VA officials had received briefings on the safety of the phenoxy herbicides, which had been used domestically with few problems. Within the broader medical community, a more mixed reaction appeared, with opinions on both sides of the issue expressed. A brief article detailing the controversy over herbicide exposure appeared in the spring of 1979. Physician Gilbert Bolger alerted physicians to some of the symptoms they might see in treating veterans, in a November 30, 1979, letter to the editor of the *Journal of the American Medical Association*. Despite the refusal of officials to acknowledge a problem, the lawsuit initiated by Reutershan had significantly expanded in the number of plaintiffs and scope of the legal case.[18]

The Agent Orange lawsuit lasted over half a decade, from 1979 to 1984, and eventually involved thousands of servicemen and scores of lawyers. Peter Schuck and Peter Sills have separately examined the Agent Orange lawsuit in detail, and my discussion of it here depends heavily on their accounts. Inaction and outright obstructionism color the entire episode, as government agencies, including the VA, the Center for Disease Control, and the Department of Defense consistently downplayed at best, actively prevaricated at worst, regarding the possible harmful effects of exposure to 2,4,5-T's dioxin contaminant and the phenoxy chemicals themselves.

In 1979 the General Accounting Office issued a report that criticized the half-hearted and ineffectual responses taken by the government. The ways the government practiced willful neglect of Vietnam veterans over Agent Orange exposure appeared to be part and parcel of the treatment these veterans received more generally. Survivors of an unpopular war, disconnected from broader society (including previous generations of soldiers), and struggling to reenter a society experiencing stagnant economic conditions, veterans still expected better treatment for their service. For many, engagement with the Agent Orange issue became an empowering act. A brief overview of the lawsuit, itself a form of protest, also provides the context for the actions taken by the VVAW, a peace group started during the war in 1967, and the cultural discourse surrounding Agent Orange toxicity and activism that emerged as the case progressed.[19]

Before his death in December 1978, Reutershan had formed an organization, Agent Orange Victims International (AOVI), and filed a lawsuit against Dow Chemical and North American Phillips Corporation. He entrusted his friend Frank McCarthy with oversight of both organizations. The attorney Reutershan had originally hired dropped out of the case; he recommended his colleague Victor Yannacone to McCarthy. Yannacone came to the case with the right mix of skills and experience. He had sued over DDT spraying on Long Island in the 1950s and worked with the newly formed Environmental Defense Fund (EDF) in the late 1960s to stop DDT spraying in Wisconsin, a precursor to the chemical being banned in 1972. His workman's compensation law practice and past involvement with what he termed "environmental law" gave Yannacone an understanding of the issues at the heart of the legal action. Yannacone filed a revised complaint in January 1979, which named more chemical manufacturers as defendants and drew immediate and worldwide attention. The class action suit prompted the plaintiffs to hire their own array of legal counsel. Yannacone also added defendants and recruited more lawyers to identify and sign up veterans potentially affected. (The greater the numbers in any class action suit, the greater the sum of the settlement, and the greater the pressure to negotiate.)

McCarthy traveled with Yannacone to find more plaintiffs even as other lawsuits began to appear. Eventually fifteen thousand plaintiffs, representing the *two and one-half million soldiers* exposed to Agent Orange in Vietnam, were combined into one suit. One important action Yannacone took was adding veterans' children who were born with birth defects, with the family of Kerry Ryan becoming the lead plaintiffs. Yannacone dealt with the challenges the case presented—scientific uncertainty and a novel legal strategy—by including the heartbreaking stories of veterans and their families and took an aggressive stance in court. In the process, he bonded more with the veterans themselves than his co-counsel. One limitation of a class action meant that Yannacone's innovative legal strategy came with restrictions and gave more power to the judge. This would have significant consequences for the outcome of the Agent Orange trial.[20]

Judge George C. Pratt wrestled with a difficult and unruly court case, exposing his judicial limitations in the process. While Yannacone respected Pratt, the judge's conventional outlook meant that defining the various terms governing the lawsuit substantially slowed its resolution. These included addressing the myriad of procedural issues, which included sorting out who was included in the plaintiffs' group, procedures for discovery, and even who could be sued. The US government potentially could have been charged as a defendant in the case, but legal precedent and veterans' reluctance to sue the military or their civilian representatives put the onus squarely on the manufacturers of Agent Orange. The chemical companies tried to claim immunity as government contractors providing necessary war material, which if it met government specifications, would have exempted the companies, along with the federal government, from legal liability.

There were two problems with this approach. One appeared to be the extent of knowledge military commanders held regarding the toxicity of the herbicides. Here the chemical companies could show that lower-level personnel, such as Alvin Young, knew about the potential harm the chemicals posed. They could not definitively show that officers at the command level knew about the potential dangers associated with the chemicals. With respect to the other difficulty, chemical manufacturers had specified what the acceptable product purity would be in their contracts with the government. In other words, industry had stipulated what product specifications would be, not the federal government. These attempts increased the state of confusion surrounding Agent Orange and its dioxin contaminant. For veterans, the case not only sought financial redress for their health problems; it served as a vindication of the soldiers and their service. They simply wanted the respect they felt they had earned and the better medical care that came with that respect.[21]

The next steps taken by government agencies expose the ways the state sought to contain what was becoming both a legal and moral dilemma. In 1981, during trial preparations, Congress ordered the VA to determine which veterans had been exposed to Agent Orange during their service in Vietnam. Over the next several months the VA repeatedly hemmed and hawed in establishing a working protocol. Eighteen months later the VA acknowledged its Agent Orange study would take another six years. The CDC lobbied to take over the Agent Orange health studies, as the agency already was studying the health effects on veterans' children; the study was transferred in October 1982.[22]

Using military records for the Ranch Hand missions, the National Academy of Sciences (NAS) had compiled a record of Agent Orange sprayings that also included the incomplete information on smaller spraying missions by helicopters and the spraying done around base perimeters with backpacks. The CDC sought to use the data, called the HERBS tapes, to identify unexposed troops, an essential control in an epidemiological study. The CDC eventually conceded that it was impossible to determine who had and who had not been exposed and focused its efforts on determining the degree of exposure, privileging acute contact over long-term chronic exposure. Even this revised protocol posed problems. Numerous accounts of empty herbicide drums being used as shower reservoirs or barbecue pits meant that it was virtually impossible to measure military exposure accurately. Agent Orange exposure was proving difficult to contain within the neat boundaries of spraying missions. An even more daunting obstacle to studying veterans' exposure-linked illnesses emerged with the next logical question. If soldiers had been affected by exposure to Agent Orange, what effects had ordinary people experienced in the everyday use of the herbicides?[23]

The widespread use and dependence on the phenoxy herbicides in the domestic market represented that millions of people had been exposed to the phenoxy herbicides. As the previous discussions of Billee Shoecraft, Ida Honorof, and Carol Van Strum have shown, there was strong resistance to the use of the chemicals in national parks, fields, and forests. These women and the grassroots anti-toxic networks they represented made the phenoxy herbicides a hotly contested issue for elected officials, government agencies, and public health programs. The evacuation of the Love Canal neighborhood of Niagara Falls in 1978 and the Missouri town of Times Beach in 1983 raised the nation's consciousness to the dangers of dioxin. Industry appears to have sacrificed 2,4,5-T (and its toxic contaminant) when Dow Chemical agreed to quit manufacturing the chemical and removed it from commercial herbicide products in 1987. The federal government's continued bungling of health studies continued the uncertainty. But the toxic legacies of Agent Orange exposure continued to come to light as veterans pursued their legal case.[24]

As the numbers of plaintiffs, and plaintiffs' lawyers, increased, tensions within the legal team mounted. Yannacone and McCarthy made numerous trips recruiting still more veterans as plaintiffs even as the legal team, what historian Peter Sills calls the consortium, paid the bills. The massive number of documents shared in the discovery phase of the trial, which included scientific papers, manufacturing records, government reports, memos, and letters, eventually tapered off. The plaintiffs continued to struggle to prepare the case for the trial.[25]

Yannacone's arrogance and risky trial strategy rubbed many of the lawyers the wrong way. Given the numbers involved, in both expenses and possible settlement, and the publicity the case received, reputations could be made or broken depending on the outcome of the case. Pratt and the legal team slowly sorted out the various obstacles to trial, while Yannacone's relationship with his legal co-counsel further soured. By 1983 the situation had deteriorated to the point that the consortium asked that Yannacone be dismissed as lead attorney. Pratt agreed and appointed two lawyers to oversee trial preparedness and the trial attorney. The appointment of new leadership changed the lawsuit. Yannacone had represented veterans' desire for moral validation and was the legal expert best versed in toxic torts. Now the lawsuit became much more like a traditional class action case with the accompanying emphasis on financial payouts to investors (people who had invested funds for the case with the expectation of a lucrative return) and the lawyers themselves. An even bigger personnel change, however, happened in October 1983 when Pratt finally recused himself from the case following his appointment to the court of appeals the previous year. The new appointed judge, Jack Weinstein, proved to be a brilliant but autocratic jurist. He demanded the case be ready to go to trial within seven months, a Herculean task by any measure.[26]

As 1984 began, the contours of the case had dramatically changed. Veterans' advocate Yannacone was no longer in charge. Weinstein had reinstated chemical companies that had been dropped and restored the federal government as plaintiffs. Both sides were working frenetically to process the mountains of information relevant to the case. Weinstein clearly wanted the case settled out of court, in part because he feared decades of appeals and wanted no case of his reversed, which had a high probability given the anticipated appeals. But a settlement denied veterans one of their goals in pursuing a lawsuit in the first place, the chance to be publicly recognized and validated with respect to their service in Vietnam. Settling the case, however, increasingly hinged on the bottom line. Veterans' desire for a bully pulpit disappeared in the calculations. Weinstein appointed procedural judges to help arrive at an acceptable settlement amount. The financiers wanted a quick settlement, as did Dow Chemical. Legal teams thought they could get more money. Monsanto, the largest supplier to the federal government of Agent Orange (most of which was heavily contaminated with dioxin) wanted to take its chances at trial. The gap in what constituted a reasonable settlement ranged from a low of $25 million from the chemical companies to the high of $700 million demanded by veterans.[27]

Keeping everyone in suspense, an agreement was reached just days before jury selection was slated to begin. As a May 1984 *New York Times* piece noted, one of the most complex civil actions, one with potentially billions of dollars at stake, was set to begin May 4. The chemical companies devised a formula to determine who would pay and how much. The veterans' legal consortium resisted an offer of $150 million, settling for a $180 million counteroffer. One of the largest, lengthiest toxic tort cases had been resolved. But not for veterans who had been contesting their exposure to the chemicals since Maude DeVictor and Paul Reutershan had made them aware of the possible link between exposure to Agent Orange and the illnesses they and their families were experiencing. One of the major veterans' groups involved in these efforts had been protesting the war since 1967.[28]

Veterans reeled from the news that they might have been exposed to hazardous chemicals during their service in Vietnam, and they began to mobilize.

For some it seemed to explain the various illnesses they had experienced or provided a reason for the birth defects their children exhibited. Yet again, the VVAW rallied. At the height of their activism, VVAW activists held major protests in 1970 and 1971, with the most prominent being the 1971 Winter Soldier Investigation in Detroit. Their most controversial action might have been Operation Dewey Canyon III, held in 1971, an action where veterans threw their service medals over the White House fence, more than a thousand of which were Purple Hearts. When Maude DeVictor, Bill Kurtis, and Paul Reutershan transformed the consciousness of Vietnam veterans with the release of *Fighting the Deadly Fog* and Reutershan's appearance on *The Today Show*, the VVAW became the voice of veterans. Agent Orange revitalized a group whose purpose had faded with the end of the war and fragmentation of the other postwar progressive social movements. Articles printed in *The Veteran*, the VVAW's official publication, reveal the major issues surrounding Agent Orange and veterans' responses. From the summer of 1978 through the settlement of the lawsuit and beyond, the VVAW provided an important service in keeping veterans informed. In the process its publication revealed veterans' perspectives as they struggled to get official recognition of their illnesses and receive proper medical care.[29]

The Veteran's first in-depth discussion of Agent Orange appeared in its Summer 1978 issue. The lead story discussed the most common symptoms of exposure. It also noted that the chemical remained in veterans' bodies for years. Initially, it seemed like another burden was placed on the backs of Vietnam veterans, although it acknowledged the "fight against this newest outrage has already begun—you can join the struggle."[30] A later article in the same issue delved more deeply into the threat. The piece recounted the current state of knowledge about Agent Orange: its composition (although it incorrectly identified 2,4-D as containing dioxin); how much was used; the destruction of the South Vietnamese environment, and the appearance of harmful effects within the Vietnamese population as early as 1970. It noted concerns about birth defects and likened the herbicides to having the same effects as thalidomide, the highly teratogenic sleep aid responsible for a host of serious birth defects throughout Europe in the late 1950s. The article described the Kurtis documentary and called out the VA's suppression of information regarding the possible health effects of Agent Orange exposure.[31]

For veterans, it was part of a pattern of neglect and abuse, and the new revelations confirmed their experiences. "This latest report about Agent Orange underlines the 'concern' of the US government for its military, for the men it sent off to do its dirty work."[32] Chicago veterans held a press conference after the documentary aired, demanding a response from the VA. At a VVAW conference, veterans developed a list of actions they wanted the VA to take. Veterans in Chicago and Milwaukee protested the VA's refusal to treat Agent Orange exposure, picketing and flooding phone lines demanding that their illnesses be classified as the result of Agent Orange exposure and their treatment would be covered by the VA. Other efforts focused on getting information out. Freedom of Information Act requests were filed to collect available information, and some veterans stormed local TV stations and insisted that *Deadly Fog* be aired. Veterans had a chance to "stick a big fist in the face of the VA and their bosses" and get medical treatment for everyone.[33]

From the group's first coverage of Agent Orange herbicides to status updates years after the settlement of the lawsuit, the VVAW's responses as published in *The Veteran* revealed some common themes. The organization's first imperative appeared to be to inform veterans about the chemicals, the illnesses, current VA procedures, and the status of the court case. The second major topic featured in the newsletter focused on veterans' responses before, during, and after the settlement of the lawsuit. The newsletter also captured the responses of ordinary vets as they sought information, shared their experiences, and grappled with how to deal with both their illnesses and the neglect of their government. The Agent Orange stories collected suggest the ways veterans' treatment and the emerging controversy over the chemical exposures proved almost impossible to contain, with concerns and protests appearing more frequently in public and cultural discourses. This prominence posed an especially vexing problem for chemical manufacturers, as it reinforced the questions about the safety of the phenoxy herbicides that environmentalists had been asking for decades. The newsletter covered the lawsuit extensively, with the editors weighing in on the problems they saw with the settlement. Finally, the veterans' experiences with Agent Orange and their poor treatment at the hands of the VA appeared repeatedly over the course of twenty years.

As 1978 progressed information about Agent Orange routinely appeared in the pages of *The Veteran*. The fall issue dealt extensively with the topic. In an article quoting Reutershan, *The Veteran* reviewed the basic facts. These included how much of the defoliant was sprayed, how much of the countryside, and how many GIs were exposed. The piece also presented more speculative information as accepted knowledge, suggesting pregnant women exposed to the herbicides had miscarried or the children born had a host of birth defects. GIs who handled the drugs experienced blisters and rashes that, while they eventually disappeared, erupted again once they returned to the United States. *The Veteran* credited Maude DeVictor with uncovering the secret that soldiers had been exposed to dangerous chemicals.

The article discussed the medical symptoms and current state of affairs. "Symptoms of Agent Orange poisoning range from numbness of toes and fingers through excessive fatigue and nervousness, a lessened sex drive, through liver cancer and skin cancer, to deformed children."[34] While veterans experienced many of these symptoms, they were not necessarily recognized by the medical profession or the VA as resulting from the exposure to Agent Orange. The article went on to note that as dioxin was stored in fat tissue, it was both difficult to detect and could be released from those tissues even years later. An inset article asked: "What Is Agent Orange?" It gave a brief history of the chemicals, acknowledging the uncertainty that pervaded the issue. It also confessed that the VVAW did not have all the information but claimed what was presented was based on confirmed facts. They noted that they "believe[d] it's important for vets to have some idea of what the stuff is so that we can better build the fight to get it tested and treated."[35] Those interested in the sources used were advised to contact the national VVAW national offices. What followed was a succinct summary of the status of 2,4-D and 2,4,5-T, including the battles over regulation and the EPA's failure to address 2,4,5-T's dioxin contaminant. The issue also called for a "National Agent Orange Day," to "bring nationwide attention to the effects of Agent Orange and the refusal of the V.A. to deal with poisoned vets."[36]

The VVAW kept veterans abreast of developments. A year after the story broke, the VVAW had begun to focus on the dioxin contaminant in Agent Orange. One article focused on the Seveso accident in Italy and the problems people exposed to dioxin had experienced. Vietnam veterans protested and demanded "that the V.A. and Defense Department notify all vets who may have been exposed, provide testing, treatment and disability payments for all who were exposed to Agent Orange."[37] The news report noted that veterans had joined with environmental groups that had been trying to get the chemical banned domestically for years. Ten years later, in his study of the Agent Orange controversy, Fred Wilcox called the domestic spraying of Agent Orange herbicides the "Vietnamization of America." The VVAW updated veterans on the status of the Reutershan family lawsuit, which sought to set up a fund to cover the costs connected to veterans' medical care. According to the article, a General Accounting Office study on Agent Orange indicated that veterans appeared to be suffering from the long-term effects of dioxin exposure and recommended that the Department of Defense map sprayed areas. The VA earned the VVAW's distrust by hiring Dow's medical director to advise the agency on illnesses related to Agent Orange. The piece also noted that Maude DeVictor had been transferred to "the V.A.'s version of Siberia."[38]

Agent Orange updates notified veterans of planned actions and current affairs in Washington. The National Veterans Task Force, of which VVAW was a member of, planned a Day of Action for Memorial Day, 1981. The task force was formed in 1979 to coordinate veterans' groups Agent Orange activities. Their 1981 conference included a number of settings where the latest scientific information was shared, strategies discussed, and updates on other groups success to get "testing, treatment and compensation for Agent Orange victims."[39] The VVAW saw the Day of Action as a means of publicizing the problem. As they noted, the absence of government support put the burden on veterans to advocate for themselves and to win support for their cause. *The Veteran* now included a new feature, "Agent Orange Shorts." The spring 1981 issue informed veterans about a change in position. The Secretary of Health and Human Services had finally conceded that more soldiers were exposed to Agent Orange than previously acknowledged. "VVAW and other vets groups ha[d] been saying, since the Agent Orange story broke in March of 1979, that the defoliant was used all over the country and that almost anyone who served in Vietnam was potentially exposed to Agent Orange."[40] A 1982 "Shorts" update charged that both chemical manufacturers and government officials must have known of Agent Orange's potential toxicity in light of the information uncovered as a part of the class action lawsuit.

Demonstrations against the VA over Agent Orange continued. In 1983 the DC chapter of VVAW picketed outside the VA. While the VVAW saw the increased coverage of Agent Orange and its dioxin contaminant in the press as a positive, it maintained that their work was not done. It recognized progress in the lawsuit and hoped that the epidemiological study ordered by Congress three years earlier would finally begin. States had already begun establishing Agent Orange/dioxin commissions. Support appeared to be in place, although legal, political, financial obstacles remained.[41]

One problem was the limited number of medical conditions that were officially recognized as symptoms of Agent Orange exposure. At the time only chloracne and soft-tissue sarcomas were included. The slow pace of the epidemiological studies proved especially frustrating, and for members of the VVAW, it was evidence of the VA's lack of commitment to treat Agent Orange exposure. The agency responded so poorly that even Congress had taken notice. Sen. Tom Daschle argued that vets should receive compensation for those conditions currently recognized. The article finished with information on what needed to be done to advance the epidemiological studies, including the creation of and participation in Agent Orange registries. Veterans were advised to be examined and placed on registries, apply for VA health care, contact the VA if suffering from skin disorders, join the class action lawsuit, and contact political representatives and urge their support for Agent Orange legislation. In an edition published a few months later, *The Veteran* recommended an "Agent Orange Bibliography" for use by local physicians. The work provided doctors with the latest dioxin research citations. Although discouraged, the VVAW continued to keep veterans informed and engaged.[42]

The VVAW consistently provided veterans with information about Agent Orange and health effects for more than a decade. In 1987 *The Veteran* reported on the failed CDC study started in 1982. According to the CDC, the failure was because they could not find enough veterans to participate in the study, "an understandable claim if the CDC were located in Nepal, perhaps."[43] The CDC's admission was expected, and it showed the ways that attempts to contain the threat of Agent Orange exposure were flawed from the beginning. The VVAW noted that the health studies required a degree of precision impossible to meet. "Most vets could simply say they were in an area that was sprayed or even how often spraying took place around base camps, along roads, etc. Records of such things were at best spotty."[44]

The VVAW discussed the results of a second, much more surprising study, in the same piece. The VA announced that more US Marines who had served in Vietnam died of lung cancer and specific lymph cancers than did their domestic counterparts. Lung cancer was as much as 58 percent higher, and there was 110 percent greater rate of non-Hodgkin's lymphoma based on a study population of fifty thousand marines and army vets who served in Vietnam between 1965 and 1973. The VVAW expressed shock at the announcement of these numbers, given the agency's long history of deflection, deflation, and discouragement of veterans with respect to "defoliant poisoning." A 1990 headline proclaimed, "Vietnam Veterans Are Still Dying from Agent Orange." The article updated veterans on what qualifications needed to be met to qualify for compensation under the settlement. It gave a specific outline of actions that veterans needed to take to challenge any VA denials, including the address of the lawyers who had undertaken the challenge to the VA compensation process. The VVAW and *The Veteran* proved a crucial means of keeping veterans informed not only about the health effects of Agent Orange, but also about the ongoing saga of the class action lawsuit undertaken to redress what veterans considered an egregious wrong.[45]

The Veteran's coverage of the Agent Orange lawsuit began with its coverage of the problem itself. Early on, the newspaper advised veterans on what they needed to do if they thought their health had been negatively affected from herbicide exposure. Veterans needed to document their military service and medical history. Although the medical exams offered by the VA were generally useless, the article explained that a record like this would help establish a date for any future claims awarded by the government. Veterans were also advised they could request a copy of the HERBS tapes to further document exposure to the herbicides, even though such information was problematic. "Those of us who stomped around the jungles know we have no way of coming up with the particular date when we saw planes spraying Agent Orange," the article noted; "Mostly we didn't even know where we were—certainly not exactly."[46] While *The Veteran* made no recommendations on legal representation, it offered aid to any veteran who wanted to join the lawsuit. Overall, veterans were given direct resources that included specific actions and contact information, as well as how to obtain even more information, such as the Agent Orange self-help pamphlet the article was based upon.[47]

As the lawsuit dragged on, *The Veteran* continued to keep vets informed on its status. The case's sheer size, both in number of litigants and the geographic area from which they came, complicated the suit. Writing in 1982 about the Supreme Court's refusal to hear an appeal on whether individual veterans could be compensated, *The Veteran*'s John Lindquist described the decision as typical of the government's attitude of "use once and then throw away!"[48] He noted that the nature of the lawsuit as a "product liability suit" meant that narrow state statutes of limitations applied to more than six thousand veterans living in states with such laws. Lindquist urged veterans to remain involved and continue fighting, and to engage in a veterans' movement. *The Veteran* kept its members informed when the trial was set to begin in 1983, and then when it was delayed for a year. Perhaps its most sustained coverage of the lawsuit came after it was settled in 1984.[49]

For veterans, the court case had always been about more than just the money. In the opinion of *The Veteran*, the May 1984 settlement omitted two vital components. The decree recognized no link between exposure to Agent Orange's dioxin contaminant and human illness. This potentially meant that the VA might not acknowledge Agent Orange exposure as a service-related disability. The second omission appeared to be the chemical companies' denial of liability. As the update put it, "And where does that leave us? Well, the corporation's answer has always been that our exposure to Agent Orange was our own fault because we didn't apply it correctly."[50] Once again, ordinary GIs got the blame just like they lost the war too.

The first published assessment condemned the chemical companies, complaining that the rise in their stock value indicated that investors thought the settlement had been a good deal for them. The attorneys that negotiated the settlement also drew the ire of veterans. They had been appointed by the judge, not hired by veterans. The piece finally admitted why a public hearing was so important. "It is our chance to speak for ourselves, to try to convince the judge what we need."[51] The next issue, an extra edition devoted entirely to the Agent Orange lawsuit, continued its criticism of the settlement, decrying the sudden out-of-court resolution of the case. Claiming that the settlement "stinks," *The Veteran* noted that the settlement's failure to assign blame meant that chemical companies and the government could put future soldiers and veterans at risk. The settlement amount was completely inadequate, and it made no provisions for research and treatment options. The edition ended with the plea that veterans act. The authors urged vets to attend scheduled public meetings and express their concerns. "This issue is much more than simply getting money to help our brothers and their families, though that is vital. The issue runs straight to who is responsible in a war for the protection of the soldier, and who should be made to pay."[52] In the eyes of the VVAW and many other veterans, fighting men suffered while chemical companies profited. Their efforts proved to be in vain.

Judge Weinstein approved the settlement in the summer of 1985 with no changes. Veterans who were partially disabled would not receive any cash payments, and veterans' children with birth defects were also excluded. The 1985 newsletter update told veterans that now was the time to unify and fight. Veterans unhappy with the settlement tried to overturn it, but the settlement was upheld by the Supreme Court in 1988. As late as 1989, veterans still had not received any of the settlement funds, although lawyers had been paid. One of the possible reasons for delay appeared to be the US Air Force's intent to use the defoliants again. In 1987 the US used the herbicides as a part of a marijuana eradication program in Guatemala and reportedly in El Salvador to kill crops and cattle in areas controlled by left-wing rebels. One article in the 1989 annual edition recounted the story of veterans' history with Agent Orange after the war. This battle, the piece noted, had gone on longer than the VVAW's protests over the war itself. Unlike the ending of the Vietnam War, veterans had failed in their protests over the Agent Orange lawsuit. "Even to-day we are still fighting against the continued use of the same chemicals, and for the just medical care and treatment for our affected children."[53] The arti-cle informed veterans about the process they needed to follow to file a claim.

An update a year later told veterans about the settlement of a related case that affected the VA's regulations on Agent Orange claims. By this point the VA and veterans had normalized the process of Agent Orange claims. Subse-quent developments saw expansions in the conditions covered and the num-bers and types of veterans eligible. Veterans' children remained mostly in-eligible for disability claims related to their fathers' (and mothers') Agent Orange exposure. Along with their vigilant attention in giving medical and legal information about Agent Orange, perhaps one of the greatest services the VVAW and *The Veteran* did was give veterans a voice, publishing their fears, anger, and frustration in dealing with the medical uncertainties, phys-ical disabilities, and emotional traumas inherent to dealing with Agent Or-ange illnesses. The group stood witness to their ongoing neglect by the gov-ernment, most particularly the VA.[54]

Problems with the VA appeared early in the Agent Orange discourse within the pages of *The Veteran*. One of the earliest issues addressing Agent Orange called for a protest to "bring nationwide attention to the effects of Agent Or-ange and the refusal of the V.A. to deal with poisoned vets."[55] The litany of complaints VVAW veterans made against the VA included its unresponsive-ness, its active deception of veterans in giving phony medical tests, and de-struction of files. These criticisms grew when the VA's health studies dragged on for years. Veterans criticized a lack of transparency as well, as the VA's working group met outside of public view. As late as 1982, health studies were still focused on determining exposure in the field. Although the fixation on mission exposure was misplaced, as soldiers and civilians were exposed in a multitude of venues, the disarray of military records further delayed the VA study. At the same time, the Air Force conducted a methodologically flawed study of Ranch Hand pilots.

The problematic nature of the Agent Orange health studies was high-lighted by studies from the EPA showing 2,4,5-T contamination of rice in Ar-kansas and Louisiana (and soil from the fields as well). Along with its poor oversight of various health studies, the VA's greatest offense might have been its continued framing of veterans' claims of ill health from Agent Orange ex-posure as psychological problems versus somatic illness. While VA chief Rob-ert Nimmo pledged to continue health studies, he characterized the issue as an "emotional" one. "I believe our responsibility [is] to relieve *unfounded* anxiety among veterans and [that this] is at least equal to our responsibility to press on for whatever the final answers there might be."[56] The 1987 VA announcement of study results showing marines who served in Vietnam dying of lung cancer and non-Hodgkin's lymphoma at much higher rates than their stateside counterparts surprised the VVAW, "given the VA's usual role in the on-going battle to shelve, squash, suppress and discourage veterans and their families on any aspect of defoliant poisoning."[57] The contentious relationship between Vietnam veterans and the VA continues to this day. One other way *The Veteran* advanced the narrative of Agent Orange hazards came in the printed stories veterans wrote about themselves.

Veterans wrote describing their experiences in Vietnam, their illnesses, and their participation in protests connected to Agent Orange. One veteran given a phony blood test by the Milwaukee VA described his interaction with one VA doctor who refused to hew to the official party line. "[The doctor] said we really got screwed and that the poison was cumulative. Even if we ate Lake Michigan fish we were adding to our poisons because PCBs and dioxin are related."[58] It appeared to be the first time this veteran had someone listen to, and validate, his concerns about Agent Orange. One article acknowledged the fight ahead as veterans dealt with new physical ailments, "physical problems that didn't exist before Vietnam and didn't exist among other people our age who hadn't been in Vietnam."[59]

Working with other veterans' groups, the VVAW recognized the need to continue to inform veterans about the problem of Agent Orange. In 1981 the organization helped plan a Memorial Day protest outside the Washington, DC, VA offices, both to publicize the issue and to hold the agency accountable. Veterans and allies wrote creative works to capture their experiences as well as protest. "Agent Orange Song" took Reutershan's famous declaration "They killed me in Vietnam and I didn't even know it" as the song's refrain, while another veteran wrote a poem to help raise consciousness. Kathy Gauthier, a veteran's wife, wrote Dow Chemical executive E. B. Barnes rebuking him for all the harm caused by Agent Orange and its dioxin contaminant. She and her husband existed in a "living Hell" along with other veterans and their families. Rewriting Henry Wadsworth Longfellow's "Paul Revere's Ride," Gauthier excoriated Dow and ended her revised poem with the claim "All the vet asks, Is to Test, Treat & compensate."[60] Although Gauthier denied that she or other veterans and their families were activists or radicals, the actions of other VVAW members and readers contradicted her words. Modifying a popular Army cadence,

Milwaukee vets chanted as their Agent Orange protest convoy approached a local Hercules Chemical plant. "Dow, Monsanto, Hercules, Diamond Shamrock if you please, Uniroyal, T.H.Ag, All you guys are a great big drag! Sound Off . . ."[61] The problem of Agent Orange expanded beyond the veterans' community as Americans began to wrestle with the other legacies of the war.

Stories about Vietnam veterans from other countries also appeared in the pages of *The Veteran*, mostly about Australian and New Zealand troops that had been exposed to the herbicides. Mobilized by events in the United States, Australian veterans began organizing to demand state recognition and aid in treating their medical conditions. Veterans' experiences in these countries mostly mirrored the treatment American Vietnam veterans received, that of government denial and delay. One difference appeared to be the role of the Australian press, which covered the case extensively and much earlier than its American counterparts.

In Australia, however, the government advanced alternative explanations for veterans' ill health. The more controversial hypothesis suggested that many of the illnesses experienced by Australian veterans were psychosomatic in nature and had been caused by the trauma of war, not exposure to toxic chemicals. While the VA appeared to have tried this line of explanation, US authorities avoided it for the most part. Veterans groups provoked the Australian government to convene a Royal Commission under the auspices of Justice Philip Evatt. In 1985, a year after the American veterans' lawsuit had been settled, the commission ruled the Australian government "not guilty" on the basis that no link between Agent Orange exposure and veterans' illnesses had been proven. As Ed Martini has documented, international cooperation extended beyond veterans' groups' collaboration but also included governments and corporations sharing information. Scientists interviewed for a 1989 conference expressed concerns. They noted that research indicated a greater degree of uncertainty than the commission acknowledged, and that a verdict of "insufficient evidence" would have been an equally acceptable decision. Instead of resolving the dilemma of Agent Orange, the commission discredited veterans, adding more acrimony to the debate. In the US one provocative criticism linked the harm done to Vietnam veterans with the wrongs of previous wars.[62]

⚜

Just as the Agent Orange issue gave new credence and importance to the VVAW, so too did it play an important part of the work done by Citizen Soldier, an organization that succeeded two others dedicated to war crimes and amnesty for draft dodgers. The group focused on issues relevant to soldiers then serving, although it also advocated for health issues related to radiation (the WWII atomic tests) and defoliation. After the Kurtis documentary aired in March 1978, Citizen Soldier held a press conference to announce a "Search and Save" (as contrasted with the standard "Search and Destroy" mission) campaign. The group set up a toll-free 800 number to "alert and identify ailing Vietnam vets." By the end of the summer, Citizen Soldier had received over three thousand phone calls and letters. Each veteran contacting Citizen Soldier received a six-page health questionnaire to be filled out by the individual. In addition to these individuals, thousands of copies were sent to various organizations—veterans' groups, unions, and counseling centers. Over a thousand surveys were returned by November, and an analysis of 536 questionnaires showed thirty-five cases of cancer, including three cases of kidney cancer, rare for this age group. In additions, seventy-seven children were born with a variety of birth defects.[63]

Citizen Soldier members Michael Uhl and Todd Ensign publicly addressed the discovery of Agent Orange's health hazards in a piece for the antiwar magazine *The Progressive* in 1978. An excerpt from their forthcoming book, *GI Guinea Pigs*, the essay told the Agent Orange origin story of Maude DeVictor's discovery and subsequent fight for benefits coverage for Agent Orange exposure. The book, published in 1980, put Agent Orange in a wider perspective, as part of a long history of military abuse of enlisted men. It specifically juxtaposed US atomic weapons testing on soldiers that produced the hydrogen bomb and the use of the phenoxy herbicides. Uhl and Ensign interviewed several veterans, among them Ranch Hand pilots and ordinary grunts, and challenged the government's version of events. Ranch Hand pilot Charley Hubbs admitted that little care had been taken to use or wear protective gear, which was confirmed by a fuel supervisor. He described soldiers being drenched while loading the herbicide into loading tanks. Men wore no gloves and often went shirtless as they worked in the heat. This contradicted the military's claim that exposure had happened only during spraying missions in the field.[64]

GI Guinea Pigs presented a searing portrait of the official military response and VA inaction in responding to the health concerns raised by veterans. By linking defoliation missions to atomic bomb testing, Uhl and Ensign sought to reveal the vulnerability of enlisted men and the uncaring nature of military and political bureaucracies. In this, they exposed only some of the military's shameful history of experimentation on military personnel. As Uhl and Ensign noted in chapter 7 of their book, military researchers tested LSD as potential "truth serum." In her history of mustard gas and WWII, historian Susan L. Smith uncovered experiments done on minority soldiers, who were exposed to the toxic chemical as a part of army experiments. More than anything, like the VVAW, Citizen Soldier wanted to arouse veterans and the general populace to demand government action. As a fundraising letter asserted, Citizen Soldier had helped inform veterans and the government about the serious conditions Vietnam veterans were experiencing, linking these conditions to Agent Orange exposure. "Therefore, Citizen Soldier stresses independent mass actions, such as rallies, marches, and demonstrations, instead of polite lobbying in the bureaucratic and legislative corridors in Washington."[65] Perhaps one of the greatest challenges the book offered to officials' narratives of safety dealt with scientific certainty.[66]

While the book's title accurately captured its subject, the use of soldiers as a means of human experimentation, it also challenged one of the most persistent and dangerous ideas embedded in the discourses around Agent Orange. Military, corporate, and administrative officials proclaimed over and over that the herbicides used in Vietnam were safe. Various agencies like the VA only reluctantly acknowledged the skin condition chloracne as evidence of dioxin poisoning and tried to make it the sole medical criterion that would be used. In GI Guinea Pigs, Uhl and Ensign indicted these state, medical, and corporate actors with the use of the guinea pig metaphor. Guinea pigs were used in the process of medical experimentation, as an important part of testing the effects of chemicals and poisons. By focusing on the multitude of ongoing and flawed health studies, Uhl and Ensign argued that the story of Agent Orange should be one of *scientific uncertainty* rather than one of certitude and assured safety. This represented an important concept in challenging conventional wisdom about the chemicals' harmless nature.

In comparing veterans exposed to atomic and chemical weapons, GI Guinea Pigs made veterans unwilling but crucial components of scientific experimentation. This contrasted with the manufactured certainty produced and broadcast by the military, chemical manufacturers, and some but not all scientists and physicians. In invoking the guinea pig metaphor, veterans (and people like Billee Shoecraft) challenged proclamations of chemical safety and instead offered a different understanding of human-environment and chemical-health interactions, one that emphasized the uncertainty of scientific knowledge regarding chemical hazards, and that argued the science was still not yet decided. In this sense these Americans were part of a much broader shift taking place in America (and I would argue worldwide), one where the benefits of the postwar chemical world and its benefits were questioned, and the Cold War military and political consensus challenged. In the narratives of toxic uncertainty presented by veterans (and environmental activists), personal stories portrayed confusion, frustration, and harm as the norm associated with Agent Orange chemicals, rather than the carefully controlled, objective process of science. It is worth noting that many of the medical problems identified by the VVAW and in *GI Guinea Pigs* have subsequently been shown to be valid, most strikingly the connection between Agent Orange exposure and cases of spina bifida in veterans' children.

Several visual representations of Vietnam veterans' plight added to the national conversation around Agent Orange herbicides. Filmmaker Jacki Ochs released her documentary *Vietnam: The Secret Agent* in 1983, at the same time the veterans' frustrations were mounting over the delayed lawsuit. The film gives a history of chemical warfare, the development of the phenoxy herbicides, and their use in South Vietnam. The film captures veterans and their families in tense confrontations with public officials as well as the usual talking heads. In an interview about the film, Ochs described the difficulties Vietnam veterans faced when they returned to the United States, but saw veterans as creating a "political movement," one independent of ideology. Ochs noted that their treatment by the government was politicizing a conservative segment of the population. "You are talking about people, a large portion of whom, had they returned from Viet Nam and been treated humanely and given good benefits, would not be anti-government. They are becoming politicized as a result of being shafted first hand."[67]

Dramatic tension built as the stories of veterans appeared on screen, culminating in veterans holding a hunger strike at the VA hospital in Wadsworth, California. Veterans called the strike in support of James Hopkins, a veteran denied treatment who then drove a truck into the lobby and shot at individuals. In the 1982 interview, Ochs confessed her shock that the action received almost no news coverage, with what little there was perpetuating false rumors put out by the VA. Veterans occupied the hospital lobby and a tent city. Eleven men started a hunger strike. The sit-in ended in June 1981 with the arrest of six strikers and the ejection of the remaining veterans, including hunger strikers. Promised a congressional hearing, the men ended their strike. Unfortunately, the multitude of congressional hearings, studies, working groups, and testimony formed a quagmire equivalent to the war itself.[68]

Agent Orange remains a missing element in almost all the fictional accounts of the Vietnam War, of which there are a significant number. It appears as an important plot point in *First Blood* (1982), the story of Vietnam veteran John Rambo, who has returned home a damaged man. Bobbie Ann Mason's 1985 novel *In Country* centers on the illnesses experienced by one of the major characters, a veteran suffering from what appears to be chloracne. Both these treatments emphasized veterans as damaged, even if that trauma took completely different forms.

Television offerings, however, tell the story of Agent Orange and veterans emphasizing not only their anger but their organizing. Working from a script written by John Sayles, Alfre Woodard starred as Maude DeVictor in *Unnatural Causes*, with John Ritter playing a composite character as a veteran dying of cancer. The movie aired in 1986, after the veterans' lawsuit had been decided. It made, as the *Washington Post* review described it, "for a compelling viewing, a three-hanky movie that drains the viewer without squeezing too hard."[69] It ended with an epilogue noting the settlement of the veterans' lawsuit and that Maude DeVictor had been fired by the VA in 1984.[70]

Maude DeVictor discussed her firing in a 2008 interview where she acknowledged the VVAW for having her back. They had supported her unreservedly. In the same interview, DeVictor articulated that an important change had occurred. Agent Orange, she said, marked an evolution in understanding. The reproductive rights of men had been recognized for the first time. This included their right to have children, healthy children. The next, and final, chapter of this story considers the activism centered around Agent Orange herbicides' "unexpected casualties," the children of those exposed.[71]

CHAPTER NINE

Unexpected Casualties

The Phenoxy Herbicides and Reproductive Harm

Under the headline "Vets Face New War at Home," an article in the *Buffalo Veteran*, the local area newsletter, featured an interview with two Love Canal residents in the summer of 1980. Both men, given pseudonyms, talked about the fears and anxieties they had for their homes and families. Exposed to Agent Orange during the Vietnam war, the men agreed that they had faced a much more difficult battle at Love Canal. Veteran Hough saw himself as a "angry, bitter, dying old man who's ready to start killing people," well aware of the chemical contamination permeating his neighborhood.[1] He ended his interview with an implicit analogy: he pointed to a dying tree in his backyard and compared it to the ones defoliated by Agent Orange in Vietnam. Jack Spencer, the other veteran, connected his family history of illness to medical conditions known to have chemical causes. Spencer thought the government had failed both Agent Orange victims and the residents of Love Canal, but his neighbors bore an especially grievous burden. The risk of combat constituted a known part of his job in Vietnam. Regarding Love Canal, he noted, "It was always my understanding that you don't take your kids into combat with you. . . . At least not in the American Army."[2] The experiences of these Vietnam veterans with another toxic disaster reveals one of the most difficult aspects of veterans' exposure to Agent Orange: their haunting fears about what it had done to their children.

A powerful moment in Jacki Ochs's documentary *The Secret Agent* comes when a woman stands and charges that Agent Orange had affected two generations, both veterans and their children. In addition to veterans, those affected include the children of American veterans, agricultural workers, and Vietnamese citizens. Given the difficulty in determining the harmful effects of Agent Orange on veterans themselves, epidemiological research on the birth defects and illnesses experienced by the next generation has been even more difficult to determine. The science remains contested, complicated by other toxic exposures (especially in the case of agricultural workers), voices dedicated to discrediting contrary results, and the political, economic, and diplomatic ramifications of declaring the phenoxy herbicides hazardous. While the scientific community mostly agrees on the toxicity (teratogenic, carcinogenic) of 2,4,5-T's dioxin contaminant, the scope of potential toxicity has expanded. Many scientists now consider the phenoxy herbicides, both 2,4-D and 2,4,5-T, to be endocrine disruptors. Chemicals in this category do not follow established toxicological expectations, particularly with respect to the axiom "the dose makes the poison." In contrast, with the endocrine disruptors the timing of exposure matters more than the quantity of the dose. Research on the endocrine disruptors began with the work of Theo Colborn in the late 1980s and continues today. In addition to this health concern, some of the most recent research on 2,4-D suggests that it also contains a dioxin contaminant, which was the reason why 2,4,5-T was eventually banned.

Various groups have responded to these unexpected casualties of the phenoxy herbicides in different ways. If the direct effects of exposure remain contested science, so much more so do the claims that the phenoxy herbicides caused reproductive trauma, like miscarriages and birth defects. But these claims also provided a compelling reason for intervention and compensation. The individuals and groups affected here lie outside the use of the chemical herbicides for the usual stated purposes. In this respect they may best show the inherent problems with domestic chemical policies that sanctioned the phenoxy herbicides in the postwar period of American history. The perceived harm caused by the chemicals to these vulnerable populations made the reproductive effects of Agent Orange herbicides an essential ingredient in protests over the phenoxy herbicides.

While acknowledging the contested nature of the science examining the phenoxy herbicides' potentially harmful reproductive effects, this examination treats the beliefs of the actors (individual, group, and state) involved as credible and reasonable. These fears certainly motivated hundreds, even thousands, of individuals to act. Three sets of children potentially affected include the children of US Vietnam veterans; the children of Vietnam, both of veterans and civilians; and the children of agricultural workers, those in California for this study. The discussion presented here focuses on these different groups of children, although many more may have been affected by the various wars the herbicides were used to fight, internationally and domestically.

Vietnam Veterans Against the War (VVAW) recognized Agent Orange's harmful reproductive effects from the beginning. Labeling it a "chemical time bomb," the VVAW informed veterans that the herbicides sprayed contained a highly concentrated form of dioxin, which they likened to thalidomide.[3] The VVAW even repeated the extreme charge that the army had intended to produce birth defects in children as a way of undermining the combatants. (Given that the herbicides were sprayed in the part of the country the United States supported, this rumor makes no sense and has no supporting documentation.) Among the long lists of wrongs committed against veterans, the possibility that the next generation might "be born deformed" appeared especially troubling.[4] A 1978 newsletter explained in much greater detail what Agent Orange was and what illnesses might result from exposure, including still births, miscarriages, and birth defects. The article used evidence gathered in Vietnam and Seveso, Italy, which had experienced an industrial accident that had exposed the population to a massive concentration of dioxin. It noted that 2,4-D was still used, along with 2,4,5-T, which had not yet been banned despite its dioxin contaminant.[5] Another veteran, one of only five, returned to Vietnam in fall 1983 for a trade unionist meeting. He described his shame upon seeing the continuing destructive effects of the war, which included defoliated forests and children with birth defects. The article bitterly noted that the Vietnamese collected the data on abortions and miscarriages while the US government blamed veterans for their and their children's medical problems.[6]

The VVAW covered the 1983 international symposium held in Vietnam on the long-term effects of Agent Orange. Researchers there reported on a study of forty thousand Vietnamese families where the father served in South Vietnam (sprayed) or North Vietnam (unsprayed). It showed the wives of South Vietnamese as having "a significantly higher rate of miscarriages, still births, and children with abnormalities than women whose husbands stayed in the North."[7] A 1989 VVAW delegation to Vietnam included a visit to a gynecological hospital and orphanages. The newsletter piece covering the trip discussed the Vietnamese research on dioxin and birth defects (through both male and female exposure) and included the publicized 1983 conference proceedings that included extensive studies on Agent Orange's effect on reproduction, including significant birth defects seen in the children of fathers who served in

South Vietnam. Writing in the 1989 annual review, Barry Romo decried the delay in dispersing the 1984 settlement funds. He noted concerns that part of the delay might be because the US military wanted to use the chemicals again, specifically in Central America. "The rashes leading to deformed children are all too familiar to those of us acquainted with Agent Orange. . . . The thought of any child dying or suffering defects because of its reintroduction in a 'Second Vietnam' holds only shame for all Americans."[8] While the herbicides' effects on children horrified Vietnam veterans and their female partners, it also gave them a powerful means of mobilizing support for their cause. One family became the face of Agent Orange's hazardous effects on veterans' children.[9]

Michael Ryan, his wife Maureen, and their daughter Kerry became one of the first and most prominent of the veterans' families voicing concerns about Agent Orange's effects on veterans' children. The Ryans were named as the lead plaintiffs in the class-action suit brought against the chemical manufacturers by veterans. They, especially Kerry, who was born with multiple birth defects, came to represent the reproductive harm done by Agent Orange. Historian Leslie Reagan examined the prominence Kerry Ryan played as a symbol of injustice in Vietnam veterans' campaigns seeking compensation for exposure to defoliating chemicals. She noted how both the Ryan family and the press stigmatized Kerry Ryan's birth defects as "not normal" and the problematic nature of characterizing disabilities in such a manner. While the Ryans did emphasize Kerry's birth defects in their public appearances and in documentary films, not all media took the same exploitive gaze. In Ochs's 1983 documentary the Ryans appeared as compelling figures, but the filmmaker minimized Kerry Ryan's birth defects. Instead, Maureen Ryan's anger emerged as one of the most striking elements of the film. Ryan directed her anger toward government officials, who she said had misled soldiers about the dangers of Agent Orange exposure even as officials denied any responsibility. She acknowledged that she thought Kerry's condition had been an act of God until she realized that her government had lied and caused her child's birth defects. Ryan ended the documentary stating that she has exposed herself by allowing the filmmakers to come into her life in the hope that others would be helped. She wanted them to know that the government could not and should not use people the way service members were treated and that chemical companies needed to be held accountable.[10]

Along with Paul Reutershan's, Kerry's name appeared in fundraising appeals for the Vietnam Veterans Relief Foundation, sponsored by the group

A FAMILY'S PRIVATE TRAGEDY—
A RAGING PUBLIC CONTROVERSY

KERRY

AGENT ORANGE AND AN AMERICAN FAMILY

DELL· 14516 ·U.S. $3.95 CAN. $4.95

Clifford Linedecker
with Michael and Maureen Ryan

Figure 11. *Kerry* tells the story of Kerry Ryan, born with more than twenty birth defects that her parents attributed to Agent Orange exposure. The family blamed her disabilities on her father Michael Ryan's exposure to Agent Orange during the Vietnam War. Copy in author's possession.

founded by Reutershan, Agent Orange Victims International (AOVI). She and other children with disabilities appeared in fundraising materials. Measuring Kerry Ryan's appeal as a symbol of harm can be seen by contrasting her family's story with another one afflicted by the consequences of Agent Orange. This family's prominence meant they loomed even larger in the American discussions about the war, as the country wrestled with what the war had meant. The story of Admiral Elmo Zumwalt Jr., the chief commander of naval forces in Vietnam, and his son Elmo Zumwalt III, who had served in the navy during Vietnam, presented the nation with a Greek tragedy. Zumwalt III died of non-Hodgkin's lymphoma, a kind of cancer increasingly connected with Agent Orange exposure, which both men blamed on Agent Orange. Both military families, the Ryans and Zumwalts, wrote books about these unexpected casualties of the war.

In their 1982 book *Kerry: Agent Orange and an American Family*, Michael and Maureen Ryan tell the story of how Agent Orange exposure profoundly affected their family. The book began with an immigrant tale with the arrival of Michael Ryan's father, Mick, and good friend Daniel O'Connor, Maureen's father. The men worked, fought for, married, and had families. They represented solid American citizens and values. Michael Ryan's service in Vietnam, his secret marriage to Maureen, and the birth of their daughter Kerry after the war made up the first half of the book. The Ryan and O'Connor families united to take care of Kerry, born with significant birth defects—as many as twenty-two by one count—requiring several major surgeries before she was five. The Ryans' story merges with the story of Agent Orange activism half-way through the book, when Maureen Ryan realizes that Kerry's medical problems may have been caused by her father's exposure to herbicides in Vietnam. The family became one of the lead plaintiffs in the veterans' Agent Orange lawsuit. In the process they also joined forces with Frank McCarthy, who led the AOVI after Paul Reutershan's death in 1978. For many veterans, the Ryans represented both the injustice done to soldiers who had served in South Vietnam and were exposed to toxic chemicals, and a cautionary tale about their futures as fathers of children born with birth defects. The trajectory of the book, starting with the family's trauma, ends on an uplifting note, with an engaged, resilient family demanding restitution from the government. Their story contrasts with another narrative about a prominent military family published just a few years later.[11]

Credited with modernizing the navy, Admiral Elmo Zumwalt Jr. was honored by *Time* with its December 1, 1970, cover. Zumwalt's job was overseeing what were known as "brown-water" naval operations in South Vietnam, an arduous task. Zumwalt Jr. had ordered the spraying of Agent Orange in the South Vietnamese delta and riverways in 1968, a command both ironic and tragic. Zumwalt's son, Elmo Zumwalt III, was exposed to the herbicides while commanding a swift-boat patrol during his service in 1969–70. He returned to North Carolina and was attending law school when his father was appointed chief of naval operations. His own son, Russell Zumwalt, experienced a condition called "sensory integration dysfunction," which he and his wife wondered might have been caused by Zumwalt III's exposure to Agent Orange. It was not until 1983 that Zumwalt III was diagnosed with stage 4 non-Hodgkin's lymphoma. His diagnosis, coupled with the veterans' lawsuit and the family's military prominence, brought the story to the public's notice. In 1986 the Zumwalts wrote a book, *My Father, My Son*, that told their family

Figure 12. Three generations of Zumwalts appeared on the back cover of My Father, My Son. The book captured the tragedy of the Zumwalt family and the conflict and continuing divisions in American society caused by the Vietnam War. Copy in author's possession.

ment in speaking out against Agent Orange, My Father, My Son centered on Zumwalt III's illness and treatment. The book served, to some extent, as an apologia for the father's decision to spray the herbicides, and as a farewell to a beloved son. It made no call to arms, and its trajectory bent toward mourning. In this story of Agent Orange, the chemical damage done appeared to doom its clan. Unlike the Ryans', the Zumwalts' saga ended on a somber note. While the Ryans committed themselves to challenging the perceived injustices facing them, the Zumwalts reflected pain, love, sadness. Both stories—the engaged and the elegiac—characterized another group living halfway around the world plagued by the reproductive harmed caused by Agent Orange.

North Vietnamese scientists and a group of international supporters charged that the defoliants caused fetal anomalies during the war. There were extreme claims made that charged chemicals had killed the newborn babies of South Vietnamese Roman Catholics. Other charges compared the suffering attributed to the chemicals as comparable to that of the people of Hiroshima and Nagasaki. Many of the charges that declared defoliants were "toxic chemicals" found a receptive audience throughout Asia. US officials dismissed these assertions as propaganda, a position challenged during the war and afterward. A 1970 Stanford University working group noted that after heavy herbicide spraying in 1967, Saigon papers printed "front-page stories of a novel and increasingly common birth defect described as 'egg-bundle-like fetus.'"[15] Other papers noted the increase of birth defects in heavily sprayed areas and speculated this was potentially linked to herbicide exposure. A report produced by a French aid society discussed the challenges of determining the herbicides' reproductive effects, focusing on the 1970 American Association for the Advancement of Science (AAAS) study led by Matthew Meselson and the 1974 National Academy of Science (NAS) report.[16]

The NAS report provided some evidence that children if not fetuses had suffered from herbicide spraying. Anthropologist Gerald Hickey interviewed approximately thirty-six informants from the South Vietnamese highlands. The group included some village chiefs and North Vietnamese defectors, and an agricultural engineer who provided the most sophisticated testimony. Everyone interviewed stated that their village's exposure had come from drift and not because they were directly sprayed. Some of the interviewees claimed a higher than normal number of children had died after herbicide spraying, although as the report noted "they were very cautious in concluding that the spraying affected childbirth."[17] One elderly woman recounted one stillborn case after a defoliation spraying. She could not identify herbicides as the cause and took care to acknowledge the high rates of stillbirths within the village. One village, Dak Tang Plun, claimed that "many children" had become ill after spraying, and at least thirty had died. Other villages reported deaths but were uncertain about the numbers. The herbicides' effects on reproduction and pregnancy remained an elusive element in official US sources. But by the late 1970s, when US Vietnam veterans became concerned about dioxin's teratogenic effects, Dr. Tồn Thất Tùng, the North Vietnamese surgeon introduced earlier, had been studying the herbicides' ill health effects for almost a decade. Tùng's initial research had focused on a possible link between herbicide exposure and increases in liver cancer. He soon broadened his attention to examine the reproductive effects of the herbicides that appeared to cause miscarriages and birth defects.[18]

In North Vietnam, Tùng represented one of the most prominent figures in publicizing the chemicals' hazardous effects. While he initially published work on the increased incidence of liver cancer, his work documenting the herbicides' deleterious effect on reproduction brought him to international attention. Tùng was a good Communist Party member, a "Labour Hero," and respected physician and intellectual. In contrast to the American media's silence on herbicides from 1970 until Reutershan, Maude DeVictor, and Bill Kurtis alerted veterans, Tùng drew attention to the problem of 2,4,5-T and its dioxin contaminant before the war officially ended. In a 1972 article, Tùng charged that dioxin had "created deformed babies" and a host of other illnesses. Anecdotal support came in the responses to a 1972 survey of almost one hundred South Vietnamese refugees. Five percent of the respondents "referred to abortions or monstrous births occurring among the animals."[19] Tùng's presence ran like a thread through all the Vietnamese research on the reproductive hazards of the phenoxy herbicides.[20]

Issues of legitimacy (and racism) haunted Vietnam's scientific studies of reproductive misfortune. Writing in a 1983 *Mother Jones* essay, author Judith Colburn interrogated Americans' hubris in their assumptions that they understood what the Vietnamese wanted, that they acted honorably in fighting the war in Vietnam. Colburn met Tùng in 1980 at a lecture at his villa in Hanoi. Tùng appeared as an erudite, renowned scholar as he spoke in his library filled with books and papers, referencing the numerous international conferences he attended. But as Colburn noted, "But Tùng is Vietnamese. Even though he has seen more Agent Orange victims than any other medical expert in the world, he is still, in the Cold War–chilled halls of the Pentagon, the Veterans' Administration and the U.N., dismissed as a propagandist."[21] Tùng's research focused on North Vietnamese soldiers exposed to Agent Orange and their children. Tùng expressed frustration as he acknowledged the need for a massive epidemiological study to gain Western attention and scientific acceptance. He conceded that given the country's ongoing postwar poverty, the likelihood of funding such studies remained remote. Other challenges faced such studies, however, as a 1982 American conference highlighted.

⁂

As with the US veterans' health studies, the difficulties of measuring Agent Orange's reproductive harm in Vietnam posed a significant methodological challenge. James and Kathleen Dwyer had visited Vietnam in January 1982, met with Tùng, and surveyed the various research studies being conducted throughout the country. In addition to Tùng's study on North Vietnamese veterans, there were at least six other ongoing studies, with at least one being done in conjunction with a Swedish university. Upon their return to the US, the Dwyers received foundation funding and organized a conference to discuss both the scientific and humanitarian aspects of studying the effects of herbicides on human reproduction. Some participants noted that while the large numbers of exposed people meant large surveys would be possible, the design and implementation of "good epidemiological studies must be *very* carefully planned."[22] Good science, they agreed, could not be hurried.

Gendered understandings of fetal risk shadowed the conference. There appeared to be two different populations to be studied. The first was composed of individuals exposed to heavy concentrations during active spraying. "While every year's delay means that fewer severely handicapped children conceived during the period of heavy spraying will be alive for study and treatment, it would be foolish to forge ahead with a flawed design."[23] Those individuals exposed to lower-level, chronic exposure represented the second group. The gathered scientists' focus on women's exposure, rather than male exposure, to the herbicides as the cause of genetic damage revealed a gendered understanding of reproductive harm. The social construction of paternity affected biological understandings of fetal harm in failing to see the links between paternal toxic exposures and its negative reproductive effects. This focus on women's exposure came despite a 1980 large-scale study that showed a greater incidence of congenital anomalies in men exposed to Agent Orange dioxin. Some of the other challenges facing researchers included finding appropriate matches for regions, siblings, and defects. Researchers expressed concern that other environmental and social factors, such as starvation, might be causative in producing birth defects. And unsurprisingly, like the war itself, the potential political ammunition of the studies divided the group.[24]

The expediency of proving harm cast doubt on the objectivity of any study done in Vietnam. One attendee at the Mt. Sinai conference expressed concern about the Vietnamese government's interest in the dioxin studies being done by the Ministry of Health, fearing that the research would be tainted. Others responded that they saw no advantage for the Vietnamese in attacking the US over the "herbicide issue," arguing that it made more sense for the country to seek common cause in trying to address the problem. Or the Vietnamese rightly feared the potential harm caused by herbicides and simply wished for medical aid. In the opinion of some participants, Vietnam wanted to join the world community, "to contribute scientifically and culturally to the development of humanity."[25] There had been attempts in the early 1970s to aid the Vietnamese scientific community, most particularly through the efforts of the group Science for Vietnam. Other researchers had noted Vietnamese scientists' pleas, Tùng most prominently, to send research teams to investigate the herbicides' effects and even establish a cooperative institute to train Vietnamese specialists to study dioxin. Dr. Arthur Galston, an attendee at the Mt. Sinai conference, had extensive contacts with both the Chinese and Vietnamese scientific communities, and advocated for cooperation. The conference ended with plans for further study, which included sending a research team to Vietnam. The Americans' efforts would be superseded by the Vietnamese, who sought to join the international scientific community by bringing it to them.[26]

⚘⚘⚘

Both the United States and the united Republic of Vietnam remained mostly silent on the issue of Agent Orange's continuing effects on the Vietnamese people in the decades after the war; this made a 1983 international symposium on the herbicides' effects especially noteworthy. Organized by the 10-80 Committee (named because it was set up in October 1980 by the Vietnamese government), the International Symposium on Herbicides and Defoliants in War, the Long-Term Effects on Man and Nature took place in Ho Chi Minh City (formerly Saigon) in January 1983.

The conference focused on the use of three chemical defoliants, Agents Orange, White, and Blue, in South Vietnam. The proceedings background information singled out Agent Orange as causing "the greatest level of continuing medical concern because of its dioxin contaminant."[27] Reading the final report reveals the major ambitions, and concerns, of the Vietnamese scientific community and government. The report's introduction emphasized the numbers of scientists and members of international organizations who had attended the conference (160) and the breadth of representation, from over twenty-one countries. It also took pains to note that the scientists came from the West as well as the East, and that the symposium was a working, dedicated gathering of scientific experts. The conference's stated goals revealed the limited status of information. The symposium sought to review and evaluate the existing scientific literature on the long-term effects of herbicides' ecological and physiological exposure, to identify future research needed, and to establish international scientific cooperation.[28]

Public dissemination of the conference in its published proceedings and media coverage complicated the Agent Orange narrative of human safety. The research presented examined the herbicides' effects on the health of humans and nature. Of the more than sixty scientific papers presented that examined the extent of defoliation activities, the topics included the long-lasting consequences of the chemicals on the natural environment and on human beings, their effects within the laboratory and field, and on groups of workers exposed to the chemicals. The working group on reproductive epidemiology suggested an increase in congenital anomalies but acknowledged the difficulty in proving some outcomes. Symposium participants reached agreement in ten areas. One key point recognized the ongoing research projects studying the phenoxy herbicides and growing scientific consensus that these chemicals might be mutagenic, carcinogenic, and/or teratogenic. Another point of agreement validated, to a certain extent, the findings of Vietnamese scientists working under difficult conditions. A related point acknowledged Vietnamese studies of the ways the herbicides negatively affected chromosomes, potentially causing congenital abnormalities and molar pregnancies.[29]

The tension between science and politics echoed the US Mt. Sinai workshop. One participant expressed his ambivalence toward the symposium, noting that the Vietnamese believed their data. They also "wanted to talk science, not politics."[30] Perhaps the most disappointing moment was the press conference with almost no one in attendance. Although immediate news coverage of the symposium proved disappointing, there was some coverage in scientific and political journals. In 1983 the *Vietnam Courier*, a weekly communist newspaper, published the symposium's summary report. The Vietnamese continued to disseminate and promote the conference proceedings even as their concerns and findings were dismissed. They also continued to study the chemicals' mutagenic and teratogenic effects. Ordinary Vietnamese continued to live with the chemicals' presence in their lives.[31]

⁂

Vietnamese women had limited birth control options in the decades after the war ended. Abortion rates in postcolonial Vietnam remain unclear, although it appears it was legal in the post-1945 Democratic Republic of Vietnam and, later, the reunited Vietnam. Women could have abortions and use birth control in the 1960s and 1970s. Vietnam's recorded abortion rate in the 1990s appeared to be some of the highest in the world, with an average of 2.5 abortions per each woman over the course of her lifetime. This number may underestimate the abortion rates, as it measured only those performed at public institutions. While it remains impossible to know the exact motives of these public-sector abortions, the economy and general poor health of citizens, in addition to personal choice, factor among the reasons. These numbers may have also made Vietnam an attractive site for reproduction studies.[32]

Chemical sterilization trials took place in Vietnam from 1989 through 1992, probably in part because of the high abortion rates and the country's status as a developing nation. Study participants were required to be at least thirty years old and have two living children. Several factors raised ethical concerns regarding the sterilization trial: the low socio-economic status of the women participating; a reported failure to enforce the study's screening criteria; the broader medical context in Vietnam, where health expenditures on mothers and children was approximately twenty cents per capita; and the limited birth control methods available to Vietnamese women (IUDs and surgical sterilization). Vietnamese women's perceptions of reproductive risk, produced by personal experience and public discourses, were not included among these concerns.[33]

Vietnamese women appeared to be acutely aware of the reproductive risks to which they may have been exposed. Fears over familial genetics and the presence of dioxin hotspots were seemingly confirmed with the introduction of pregnancy diagnostic technologies in the late 1980s. So, while virtually impossible to measure, it appears reasonable to consider the way women's reproductive anxieties may have led to abortions and sterilization after the war. Some women may have followed the path taken by Mrs. Ha, a woman living in the northern province of Thai Binh.[34] Ha's husband had served as a special forces soldier in the south, Tay Ninh, an area that was heavily sprayed with defoliants. Ha had borne a severely deformed fetus, a mentally disabled son, and a daughter with epilepsy. She had chosen to be sterilized rather than have any more children. Her decision appears especially heart-rending given the importance childbearing has on women's status in Vietnam.

Parents' desire to have a healthy child led some families to continue having children, and because contemporary public attitudes condemned abortion, resulted in their having three or more children with disabilities. Yet the public discourse surrounding voluntary abortions considered procedures done for fetal defects as fundamentally different. I should note these attitudes reflect opinions as surveyed in the late 1990s. Like Ha, one young woman ended her pregnancy after finding out the fetus was hydrocephalic. She consulted with her extended family as well as her in-laws. No evidence exists to measure earlier attitudes about abortions done in the 1970s and 1980s. The understandings expressed at this later time suggest that acceptance of abortion for fetal birth defects and exposure to Agent Orange may have justified earlier generations' decision to end their pregnancies. A 1996 documentary, Tran Van Thuy's *A Story from the Corner of the Park*, made Agent Orange's reproductive effects visible and an acknowledged part of the public's consciousness. Mass media stories on the struggles of Agent Orange–affected families reinforced the narrative of suffering. This increased awareness of what Vietnamese society viewed as "Agent Orange victims" also affected an intertwined public discourse of protest in the extensive visual and artistic representations of fetuses and babies with severe deformities.[35]

An exhibit of fetuses with severe deformities and fetal remains contained in glass jars appears in the Museum of American War Crimes opened in 1975. The exhibit materials provided the content of some of the more disturbing photographic images of "Agent Orange babies." Like the My Lai exhibit, the Agent Orange fetuses were intended as propaganda. Vietnamese American novelist Viet Thanh Nguyen described it as "far from beautiful," seeing it as proof of the nation's "poverty of memory found in poor countries" and setting a mood of "shabbiness and sadness."[36] The glass jars, as one historian described it, were displayed "very clinically, devoid of emotions."[37] Presented in this way, the materials act as a semblance of scientific evidence, reinforcing the ongoing studies of reproductive risk undertaken by Vietnamese physicians.[38]

The photographs of these bottled fetuses, along with images of children with varying kinds of birth defects, pervade the visual discourse of Agent Orange and its reproductive harm in Vietnam. But photographs, perhaps even more than film, have captured the war and its experiences. Two sets of photographs embody Agent Orange herbicides, those of defoliated landscapes and of fetuses and children with deformities. These images, in the words of one historian, "fix and disseminate memories of the past."[39] One of the first published photographs appears to be a United Press International (UPI) photograph taken by Frances Starner in 1980. The photo shows a young woman holding a child with a visible harelip. The caption reads:

"A mother holds her small daughter at a Hanoi hospital where the child is being treated for birth defects. The child's father, a North Vietnamese soldier, was exposed to Agent Orange defoliant sprayed by US forces during the Vietnam War. Vietnamese doctors believe the chemical's aftereffects are responsible for many malformed babies."[40] So, one of the earliest images of the children many Vietnamese referred to as "Agent Orange babies" connected

Figure 13. This 1980 photograph taken by Frances Starner of a Vietnamese mother and child with a birth defect supported understandings of the phenoxy herbicides' toxicity. Courtesy of Bettman Collection, Getty Images.

birth defects with Agent Orange exposure through its image and text, noting the father's exposure as causative.

In the absence of state mediations, graphic and disturbing images of fetal remains and children with severe disabilities dominated. The images were distributed more widely after the normalization of relations in 1995 and related demands for Agent Orange reparations. By the early 2000s, photographs of Vietnamese children with disabilities inundated the visual discourses. In 2003 photojournalist Philip Jones Griffiths released an oversized collection of photographs entitled *Agent Orange: "Collateral Damage" in Vietnam*. A photographic coda to his award-winning war coverage, the book shows the next generations of Vietnamese and Cambodians harmed by the chemical herbicides. Griffiths includes a series of photographs taken of the contents of a locked room in a Ho Chi Minh City hospital. The photos show numerous abnormal fetuses, what Griffiths calls the "most inhuman of human remains."[41]

Over 140 pages follow, each page filled with images of fetuses, birth defects, and people with varying degrees of disability, what Griffiths calls the "human toll" of Agent Orange.[42]

The black-and-white photographs are stark, disturbing, and arresting in their simplicity. A 2007 *Vanity Fair* feature prompted by a 2007 Vietnam Association of Victims of Agent Orange (VAVA) lawsuit sent Christopher Hitchens and photographer James Nachtwey to Vietnam. Hitchens acknowledged the power of the photographic testaments. "To be writing these words is, for me, to undergo the severest test of my core belief—that sentences can be more powerful than pictures. . . . Unless you see the landscape of ecocide, or meet the eyes of its victims, you will quite simply have no idea. I am content, just for once . . . to be occupying the space between pictures."[43] While photographs became one way of challenging official US narratives of safety, literary artists have been equally explicit about Agent Orange's toxic legacy.[44]

In 2010 an anthology that confronted the issue of Agent Orange through art was published. In *Family of Fallen Leaves: Stories of Agent Orange by Vietnamese Writers*, literary artists of that nation identified and explored the ongoing problems they perceived as a toxic legacy of the herbicides. The collection's stories deal with families, the interconnectedness of nature, and defoliation. One of the volume's editors notes, "Their work allows readers to think, live, and suffer vicariously as a person exposed and to identify with that person and establish a deeply felt human connection."[45] Using stories, the narratives embody the ongoing efforts of the Vietnamese people to receive reparations from the United States government for the spraying of more than 20 million gallons of defoliating herbicides over 12 percent of the southern countryside. Many of the tales directly confront the reproductive misadventures caused by Agent Orange, a few in graphic and haunting ways. At the same time, they present a powerful call to action. The next paragraph discusses two stories in greater detail.

Reproduction follows an intimate act. The first-person narrator, Sao, in Suong Nyet Minh's 2004 story "Thirteen Harbors," begins her tale describing her efforts to find her husband a new bride. The explanation of how she got to this point unfolds as a series of brutal, horrific losses as four of her pregnancies end in what she calls "pieces of red meat," and with "even greater terrors" produced during her fifth.[46] Her husband and her mother-in-law convince Sao that a new wife (and a new husband) may be able to have healthy children, leading to the dissolution of the marriage and to her husband's new wife. Unlike some of the other stories, "Thirteen Harbors" explicitly connects the couple's misfortunes with Agent Orange exposure, even mentioning dioxin. The story ends in tragedy, as the new bride delivers "eleven pieces of red meat," and Sao returns to her husband's household to care for him and his ill mother.[47]

In contrast to the deeply intimate tone used by Minh and others, the collection ends with a nonfiction essay by Minh Chuyen, "Le Cao Dai and the Agent Orange Sufferers." Chuyen reports on Dr. Le Cao Dai, who served in Central Highlands military hospital during the war and was exposed to Agent Orange. He was ill after the war, a condition he attributed to Agent Orange. Dai and his wife lost three children, all of whom "were deformed, and finally they died."[48] For more than thirty years, Dai studied Agent Orange and met with numerous families believed to be suffering from the physical and emotional effects of dealing with dioxin contamination. In their travels around the countryside, Dai and Chuyen met James Zumwalt, the youngest son of Admiral Zumwalt. He told them that he represented his father, too frail and ill to come to Vietnam. "He said that the most fatal mistake in his life was to command the spraying of toxic herbicides in South Viet Nam."[49] Chuyen calls out Americans on their responsibility to address the "pain and destruction" they caused even as she praises the service of Dai. The "stories of Agent Orange by Vietnamese writers" ends by speaking truth to power. The final group of unexpected casualties discussed here receive no such moment.

Photographic and literary portraits of suffering haunt the public discourses centered on Agent Orange. Children play a significant, if not the most prominent, role in these artistic representations, not just because they are the most affected, but because they may gather the most sympathy. Just as American Vietnam veterans put their children front and center, other activists have used images of children to raise awareness (and funds) for their afflictions. The first poster children appeared in March of Dimes literature to educate Americans about polio and its devastating effects. By midcentury, children with cancer also appeared on posters and in flyers as advocates tried to change minds and gain support for research and funding. Children appeared in discourses of chemical harm, as demonstrated by the Love Canal story. While not everyone agreed with the use of the children's photographs, it appeared relatively unquestioned as a means of raising awareness of severe medical problems. With the exception of Love Canal, children remain conspicuously absent within discourses of environmental justice. This is especially true for the next group discussed, the children of agricultural workers exposed to the phenoxy herbicides and other agricultural chemicals.[50]

In 1987 many of California's Central Valley agricultural workers lived in McFarland, located in the San Joaquin Valley, where crops of cotton, grapes, almonds, and kiwi fruit were grown. Almost a third of the city's more than six thousand residents were Latino By this point in time, residents had been wrestling with a horrifying reality for over two years. Children in McFarland were dying of a multitude of cancers in numbers and kinds disproportionate to the population numbers, known in epidemiological terms as a cancer cluster. Along with the nearby town Earlimart, a Latino farming community, McFarland residents had documented the unusual number of cancer cases before any state medical officials. The realization set off a series of actions, with state health officials launching studies and labor and grassroots activists protesting the use of pesticides in the San Joaquin Valley. Invariably the two would clash, with families caught in the middle, uncertain and scared about what they saw happening to their children, and angry that it might be the result of environmental exposures.[51]

Like the United Farm Workers (UFW) protests discussed earlier, the case of McFarland's cancer clusters fits imperfectly within various groups' protests about Agent Orange herbicides, for several reasons. Perhaps the most significant is that phenoxy herbicides *harm* most of the crops grown in the San Joaquin Valley—such as grapes, cotton, alfalfa, tomatoes—and the herbicides would not have been used on them. The phenoxy herbicides are used on asparagus, however; Stockton produces most of the US domestic crop. By 1985 use of 2,4,5-T had been discontinued. Other pesticides would have been used on these crops, including much more toxic chemicals, like the organophosphates, presenting another complicating factor. Many of these other chemicals show carcinogenic properties. As agricultural workers experienced less dangerous workplace accidents because of mechanization, their chemical exposures increased with the pervasive use of pesticides. The nation's four to five million agricultural workers were routinely exposed to multiple hazardous chemicals with their synergistic effects unknown.

But research measuring the presence of chemicals in the Sacramento–San Joaquin Delta in the 1990s noted a complicated pattern of distribution for 2,4-D. While it appears most heavy in samples taken from the Sacramento River sites, the chemical still showed significant concentrations in the San Joaquin river basin. The study failed to explain the source of the chemical. Given the chemicals' ubiquitous presence, and the fact that later health studies included dioxin monitoring, some exposure to the phenoxy herbicides would be reasonable. Examining reactions within the community and their allies follows the pattern seen in the protests over Agent Orange herbicides while also encapsulating the toxic uncertainties of hazardous chemical exposure.[52]

Agricultural workers faced a host of precarious situations, with the scientific uncertainty surrounding chemical exposure as just one, although significant, of many. Local growers violated a multitude of regulations and laws imposed on the workers, including the age of workers, the hours worked, the proper and improper conditions, and other safety issues. Chemical poisoning, however, increasingly rose to the top of the list of workers' concerns. In 1980 Guadalupe Rivera testified before an EPA-sponsored panel on these kinds of problems. At just nine years old, the girl worked with her family as part of a cauliflower harvesting crew. The workers were paid by the row, and Rivera along with other children helped pick. Rivera and the rest of the workers became sick. "My stomach began to hurt and my eyes felt like they were jumping. . . . I saw everybody getting sick. They were vomiting and holding their stomachs."[53] While the UFW had campaigned against pesticides as a part of their union organizing, from the mid-1960s through the early 1970s, agricultural chemicals had remained a significant problem. Growers avoided the health consequences to agricultural workers, predominantly Mexican migrant and Mexican American citizens, by adopting a racialized understanding of fieldwork. Illness happened because of workers' physical limitations, not because of hazardous environmental conditions. Pesticide poisoning episodes increased by approximately 30 percent from 1970 to 1980.[54]

In cases of pesticide poisoning, state agencies compounded the problem, particularly with respect to public health. Several conceptual constraints that emerged over the course of the twentieth century influenced these difficulties. Traditionally, public health departments focused on urban ills, and it took a while to shift that focus to rural concerns. At the same time, the division of occupational health received little attention or funding from public health agencies. The emphasis on laboratory testing also narrowed public health officials' vision. As historian Linda Nash described it, "the 'field' never conforms to the idealized space of the laboratory."[55]

Chemical exposures within this setting also frustrated occupational health toxicologists because of their seemingly random nature. Outbreaks of illnesses failed to correlate with the concentration or time of exposure. Toxicologists' focus on mortality as a measure of toxicity also obscured the kinds of illnesses often reported by field workers, which meant the organophosphates like parathion and malathion were identified as toxic, but other chemicals like the phenoxy herbicides eluded scrutiny. Studies as far back as the 1970s had begun to challenge the accepted relationships between dose and toxicity. Agencies on the federal and state levels also failed to communicate with one another, as when the USDA's research ignored health studies being done on agricultural workers. The research on chemical toxicity done by both public sector and industry scientists also showed significant gaps. A 1984 National Research Council review of toxicity data showed that there was no information on 38 percent of the chemicals in use. Environmental factors muddled health studies as well.[56]

The geology of the Central and San Joaquin Valleys showed important differences in soil composition and water aquifers. Nash noted that these differences profoundly affected the migration of hazardous chemicals and made tracking chemical pathways especially difficult. Less-compact soil and shallow groundwater meant that the eastern side of the region was more vulnerable to migrating chemicals than the western. Human use and intervention, which increased with agricultural modernization, affected water flow, increased chemical contamination, and increased numbers of poorly constructed irrigation wells; all meant that chemical contamination happened more easily. Worse yet, the spread of pollutants depends on the combination of the wells used and when. The third way chemicals migrated, via air transmission, proved even more difficult to measure than soil and water contamination. Local conditions included a thick ground mist called a tule fog, which was shown to effectively carry pesticide residues far afield even as the mist increased the concentration of the chemicals. Water contamination presented the greatest challenge to California public officials, as chemical contamination extended throughout the Central Valley. One way to measure the magnitude of the contamination can be seen in the California State Water Control Board's (CSWCB) decision to close more than a hundred contaminated wells in the summer of 1985. It was within this context that the cancer clusters among McFarland's children were recognized and in which people acted.[57]

McFarland's mothers continued to play a key role after helping to identify the initial problem. Connie Rosales, Tina Bravo, Rosemary Esparza, and others attended public meetings—town council, school board, water control—and challenged elected officials and city planners regarding the presence of hazardous chemicals and community safety. Like the women roused to action by the 1978 Love Canal chemical disaster, they were initially dismissed as "hysterical housewives."[58] As with other mothers protesting the use of chemicals, the mourning and anger of McFarland's mothers made a compelling story, one told repeatedly in the national headlines. Here, Connie Rosales

Figure 14. The McFarland community turned out en masse to support the Bravo family at their son Mario's funeral. His death occurred with a number of others in an unexplained cancer cluster. The community marched in a processional after the funeral. Courtesy of John Harte © 1987.

charged that the town's children were "canaries," unknowingly exposed in the fields and their homes, what Rosales called "human laboratories," to carcinogenic and teratogenic chemicals.[59] Frustrated by the town's reluctance to address the problem, the women pressured the county health department to intervene. After Kern County public health officials initially rejected the idea of a cancer cluster, Rosales, whose child had cancer, contacted the Mexican American Political Association (MAPA); eventually her plea drew the attention of an ambitious state senator.[60]

The mothers' persistence drew increased political and medical attention, but to no avail. In 1985 Art Torres, chair of the California Senate Committee on Toxic and Public Safety Management, decided to hold public hearings in McFarland. Residents gave moving testimony, and the media coverage prompted the county health department to investigate. McFarland's water, air, and soil were all tested but revealed chemical levels within acceptable limits. Attempts to continue epidemiological studies stalled as Kern County public health officials decried a lack of funding and expertise. At the same time, Gov, George Deukmejian effectively limited citizens' rights to challenge pesticide use or take legal action. The death of Mario Bravo in late 1987 marked a turning point in the town's consciousness. Almost two hundred people from

McFarland and nearby Delano attended his funeral, and it was covered by Southern California television stations. Bravo's mother, Tina, gave permission for those attending to form a processional and march through town to honor her son. The UFW joined the march, prominently displaying the union flag. Their action ignited Tina Bravo's anger, who demanded they put the flags away. The confrontation marked the beginning of a contentious relationship between the families, the union, and the medical experts.[61]

Even before Mario Bravo's death, McFarland's cancer clusters had become intertwined with the UFW's broader campaign against pesticides. As geographer Laura Pulido has noted, working-class and nonwhite people respond to "environmental issues within the context of material and political inequality."[62] By 1984 UFW leader César Chávez had called for another grape boycott. This labor action followed the well-known 1965–70 strike, which ended when consumers boycotted nonunion grapes, and the Monitor 4 pesticide scandal was uncovered by Ida Honorof in 1972. In its earliest incarnation, this new boycott was not focused on pesticide exposure and residue. In his renewed call for a boycott, Chávez pointed to McFarland and other farming communities like Valdez that saw children's cancer rates spike to four hundred times the national average. The UFW had identified several problems with the pesticides used in California's agricultural industry. These included the use of illegal hazardous chemicals (like DDT), or the use of pesticides with unknown safety profiles, or unreliable safety records. Workers were exposed not only to the chemicals sprayed in the fields where they worked but to drifting chemicals sprayed on other fields. They and their families were exposed again when airborne chemicals drifted over their homes and communities. The UFW led a series of marches in September 1985 to commemorate the twentieth anniversary of the original grape boycott and to mobilize workers, communities, and consumers to support the newly launched one.[63]

The UFW and its progressive partners sharpened the campaign's focus on the role of pesticides in harming workers, consumers, and their families. Writing in *The Witness*, Pat Hoffman described Bravo's funeral, connecting the tragic personal story with the broader goals of the UFW to ban hazardous chemicals. "Consumers should feel grateful for organizational efforts such as the United Farm Workers' drive to ban several of the most dangerous pesticides and to limit the use of others."[64] Five months after Bravo's death, Chávez told Southwestern College students about the "Wrath of Grapes" campaign, an obvious play on John Steinbeck's *Grapes of Wrath*. He explained that the boycott focused on table grapes because growers sprayed more chemicals on them. Noting that consumers had more power than growers, Chávez "said the movement would be successful if 7 percent of consumers boycotted grapes," which meant an increase of more than 5 percent.[65] The boycott drew further attention to McFarland while gaining influential supporters for the labor/consumer action.[66]

Chávez united previous UFW supporters with powerful national groups. The UFW's February 1988 *Food and Justice* newsletter heralded the endorsement of the grape boycott by consumer advocate Ralph Nader and the leaders of fifteen other advocacy groups. In addition to calling upon consumers to join the boycott, Nader wrote to more than thirty grocery chains demanding immediate action in removing California grapes from their shelves, and more long-term support in calling for the testing of produce for residues and pressuring Congress for better worker and consumer regulations on pesticide use. A host of other progressive groups, including Friends of the Earth and Environmental Action, signaled their support by signing the boycott pledge. Among the supporters was the American Public Health Association. In July 1988, while campaigning in California, presidential candidate Jesse Jackson visited McFarland and marched with Chávez and families to draw attention to the continuing problem.[67]

A thirteen-minute video, *The Wrath of Grapes*, was to be a key part of the UFW campaign. Narrated by actor Mike Farrell, the film forged a link between the UFW and field workers, pesticide use, affected workers, children with birth defects, and cancer, and concluded with the grape boycott. The brief film also highlighted a press conference with a constellation of Hollywood stars, including Martin Sheen, Edward James Olmos, Lou Diamond Phillips, Mike Farrell, and Charles Haid. More interesting than their celebrity status, three of the stars—Olmos, Phillips, and Haid—emphasized the chemicals' threat to human life, and especially to the lives of children. The film proved remarkably effective, gaining support among 90 percent of those who watched the documentary. Viewers may have been convinced by a point made repeatedly in the film. McFarland mother Rosales, along with the other mothers interviewed, and with longtime UFW medical consultant Marion Moses, emphasized the ways children figured as canaries and guinea pigs in revealing pesticides' toxicities, paying the price with their deaths. In later interviews, Rosales disavowed her appearance. In 1989 a California judge issued an injunction to stop showings of the video based on claims the families interviewed had not known it would support the grape boycott. Rosales's action highlighted the escalating conflicts among the various actors.[68]

McFarland's cancer cluster became the center of an intense battle, fueled in part by the continued failure of local public health officials to determine the community's chemical culprits. Rosales, as she put it, "flipped out" when she realized the video she appeared in supported the grape boycott.[69] She did not keep her displeasure to herself, speaking out against the UFW and its partner, the National Farm Workers Ministry, in media interviews and public meetings. She charged that the UFW had exploited McFarland's children. The union had sought to profit from the community's tragedy in its labor battles. Rosales expressed anger at both sides. "Those two forces, the union and the growers, have been at it for so long, and we're in the middle."[70]

Dr. Thomas Lazar actually switched sides, leaving his position in the Kern County public health department to take a full-time position with the UFW. Lazar claimed that Rosales had aligned herself with the Valley Action Network, a group linked to a local anti-union grower. Lazar himself earned the wrath of his former colleagues and some community members when he challenged the public health report's finding in general, and in specific that arsenic contamination had not been the cause of McFarland's cancers. The problem, however, appeared to be that the kinds of cancers typically seen with arsenic poisoning were not the kind seen in the McFarland cancer cluster. Noted epidemiologist Beverly Paigen, who had worked with families at Love Canal, also expressed the opinion that the science promoted by the UFW didn't add up. "I always liked the UFW, . . . but I think they really are using McFarland."[71] The UFW's environmental allies eventually addressed the problem of water contamination with legislation that mostly ignored the "unexpected casualties" of chemical contamination. Instead, illness, politics, old feuds, and scientific uncertainty all made for a toxic fog that obscured what had caused children's cancers in McFarland and other rural towns like it and offered no relief or solace.

The haunting fears the phenoxy herbicides engendered in countless families may be the most lasting and tragic legacy of the chemicals. These "unexpected casualties" capture the futility of Cold War chemical policies that played out in jungles, cities and suburbs, and rural towns in the United States and Vietnam. The next generation had joined the previous in wrestling with the uncertainty of hazardous chemical exposure, showing the magnitude of the continued narrative of safety perpetuated by scientists, industry, and government.

Conclusion

The Dissenters:
Citizens Protest Chemical Herbicides

Responding to an interviewer's identification of her as a "female Ralph Nader," consumer advocate Ida Honorof pointed out that while he did good work, Nader had a "whole staff of people" working with him, while she mostly worked alone. She acknowledged receiving support but pointed to the publication of her bi-monthly newsletter as an example of what she meant. During the twenty-three-year span she had been publishing *A Report to the Consumer*, Honorof had filled a variety of roles: writer, editor, and publisher. She wrote it the way she thought people should get to read it. So, while she respected Nader and appreciated the work that he did, she disagreed with being labeled a female Ralph Nader, or being called Lady Nader. "I'm Ida Honorof," she proclaimed with pride, "I'm Ida not Ralph."[1]

One of the major questions this story seeks to answer deals with the *why* of chemical herbicides. Agent Orange herbicides appear as a fundamental part of the postwar period as an unrecognized means of challenging the Cold War political consensus. The phenoxy herbicides saw widespread use in the 1940s and 1950s in urban lots, agricultural fields, and suburban lawns for a variety of purposes. Doubts about the necessity of aerial spraying and growing concerns about safety represented cracks in the complacency surrounding the use of the phenoxy herbicides. With the advent of the United States' formal involvement in South Vietnam, President John F. Kennedy's military advisers considered the herbicides important tools in their counterinsurgency efforts. The war made the issue of chemical spraying visible to the world. The 1970s represented a key decade in these various anti-toxic campaigns, as domestic use of, and protests over, the chemicals multiplied.

The herbicides both represented and facilitated the American air war in Vietnam, a symbol of US technological superiority. Advisers wanted clearly demarcated boundaries, and then clearly identified enemy routes, and finally cleared landscapes to prevent attacks. The herbicides also played an important role in the psychological war and in crop destruction, which sought to deny food to enemy combatants. Herbicide spraying helped contain what was perceived as an increasingly resistant peasant population. In this sense, the herbicides allowed for physical and ideological attacks against resistant landscapes and people. In addition, the chemicals were used at home to increase agricultural production—an understudied part of the Cold War—and in fire prevention, one of the potential weapons perfected in the wars of the twentieth century.

The various individuals and groups challenging the use of Agent Orange herbicides also challenged the political consensus. The conflicts exposed gaps in citizen understanding of corporate and government obligations. People wanted corporations to take responsibility for the products they made and sold. They expected elected officials and government employees to be responsive to citizens and not corporate interests. As the individuals exposed to the phenoxy herbicides gained experiential knowledge of the chemicals, new expectations of industry and government arose. The indifference and hubris of those in power increasingly angered ordinary people who perceived they had been harmed through the actions of politicians and corporate leaders who lived remote from harm. More distressing were the ways people were *unknowingly* exposed to the chemicals that were sprayed on recreational lands, contaminated water supplies, and entered the food chain. Citizens expected a government responsive to their concerns about safety.

With the heightened visibility wartime use gave to the phenoxy herbicides, several groups began protesting their use as defoliants and in crop destruction. The work of American biologists made fundamental contributions in the critiques offered when they questioned the long-term effects of herbicide spraying on the natural environment. These scientists, all men, aroused the conscience of their professional society and empowered the international community of scientists to challenge the US with evidence of defoliation. Concerns about the herbicides' effects were raised by other groups during the war, most prominently the religious and academic communities.

The Catholic Left judged corporate America as complicit in supporting an unjust war, and both Catholics and other religious denominations protested the manufacture and use of 2,4-D and 2,4,5-T. Students condemned their universities' cooperation with both government and corporations in providing knowledge and materials to undertake the war. They included chemical herbicides as a part of their napalm protests. Domestic protests showed an awareness of these wartime issues.

The examination of three women in the western United States allows for a consideration of why the herbicides were sprayed and the kinds of campaigns mounted against their use. In Arizona the Forest Service sprayed 2,4-D and 2,4,5-T as part of a program designed to increase water supplies. Billee Shoecraft mobilized the community of Globe, which demanded that aerial spraying of the herbicides stop. Community members also sued. While Shoecraft died before the lawsuit was settled, her book about the protests inspired others. Consumer advocate Ida Honorof broadcast a weekly radio program and published a bi-monthly newsletter to keep Californians informed about environmental issues. Honorof routinely addressed the problems of chemical herbicides. In California, massive spraying campaigns were used for a variety of purposes that included removing brush to lessen the chance of wildfires, to increase crop yields and efficiency, and to control brush in public forests. Honorof routinely spoke at public hearings and to community groups, as well as writing and publishing a newsletter, as a means of educating the public about toxic chemicals. She sought to expose the cozy relations between state officials, university scientists, and chemical company representatives. Honorof praised one Oregon-based grassroots group in her newsletter.

For Carol Van Strum, the accidental spraying and subsequent illness of her children motivated her to research the phenoxy herbicides. She helped form a group that challenged the Forest Service's herbicide spraying program. Used to help control valuable timber stands and keep roadsides clear, the phenoxy herbicides were sprayed on thousands of acres every spring. After Van Strum's group won their lawsuit, a much more serious challenge to herbicide use appeared. Bonnie Hill and a group of women from Alsea, Oregon, sent a letter to the EPA in the spring of 1978. It presented information correlating spring herbicide spraying to miscarriages the women had suffered. The EPA suspended almost all uses of 2,4,5-T over concerns of the dioxin contaminant. The initial ban led to a permanent ban by 1985. The events in Oregon strengthened environmental battles later known as the "herbicide wars."

Tensions between longtime farmers and ranchers and counterculture newcomers led to political conflicts centered on herbicide spraying in northern California. Logging companies routinely sprayed private lands, while the Forest Service sprayed public timber stands. While the illicit cultivation of marijuana crops represented a significant economic activity in the region, residents also displayed a new consciousness of risk and the human-nature relationship. They sought to stop aerial herbicide spraying to prevent illness with an awareness of the chemicals' potential for harm. Another group newly conscious of risk, Vietnam veterans expected the government to treat illnesses they were convinced were caused by exposure to Agent Orange. Vietnam veterans brought a lawsuit seeking reparations and acknowledgment of the harm that had been done to them. Of special concern was the perceived threat to their children. Veterans' children, children born in Vietnam, and the children of agricultural workers all represented the "unexpected casualties" of herbicide use. Activists and scientists tried to hold corporate and state actors accountable, mostly to no avail.

Even as veterans challenged the decision of the Agent Orange settlement and the VA's treatment policies, perspectives had started to change. In a major shift in policy, in 1990 the secretary of veterans affairs announced a decision that soft tissue sarcomas (STS) would now be a covered condition for Vietnam veterans. STS affected tissues in the body like lymph and blood vessels, muscle and fat, and connective tissue. These types of cancer appeared rarely in the general population. The secretary's decision was based largely on the recommendations made by a Veterans' Advisory Committee on Environmental Hazards, which reviewed more than eighty scientific articles that showed a statistical link between exposure to dioxin-herbicides and STS. Of note, the advisory committee "concluded that it was at least as likely as not that such an association existed."[2] The committee acknowledged the "strongly compelling" work done in Sweden. Congress passed the Agent Orange Act of 1991 with the leadership of elected officials like Sen. Tom Daschle (D-North Dakota). The law provided for permanent disability benefits for veterans with STS or non-Hodgkin's lymphoma and chloracne. These revised policies and legislation encouraged veterans to seek benefits for other medical conditions that had previously been denied.[3]

In another key piece of legislation, the Veterans Claims Assistance Act of 2000 removed the standard of a "well-grounded claim," while revising the VA's obligation to assist. By 2017 the VA recognized fourteen conditions as related to Agent Orange exposure, for which veterans could receive benefits.

Notably, both the chronology and geography of exposure had expanded for coverage dating back to 1962 and up to 1986 for military personnel, including reserves, who served in Vietnam and Korea, or on refitted C-123s used by Air Force Reserves. The health conditions of several groups, however, remain unrecognized, including female veterans, the children of male US Vietnam veterans (excluding cases of spina bifida), and the people of Vietnam.[4]

The environmental harm wrought by the United States' defoliation campaign appears written upon the Vietnamese landscape. The defoliation spraying program eradicated between 26 and 61 million square yards of timber resources. The areas that were repeatedly sprayed, primarily in the Mekong Delta, have seen regrowth of the mangrove forests after intensive cultivation efforts of the Vietnamese government. Dioxin contamination has presented a more significant, if less visible, problem. Jeanne and Steven Stellmans' further research on dioxin in Vietnam revealed that much greater quantities of the chemical contaminant were sprayed over the countryside than previously thought, increasing the likelihood of health problems in both the Vietnamese and American populations exposed during and after the war. The Stellmans' research and conclusions were bolstered by the findings of a Canadian firm hired by the Vietnamese 10-80 Division of the Vietnamese Ministry of Health in 1992. In 2007 Hatfield Consultants released their most recent findings on dioxin contamination in Vietnam. The report concluded that while the overall levels of dioxin contamination appeared lower than in industrialized countries, there were twenty-eight dioxin hotspots associated with US military bases, particularly those where Ranch Hand Operations took place. The epidemiological and toxicological evidence gathered from the 1990s onward helped change both consciousness of elected officials and strengthened the credibility of Vietnam veterans and Vietnamese citizens' claims of illness and reproductive harm.[5]

People protesting toxic chemicals commonly get dismissed as "NIMBYs"— Not In My Back Yard. Their protests are portrayed as selfish and elitist. Yet the reality is quite different. Instead of individuals whining, many protesters organize the community to make their concerns visible, and hazardous chemical landfills are almost exclusively sited in poor neighborhoods amidst communities of color. The questions raised about the safety of the chemicals used and how they are disposed of engage people and provoke citizen protests. The various groups of citizens I write about here were challenging state and scientific authority, demonstrating a changed environmental and health consciousness, and acting on new understandings of the human-nature relationship. Despite official indifference or even hostility, little money, corporate harassment, and state complicity, the individuals and groups protesting the phenoxy herbicides spoke up. They challenged the military-industrial complex. They demanded more of their government and corporate America. They cared. They persisted.

Notes

Introduction

1. For more on modernity and risk society, see David Harvey, *The Condition of Post Modernity: An Enquiry into the Origins of Cultural Change* (Cambridge, MA: B. Blackwell, 1980); Anthony Giddens, *Consequences of Modernity* (Stanford: Stanford University Press, 1990); and Ulrich Beck, *Risk Society: Toward a New Modernity*, translated by Mark Ritter (London: Sage, 1992).

2. I use the term "anti-toxic/s" as more representative than the commonly used term Not In My Backyard (NIMBY), which was often used to dismiss these protests. "Antitoxic" with no hyphen, as defined by *Merriam-Webster's Collegiate Dictionary*, refers to something that counteracts toxins.

3. See Chad Montrie's *The Myth of "Silent Spring": Rethinking the Origins of American Environmentalism* (Oakland: University of California Press, 2018) for a discussion of the problematic nature in the narrow definitions of what is considered environmentalism.

4. A more explicit discussion of anti-toxic and environmental justice literature appears later in this introduction.

5. The scholarship on scientific uncertainty abounds. Recent and/or influential works examining this topic include Naomi Oreskes and Erik M. Conway's *Merchants of Doubt: How a Handful of Scientists Obscured the Truth on Issues from Tobacco Smoke to Global Warming* (London: Bloomsbury, 2012); Nancy Langston, *Toxic Bodies: Hormone Disruptors and the Legacy of DES* (New Haven: Yale University Press, 2011); Sarah Vogel, *Is It Safe? BPA and the Struggle to Define the Safety of Chemicals* (Berkeley: University of California Press, 2013); and specifically for Agent Orange, Edwin Martini, *Agent Orange: History, Science, and the Politics of Uncertainty* (Amherst: University of Massachusetts Press, 2012). Frederick Rowe Davis's history of the science of toxicology, *Banned: A History of Pesticides and the Science of Toxicology* (New Haven: Yale University Press, 2014), provides an excellent study of the development of key concepts. See "Interview with Dr. Linda Birnbaum about Dioxin," April 19, 2004, available on YouTube at https://www.youtube.com/watch?v=c65cTmsWkxw.

6. Rob Nixon, *Slow Violence and the Environmentalism of the Poor* (Cambridge: Harvard University Press, 2011).

7. Arthur M. Schlesinger, *The Vital Center: The Politics of Freedom* (Cambridge, MA: Riverside, 1949). Excellent examples of these state-sponsored projects would be the campaigns against various pests, such as the gypsy moth and especially fire ants. See Joshua Blu Buhs, *The Fire Ant Wars: Nature, Science, and Public Policy in Twentieth-Century America* (Chicago: University of Chicago Press, 2004); and Pete Daniel, *Toxic Drift: Pesticides and Health in the Post–World War II South* (Baton Rouge: Louisiana State University Press, 2005). Buhs charts the ways the fire ant campaign was imbued with Cold War ideology with an emphasis on the *red* fire ant as a surrogate for Communism.

8. Gale E. Peterson, "The Discovery and Development of 2,4-D," *Agricultural History* 41, no. 3 (1967): 243–54; Nicolas Rasmussen, "Plant Hormones in War and Peace: Science, Industry, and Government in the Development of Herbicides in 1940s America," *Isis* 92, no. 2 (2001): 291–316; Daniel, *Toxic Drift*; J. L. Anderson, *Industrializing the Corn Belt: Agriculture, Technology, and Environment, 1945–1972* (DeKalb: Northern Illinois University Press, 2009).

9. James Whorton, *Before "Silent Spring": Pesticides and Public Health in Pre-DDT America* (Princeton: Princeton University Press, 1974); Thomas Dunlap, *DDT: Scientists, Citizens, and Public Policy* (Princeton: Princeton University Press, 1981); Edmund Russell, *War and Nature: Fighting Humans and Insects with Chemicals from World War I to "Silent Spring"* (Cambridge: Cambridge University Press, 2001); Buhs, *Fire Ant Wars*; Vogel, *Is It Safe?*; Davis, *Banned*; Michelle Mart, *Pesticides, a Love Story: America's Enduring Embrace of Dangerous Chemicals* (Lawrence: University of Kansas Press, 2015).

10. Gerald Markowitz and David Rosner have published numerous studies on occupational risk and industrial pollution. Their most recent work, *Deceit and Denial: The Deadly Politics of Industrial Pollution* (Berkeley: University of California Press, 2002), examines the case of polyvinyl chloride contamination. Christopher Sellers, *Hazards of the Job: From Industrial Disease to Environmental Health Science* (Chapel Hill: University of North Carolina Press, 1997), and *Dangerous Trade Histories of Industrial Hazard across a Globalizing World* (Philadelphia: Temple University Press, 2011); Christian Warren, *Brush with Death: A Social History of Lead Poisoning* (Baltimore: Johns Hopkins University Press, 2000); Nancy Langston, *Toxic Bodies: Hormone Disruptors and the Legacy of DES* (New Haven: Yale University Press, 2010); David Kinkela, *DDT and the American Century: Global Health, Environmental Politics, and the Pesticide That Changed the World* (Chapel Hill: University of North Carolina Press, 2011); Jennifer Thomson, *The Wild and the Toxic: American Environmentalism and the Politics of Health* (Chapel Hill: University of North Carolina Press, 2019).

11. Robert D. Bullard, *Dumping in Dixie: Race, Class, and Environmental Quality* (Boulder, CO: Westview Press, 1990); Andrew Hurley, *Environmental Inequalities: Class, Race, and Industrial Pollution in Gary, Indiana, 1945–1980* (Chapel Hill: University of North Carolina Press, 1995); Eileen M. McGurty, *Transforming Environmentalism:*

Warren County, PCBS, and the Origins of Environmental Justice (New Brunswick, NJ: Rutgers University Press, 2006); Elizabeth D. Blum, *Love Canal Revisited: Race, Class, and Gender in Environmental Activism* (Lawrence: University of Kansas Press, 2008); Kate Brown, *Plutopia: Nuclear Families, Atomic Cities, and the Great Soviet and American Plutonium Disasters* (Oxford: Oxford University Press, 2013); and Ellen Griffith Spears, *Baptized in PCBs: Race, Pollution, and Justice in an All-American Town* (Chapel Hill: University of North Carolina Press, 2014).

12. Fred Wilcox, *Waiting for an Army to Die: The Tragedy of Agent Orange* (New York: Vintage Books, 1983); Peter Schuck, *Agent Orange on Trial: Mass Toxic Disasters in the Courts* (Cambridge: Belknap Press of Harvard University Press, 1986); Wilbur Scott, *The Politics of Readjustment: Vietnam Veterans since the War* (New York: Aldine de Gruyter, 1993); David Zierler, *The Invention of Ecocide: Agent Orange, Vietnam, and the Scientists Who Changed the Way We Think about the Environment* (Athens: University of Georgia Press, 2011); Edwin A. Martini, *Agent Orange: History, Science, and the Politics of Uncertainty* (Amherst: University of Massachusetts Press, 2012); and Peter Sills, *Toxic War: The Story of Agent Orange* (Nashville: Vanderbilt University Press, 2014).

13. Mark Lytle emphasizes the radical nature of Carson's work in his biography *The Gentle Subversive: Rachel Carson, "Silent Spring," and the Rise of the Environmental Movement* (New York: Oxford University Press, 2007).

14. "Science News," *Science*, n.s., 95, no. 2471 (May 8, 1942): 11.

15. This idea taken from the title and subject of Peter Taylor's book *Unruly Complexity: Ecology, Interpretation, Engagement* (Chicago: University of Chicago Press, 2005).

Chapter One

1. Anthony Ripley, "Napalm Protests Worrying Dow, though Company Is Unhurt," *New York Times*, December 11, 1967, 2.

2. For some standards of historical literature on the Kennedy administration and US involvement in South Vietnam, see David Milne, *America's Rasputin: Walt Rostow and the Vietnam War* (New York: Hill and Wang, 2009); Michael Lind, *Vietnam: The Necessary War: A Reinterpretation of America's Most Disastrous Military Conflict* (New York: Basic Books, 2002); Thomas Paterson, *Kennedy's Quest for Victory: American Foreign Policy, 1961–1963* (New York: Oxford University Press, 1989). One other issue that needs to be addressed is the changed naming of the country of Vietnam. In the immediate postwar period, the geographic region we now recognize as the Socialist Republic of Vietnam was known as Indochina, a French colonial taxonomy. With the withdrawal of the French and creation of two states, the US supported the nominally democratic southern provinces, known then as South Viet Nam, and many of my sources use this nomenclature. In the interests of consistency, and to hopefully reduce confusion, this work uses the historical naming only in direct quotes of sources. Otherwise, the text refers to the geographic regions as South and North Vietnam.

3. W. W. Dennis and W. M. Dennis, "Chemicals: A Growth Industry," in special

issue "Science and Security," *Analysts Journal* 8, no. 2 (March 1952): 61–65; Jeffrey Meikle, *American Plastic: A Cultural History* (New Brunswick, NJ: Rutgers University Press, 1995), 177, 215, 216; and Gerald Markowitz and David Rosner, *Deceit and Denial: The Deadly Politics of Industrial Pollution* (Berkeley: University of California Press, 2002), 139–50.

4. Robert E. Burk, "What Chemistry Is Doing to Us and for Us," *Scientific Monthly* 49, no. 6 (December 1939): 492.

5. E. N. Brandt, *Growth Company: Dow Chemical's First Century* (East Lansing: Michigan State University Press, 1997), 247–50; Marvin B. Lieberman, "Magnesium Industry in Transition," *Journal of Industrial Organization* 19 (June 2001): 3; and "Obituary," *New York Times*, April 3, 1949, E2.

6. Eldridge Haynes, "Industrial Production and Manufacturing," *Journal of Marketing* 8, no. 1 (July 1943): 16.

7. Dan J. Forrestal, *Faith, Hope, and $5000: The Story of Monsanto: The Trials and Triumphs of the First 75 Years* (New York: Simon and Schuster, 1977), 105.

8. "Big Growth by '75 Seen in Chemicals," *New York Times*, May 16, 1953, 23.

9. Alfred D. Chandler Jr., *Shaping the Industrial Century: The Remarkable Story of the Evolution of the Modern Chemical and Pharmaceutical Industries* (Cambridge: Harvard University Press, 2005), 3–5, 21–28; Meikle, *American Plastic*, 149; Markowitz and Rosner, *Deceit and Denial*, 139–45; Linda Lear, "Bombshell in Beltsville: The USDA and the Challenge of 'Silent Spring,'" in special issue "History of Agriculture and the Environment," *Agricultural History* 66, no. 2 (Spring 1992): 155–60; Jerry Harrington, "The Midwest Agricultural Chemical Association: A Regional Study of an Industry on the Defensive," *Agricultural History* 70, no. 2 (Spring 1996): 417, 418.

10. Brandt, *Growth Company*, xi.

11. Brandt, *Growth Company*, 114, 115; "Researcher Is Honored by Chemical Industry," *New York Times*, November 6, 1955, 82.

12. "News of Food: Styron Boxes, First Made for Novelty, Used Widely in Home Refrigerators," *New York Times*, January 31, 1948, 23.

13. Chandler, *Shaping the Industrial Century*, 22, 23.

14. The three key articles claiming British discovery all appeared in a 1945 issue of *Nature*. See G. E. Blackman, "Plant-Growth Substances as Selective Weed-Killers: A Comparison of Certain Plant-Growth Substances with Other Selective Herbicides," *Nature* 155, no. 3939 (1945): 500, 501; P. S. Nutman, H. G. Thornton, and J. H. Quastel, "Plant-Growth Substances as Selective Weed-Killers: Inhibition of Plant Growth by 2: 4-Dichlorophenoxyacetic Acid and Other Plant-Growth Substances," *Nature* 155, no. 3939 (1945): 498–500; and R. E. Slade, W. G. Templeman, and W. A. Sexton, "Plant-Growth Substances as Selective Weed-Killers: Differential Effect of Plant-Growth Substances on Plant Species," *Nature* 155, no. 3939 (1945), 497, 498.

15. Gale E. Peterson, "The Discovery and Development of 2,4-D," *Agricultural History* 41, no. 3 (July 1967): 243; E. Holmes, "The Role of Industrial Research and

Development in Weed Control in Europe," *Weeds* 6, no. 3 (July 1, 1958): 245; and H. E. Thompson, Carl P. Swanson, and A. G. Norman, "New Growth-Regulating Compounds," *Botanical Gazette* 107, no. 4 (June 1, 1946): 476.

16. "Science News," *Science—Supplement* 95, no. 2471 (May 8, 1942): 11; Charles L. Hamner and H. B. Tukey, "The Herbicidal Action of 2,4 Dichlorophenoxyacetic and 2,4,5 Trichlorophenoxyacetic Acid on Bindweed," *Science*, n.s., 100, no. 2590 (1944), 154, 155; J. M. Beal, "Further Observations on the Telemorphic Effects of Certain Growth-Regulating Substances," *Botanical Gazette* 106, no. 2 (December 1, 1944): 165; and Paul C. Marth and John W. Mitchell, "2,4-Dichlorophenoxyacetic Acid as a Differential Herbicide," *Botanical Gazette* 106, no. 2 (December 1, 1944): 224.

17. "Science and Life in the World," *Science*, n.s., 103, no. 2683 (May 31, 1946): 662, 663.

18. "Plant Growth Regulators," *Science*, n.s., 103, no. 2677 (April 19, 1946): 469; "Science and Life in the World," 662, 663.

19. E. John Russell, *A History of Agricultural Science in Great Britain: 1620–1954* (London: George Allen & Unwin, 1966), 441–43.

20. Hamner and Tukey, "Herbicidal Action of 2,4 Dichlorophenoxyacetic and 2,4,5 Trichlorophenoxyacetic Acid on Bindweed," 154, 155; J. H. Quastel, "2,4-Dichlorophenoxyacetic Acid (2,4-D) as a Selective Herbicide," *Agricultural Control Chemicals*, Collected Papers from the Symposia on Economic Poisons (Washington, DC: American Chemical Society, Division of Agricultural and Food Chemistry, 1950), 244–46; Nicolas Rasmussen, "Plant Hormones in War and Peace: Science, Industry, and Government in the Development of Herbicides in 1940s America," *Isis* 92, no. 2 (July 1, 2001): 299.

21. Nutman et al., "Plant-Growth Substances as Selective Weed-Killers," 498, 488.

22. Nutman et al., "Plant-Growth Substances as Selective Weed-Killers," 499.

23. Hamner and Tukey, "Herbicidal Action of 2,4 Dichlorophenoxyacetic and 2,4,5 Trichlorophenoxyacetic Acid on Bindweed," 154, 155; Beal, "Further Observations on the Telemorphic Effects of Certain Growth-Regulating Substances," 167–77; Marth, "2,4-Dichlorophenoxyacetic Acid as a Differential Herbicide," 224–31; Blackman, "Plant-Growth Substances as Selective Weed-Killers," 500, 501; A. S. Crafts, "Selectivity of Herbicides," *Plant Physiology* 21, no. 3 (1946): 346–55.

24. Gregg Mittman, *Breathing Space: How Allergies Shape Our Lives and Landscapes* (New Haven: Yale University Press, 2007), 78.

25. Zachary J. S. Falck, *Weeds: An Environmental History of Metropolitan America* (Pittsburgh: University of Pittsburgh Press, 2010), 96–110.

26. Falck, *Weeds*, 102, 103; Mittman, *Breathing Space*, 81, 82.

27. George S. Avery Jr., "Death to Weeds! Powerful Weapons in an Endless Battle Are the New Synthetic Plant Hormones," *New York Times*, July 28, 1946, X13.

28. "Boy Scouts to Aid Ragweed Campaign," *New York Times*, August 11, 1946, 35; Falck, *Weeds*, 103; and Mittman, *Breathing Space*, 85.

29. Mittman, *Breathing Space*, 86.

30. Falck, *Weeds*, 113, 114.

31. Peterson, "Discovery and Development of 2,4-D," 251.

32. J. L. Anderson, *Industrializing the Corn Belt: Agriculture, Technology, and Environment, 1945–1972* (DeKalb: Northern Illinois University Press, 2009), 37.

33. Peterson, "Discovery and Development of 2,4-D," 252.

34. Harrington, "Midwest Agricultural Chemical Association," 420.

35. David Vail focuses on this shift in his book *Chemical Lands: Pesticides, Aerial Spraying, and Health in North America's Grasslands since 1945* (Tuscaloosa: University of Alabama Press, 2018), especially in chs. 1 and 2.

36. J. L. Anderson, "War on Weeds: Iowa Farmers and Growth-Regulator Herbicides," *Technology and Culture* 46, no. 4 (October 9, 2005): 733.

37. Anderson, *Industrializing the Corn Belt*, 33.

38. Pete Daniel, *Toxic Drift: Pesticides and Health in the Post–World War II South* (Baton Rouge: Louisiana State University Press, 2005), 51.

39. This quote appears in David D. Vail, "'Kill That Thistle': Rogue Sprayers, Bootlegged Chemicals, Wicked Weeds, and the Kansas Chemical Laws, 1945–1980," *Kansas History* 32, no. 2 (2012): 117.

40. Anderson, "War on Weeds," 730, 731; Daniel, *Toxic Drift*, 50–52.

41. Virginia Scott Jenkins, *The Lawn: The History of an American Obsession* (Washington, DC: Smithsonian Institution Press, 1994), 99–102.

42. Jenkins, *Lawn*, 100; Paul Robbins, *Lawn People: How Grasses, Weeds, and Chemicals Make Us Who We Are* (Philadelphia: Temple University Press, 2007) 36, 38, 45; Steinberg, *American Green: The Obsessive Quest for the Perfect Lawn* (New York: W. W. Norton, 2006), 24.

43. Steinberg, *American Green*, 51.

44. For the postwar American housing boom, see Kenneth Jackson, *Crab Grass Frontier: The Suburbanization of the United States* (New York: Oxford University Press, 1987); Adam Rome, *The Bulldozer in the Countryside: Suburban Sprawl and the Rise of American Environmentalism* (New York: Cambridge University Press, 2001); and Ted Steinberg, "Lawn and Landscape in World Context, 1945–2000," *OAH Magazine of History* 19, no. 6 (November 1, 2005): 62–64.

45. Jenkins, *Lawn*, 99.

46. Steinberg, *American Green*, 46.

47. For the most significant work done on the idea of "domestic containment," see Elaine Tyler May, *Homeward Bound: American Families in the Cold War Era* (New York: Basic Books, 1988).

48. May, *Homeward Bound*, 142.

49. For more on postwar middle-class anxiety and literature as a sociological study of it, see Catherine Jurca, "The Sanctimonious Suburbanite: Sloan Wilson's *The Man in the Gray Flannel Suit*," *American Literary History* 11, no. 1 (Spring 1999); regarding

the home as a site of domestic containment and Cold War defense, see May, *Homeward Bound*, 10–15, 143–62; Sarah A. Lichtman, "Do-It-Yourself Security: Safety, Gender, and the Home Fallout Shelter in Cold War America," *Journal of Design History* 32, no. 1 (April 2006): 39–45.

50. Sloan Wilson, *The Man in the Gray Flannel Suit* (New York: First Four Walls Eight Windows, 2002; orig. published 1955), 1.

51. William H. Whyte Jr., *The Organization Man* (New York: Simon and Schuster, 1956), 380.

52. Whyte, *Organization Man*, 372.

53. Whyte, *Organization Man*, 296.

54. *Time* 64, no. 5, cover (August 2, 1954); Lichtman, "Do-It-Yourself Security," 45.

55. David Snell, "Snake in the Crab Grass," *Life* 53, no. 15 (October 12, 1962), 23, 25; Steinberg, *American Green*, 54–58.

56. See Rome, *Bulldozer in the Countryside*; Christopher Sellers, *Crabgrass Crucible: Suburban Nature and the Rise of Environmentalism in Twentieth-Century America* (Chapel Hill: University of North Carolina Press, 2012); Markowitz and Rosner, *Deceit and Denial*.

57. Elaine Woo, "Dagmar Wilson Dies at 94; Organizer of Women's Disarmament Protesters," *Los Angeles Times*, January 30, 2011.

58. See Rome, *Bulldozer in the Countryside*, and "'Give Earth a Chance': The Environmental Movement and the Sixties," *Journal of American History* 90, no. 2 (September 2003); Terrianne K. Schulte, "Citizen Experts: The League of Women Voters and Environmental Activism," *Frontiers* 30, no. 3 (September 2009): 1–29.

59. For a detailed discussion of the cases, see Christopher Sellers, "Body, Place and the State: The Making of an 'Environmentalist' Imaginary in the Post–World War II U.S.," *Radical History Review* 74 (1999): 31–64.

60. Sellers, "Body, Place and the State," 50–57. Thomas Dunlap discusses Americans' changed chemical consciousness in *DDT, "Silent Spring," and the Rise of Environmentalism: Classic Texts* (Seattle: University of Washington Press, 2008); and Charles F. Wurster tells about the second Long Island case and other significant DDT cases in the emergence of the Environmental Defense Fund in his memoir, *DDT Wars: Rescuing Our National Bird, Preventing Cancer, and Creating the Environmental Defense Fund* (New York: Oxford University Press, 2015).

61. Maril Hazlett, "Voices from the Spring: *Silent Spring* and the Ecological Turn in American Health," in *Seeing Nature through Gender*, ed. Virginia Scharff (Lawrence: University Press of Kansas, 2003), 111–16, 119.

62. Sellers, "Body, Place and the State," 47–50. For a work that problematizes "flexible bodies," see Emily Martin, *Flexible Bodies* (New York: Beacon Press, 1995).

63. Bui Thi Phuong-Lan, "When the Forest Became the Enemy and the Legacy of American Herbicidal Warfare in Vietnam," PhD diss., Harvard University, Cambridge, Massachusetts, 2003, ch. 2, 66–116.

64. Simon Schama, *Landscape and Memory* (New York: Vintage Books, 1995), 61–120.

65. David Arnold, *The Tropics and the Traveling Gaze: India, Landscape, and Science, 1800–1856* (Delhi: Permanent Black, 2005), 35.

66. David Arnold, *The Problem of Nature: Environment, Culture, and European Expansion* (Oxford: Blackwell, 1996), 152.

67. Greg Bankoff, "A Question of Breeding: Zootechny and Colonial Attitudes toward the Tropical Environment in the Late Nineteenth-Century Philippines," *Journal of Asian Studies* 60, no. 2 (May 2001): 432, 433.

68. Stephen Frenkel, "Society Jungle Stories: North American Representations of Tropical Panama," in special issue "Latin American Geography," *Geographical Review* 86, no. 3 (July 1996): 317–33.

69. David Zierler, *The Invention of Ecocide: Agent Orange, Vietnam, and the Scientists Who Changed the Way We Think about the Environment* (Athens: University of Georgia Press, 2011), 50–60.

70. David Biggs, "Managing a Rebel Landscape: Conservation, Pioneers, and the Revolutionary Past in the U Minh Forest, Vietnam," *Environmental History* 10, no. 3 (July 2005): 466–71.

71. Bui, "When the Forest," 26, 27.

72. Bui, "When the Forest," 12; Milne, *America's Rasputin*, 105–9, 126.

73. Edwin A. Martini, *Agent Orange: History, Science, and the Politics of Uncertainty* (Amherst: University of Massachusetts Press, 2012), 54; Bui, "When the Forest," 27, 31–33.

74. Pierre Brocheux, *Indochina: An Ambiguous Colonization, 1858–1954* (Berkeley: University of California Press, 2009), 6, 200–205, 250–52; Huynh Quang Nhuong, *The Land I Lost: Adventures of a Boy in Vietnam* (New York: Harper Collins, 1982), 11, 19; James C. Scott, *The Art of Not Being Governed: An Anarchist History of Upland Southeast Asia* (New Haven: Yale University Press, 2010), 190–94; quote from Nancy Lee Peluso and Peter Vandergeest, "Political Ecologies of War and Forests: Counterinsurgencies and the Making of National Natures," *Annals of the Association of American Geographers* 101, no. 3 (2011): 587.

75. Interview with John Hodgin, February 3, 2003, John Hodgin Collection, TTU Vietnam Archive and Archive, Texas Tech University, http://www.vietnam.ttu.edu /virtualarchive, hereafter cited as the TTU Vietnam Archive.

76. Interview with Charles Hubbs, n.d., Charlie Hubbs Collection, TTU Vietnam Archive.

77. Interview with John Spey, October 4, 2000, John Spey Collection, TTU Vietnam Archive.

78. Hodgin interview.

79. Interview with Ralph Dresser, January 25, 2002, 47, Ralph Dresser Collection, TTU Vietnam Archive.

80. Interview with Robert Turk, n.d., Robert Turk Collection, TTU Vietnam Archive.

81. Chemical Warfare Charges, April 11, 1963, Folder 11, Box 02, Douglas Pike Collection: Unit 03—Technology, TTU Vietnam Archive.

82. Interview with Larry Wasserman, December 2, 2002, Larry Wasserman Collection, TTU Vietnam Archive.

83. Jay Cravens, *A Well Worn Path* (Huntington, WV: University Editions, 1994), 340.

84. Cravens, *Well Worn Path*, 229.

85. February 1966 interview with twenty-year-old male, Folder 112, "Interview 1, Studies of the National Liberation Front of South Vietnam, March 1972," conducted by Rand Corporation, Box 13, Alvin L. Young Collection on Agent Orange, Special Collections, National Agricultural Library, Beltsville, Maryland, hereafter cited as the Young Collection.

Chapter Two

1. "5 Priests among 9 Seized in Dow Protest in Capital," March 23, 1969, 18; news clipping, Folder 104, "Napalm, Misc. 1969," Box 1, "Napalm," Dow Chemical Company Collection, Archives, Chemical Heritage Foundation, Philadelphia, Pennsylvania, hereafter cited as Dow Collection.

2. "An Open Letter to American Corporations," Folder 104, "Napalm, Misc. 1969," Box 1, "Napalm," Dow Collection.

3. "5 Priests among 9 Seized in Dow Protest in Capital," March 23, 1969, 18; news clipping, Folder 104, "Napalm, Misc. 1969," Box 1, "Napalm," Dow Collection.

4. "The Nine," undated typewritten manuscript, Folder 104, "Napalm, Misc. 1969," Box 1, "Napalm," Dow Collection.

5. Anne Klejment, "The Spirituality of Dorothy Day's Pacifism," in special issue "War and Peace," *U.S. Catholic Historian* 27, no. 2 (Spring 2009): 1; Joseph G. Morgan, "A Change of Course: American Catholics, Anticommunism, and the Vietnam War," in special issue "Catholic Anticommunism," *U.S. Catholic Historian* 22, no. 4 (Fall 2004): 128–30.

6. James Finn, "American Catholics and Social Movements," in *Contemporary Catholicism in the United States*, ed. Philip Gleason, 133–45 (South Bend: University of Notre Dame, 1969).

7. Morgan, "Change of Course," 117, 118; Penelope Adams Moon, "'Peace on Earth: Peace in Vietnam': The Catholic Peace Fellowship and Antiwar Witness, 1964–1976," *Journal of Social History* 36, no. 4 (July 1, 2003): 1033–57.

8. Thomas Merton, *Faith and Violence: Christian Teaching and Christian Practice* (Notre Dame, IN: University of Notre Dame Press, 1994), 1–13.

9. Harriet Gross, "Jane Kennedy: Making History through Moral Protest," in special issue "Women's Oral History," *Frontiers: A Journal of Women Studies* 2, no. 2 (Summer 1977): 73; Shawn Francis Peters, *The Catonsville Nine: A Story of Faith and Resistance in the Vietnam Era* (New York: Oxford University Press, 2012); see Marian Mollin,

"Communities of Resistance: Women and the Catholic Left of the Late 1960s," *Oral History Review* 31, no. 2 (2004): 29–51, for more on the gender politics of the Catholic Left and its antiwar protests.

10. "The Nine," File No. 00104, "Napalm, Misc. 1969," "Napalm," Box 1, Dow Collection.

11. John M. Goshko, "Two Priests and Nun Embrace Red Revolution," *Milwaukee Journal*, January 25, 1968, 1, http://news.google.com/newspapers; George Mische, "Inattention to Accuracy about 'Catonsville Nine' Distorts History," *National Catholic Reporter*, May 17, 2013, http://ncronline.org.

12. "The Nine," File No. 00104, "Napalm, Misc. 1969," Box 1, "Napalm," Dow Collection.

13. Charles Meconis, *With Clumsy Grace: The American Catholic Left, 1961–1975* (New York: Seabury Press, 1979), 41.

14. Meconis, *With Clumsy Grace*, 41–43.

15. Meconis, *With Clumsy Grace*, 61; "An Open Letter to the Corporations of America," File No. 00104, "Napalm, Misc, 1969" Box 1, Dow Collection.

16. D.C. Nine Defense Committee, "1968 Dow Annual Report—Support the D.C. Nine," File No. 00104 "Napalm Misc. 1969," Dow Collection.

17. D.C. Nine Defense Committee, "1968 Dow Annual Report—Support the D.C. Nine," File No. 00104 "Napalm Misc. 1969," Dow Collection.

18. Meconis, *With Clumsy Grace*, 36.

19. Meconis, *With Clumsy Grace*, 52.

20. Judith Coburn, "Waiting for the Verdict," *Village Voice*, February 19, 1970, 9, http://news.google.com/newspapers.

21. Meconis, *With Clumsy Grace*, 61; Coburn, "Waiting for the Verdict," 9.

22. Coburn, "Waiting for the Verdict," 9.

23. Coburn, "Waiting for the Verdict," 64.

24. Untitled Dow report, File No. 00102, "Napalm Misc. 1968," Dow Collection.

25. "Vandals Visit Dow Building," *Brazosport Facts*, November 9, 1969, 16; Joseph Hanlon, "Anti-War Protestors Erase 1,000 Dow Tapes," *Computerworld* 3, no. 48 (December 3, 1969): 1.

26. Beaver 55, "Statement," File No. 00102, "Napalm Misc. 1968," Dow Collection.

27. Gross, "Jane Kennedy," 73–75.

28. Letter from Studs Terkel, Sid Lens, and Annette Nussbaum, "Re: Justice for Jane Kennedy Committee, August 5, 1974," Folder 024, Box 24, Social Movements Collection, TTU Vietnam Archive and Archive, Texas Tech University, http://www.vietnam.ttu.edu/virtualarchive, hereafter cited as the TTU Vietnam Archive.

29. Gross, "Jane Kennedy," 78.

30. Gross, "Jane Kennedy," 80, emphasis in the original.

31. Mitchell K. Hall, *Because of Their Faith: CALCAV and Religious Opposition to the Vietnam War* (New York: Columbia University Press, 1990), 5–67, 67.

32. Hall, *Faith*, 5–67.

33. CALCAV, "A Statement of Purpose Issued by Clergy and Laymen Concerned about Vietnam upon Announcing Its Activities at the Annual Stock Holders Meeting of Dow Chemical Company," April 30, 1969, File 00104, "Napalm Misc. 1969," Dow Collection, hereafter cited as CALCAV Statement.

34. CALCAV Statement, Dow Collection.

35. E. N. Brandt, "Napalm Program: Religious Community Activities" memo, September 24, 1969, File 00104, "Napalm Misc. 1969," Dow Collection.

36. Brandt memo.

37. Hall, *Faith*, 127–29.

38. Michael Aaron Dennis, "'Our First Line of Defense": Two University Laboratories in the Postwar American State, *Isis* 85, no. 3 (September, 1994): 427, 428; Roger L. Geiger, "Science, Universities, and National Defense, 1945–1970," in special issue "Science after '40," *Osiris*, 2nd ser., 7 (1992): 27; Jonathan Goldstein, "Agent Orange on Campus: The Summit-Spicerack Controversy at the University of Pennsylvania, 1965–1967," in *Sights on the Sixties*, ed. Barbara Tischler (New Brunswick, NJ: Rutgers University Press, 1992), 46–48. The Department of Agriculture and National Institutes of Health were the other two primary government agencies funding university research.

39. Robert Neer, *Napalm: An American Biography* (Cambridge: Belknap Press of Harvard University Press, 2013), 7–11; Jonathan Goldstein, "Agent Orange," 46–48.

40. Goldstein, "Agent Orange," 47.

41. Elinor Langer, "Chemical and Biological Warfare (I): The Research Program," *Science*, n.s., 155, no. 3759 (January 13, 1967): 174.

42. Goldstein, "Agent Orange," 47.

43. M. S. Handler, "New Magazine on Vietnam War Aims to Keep Scholars Informed," November 14, 1965, *New York Times*, 7.

44. Carol Brightman, "The 'Weed Killers'—A Final Word," *Viet-Report* 2, no. 7 (October 1966): 3.

45. Langer, "Chemical and Biological Warfare (I)," 174.

46. "ICR (Institute for Cooperative Research), Projects Spicerack and Summit Research Satirized by Student Newspaper [*Daily Pennsylvanian*] [graphic]," 1967 February, Groups and Events, UARC20080613004, University Archives Digital Image Collection, University Archives, University of Pennsylvania, http://hdl.library.upenn.edu/1017/d/archives/20080613004.

47. Goldstein, "Agent Orange," 53–60.

48. R. J. Williams, "Attitudes toward the War, the Use of Chemical Weapons, and the Dow Chemical Company," March 1968, 4, Folder 00103, Box 1 Napalm, Dow Collection.

49. David Maraniss, *They Marched into Sunlight: War and Peace in Vietnam and America, October 1967* (New York: Simon and Schuster, 2003), 233–42; R. J. Williams, "Attitudes toward the War, the Use of Chemical Weapons, and the Dow Chemical Company," March 1968, 1–4, 9, Dow Collection; "War Protest," November 27, 1967, Folder 05, Box 07, Douglas Pike Collection: Unit 03—Antiwar Activities, TTU Vietnam Archive.

50. "Hand-Out for Passerby: What Are We Trying to Say?," n.d., Folder 45, Box 18, Social Movements Collection, TTU Vietnam Archive.

51. "Organizing Case Study: Duke and Dow," n.d., 6, Folder 45, Box 18, Social Movements Collection, TTU Vietnam Archive.

52. "22 Scientists Bid Johnson Bar Chemical Weapons in Vietnam," *New York Times*, September 20, 1966, 3.

53. Joe McPherson, "Memo: A Posture and Program for These Times," October 25, 1967, Folder "File No. Napalm 00100, Misc. 1966–1967," Box 1 Napalm, Dow Collection.

54. J. B. Neilands, G. H. Orians, E. W. Pfeiffer, Alje Vennema, and Arthur H. Westing, *Harvest of Death: Chemical Warfare in Vietnam and Cambodia* (New York: Free Press, 1972), 118.

55. "Defoliation in Viet-Nam," Dow Chemical document, ca. January 1967, Folder "File No. Napalm 00112, Harbicides [sic] in Vietnam," Box 1 Napalm, Dow Collection.

56. "Airplane Can Help Clear Brush," newspaper clipping, unknown newspaper, Dow files; "Aerial Defoliation Used to Open New Lands," *Hawaii Business and Industry*, October 1967, 41–43; "Defoliation in Viet-Nam," Dow Chemical document, ca. January 1967; "The Value of Tordon in Expanding World Food Production," Tordon Fact Sheet; "Memo on Meeting with Caterpillar Tractor Company," November 16, 1967; "What U.S. Companies Are Doing Abroad: Jungle-Clearing Job," *U.S. News & Report*, November 27, 1967, clippings file; P. S. Motooka, D. F. Saiki, D. L. Plucknett, O. R. Younge, and R. E. Daehler, "Control of Hawaiian Jungle with Aerially Applied Herbicide," *Down to Earth* 23, no. 1 (Summer 1967): 18–22; all documents found in Folder "File No. Napalm 00112, Harbicides [sic] in Vietnam," Box 1 Napalm, Dow Collection.

57. "War Use of Chemicals Alarms Science Group," *Milwaukee Journal*, December 31, 1966, newspaper clipping, Folder "File No. Napalm 00112, Harbicides [sic] in Vietnam," Box 1 Napalm, Dow Collection.

58. Elinor Langer, "National Teach-In: Professors Debating Viet Nam, Question Role of Scholarship in Policy-Making," *Science* 148, no. 3673 (May 21, 1965): 1075–77; Report, "Government Sales," c. 1969, Folder 00104, Box 1 Napalm, Dow Collection; Anderson, *Industrializing the Corn Belt*, 46; E. N. Brandt, *Growth Company: Dow Chemical's First Century* (East Lansing: Michigan State University Press, 1997), 532.

59. Jessica Wang, *American Science in an Age of Anxiety: Scientists, Anticommunism, and the Cold War* (Chapel Hill: University of North Carolina Press, 1998), 85–145; Kelly Moore, *Disrupting Science: Social Movements, American Scientists, and the Politics of the Military, 1945–1975* (Princeton: Princeton University Press, 2008), 190–202; Charles Thorpe, *Oppenheimer: The Tragic Intellect* (Chicago: University of Chicago Press, 2006), 200–243;

60. For a focused examination of these scientists and their influence on US military policy, see David Zierler, *The Invention of Ecocide*; Bui, "When the Forest"; and Martini, *Agent Orange*.

61. Brandt, *Growth Company*, 108, 335; also see Zierler, Martini, and Moore.

62. All quotes taken from Bryce Nelson, "Herbicides in Vietnam: AAAS Board Seeks Field Study," *Science*, n.s., 163 (January 3, 1969): 58.

63. Neilands et al., *Harvest of Death*, 118; Robert Reinhold, "Scientists Call for a Ban on 2 Vietnam Defoliants," *New York Times*, December 31, 1969, 10; [Walter Sullivan, "Zoologist, Back from Vietnam, Notes Defoliants' Value"], April 4, 1969, Folder 13, Box 02, Douglas Pike Collection: Unit 03—Technology, TTU Vietnam Archive; Zierler, *Invention of Ecocide*, 119–21; Peter Sills, *Toxic War: The Story of Agent Orange* (Nashville: Vanderbilt University Press, 2014), 109–12.

64. Matthew Meselson et al., "Hearing: Background Material Relevant to Presentations at the 1970 Annual Meeting of the AAAS," War Related Civilian Problems in Indochina, Part 1: Vietnam, Hearings before the Subcommittee to Investigate Problems Connected with Refugees and Escapees, Committee on the Judiciary, U.S. Senate, 92nd Congress, 1st Session, 21 April 1971. Reprinted in the *Congressional Record* 118, no. 32 (March 3, 1972), S3226-S3233, Item ID# 900, Box 40, Series III Subseries I, Young Collection, http://specialcollections.nal.usda.gov/agentorange/agent-orange-item-id-900.

65. Arthur Galston, "Some Implications of the Widespread Use of Herbicides," *BioScience* 21, no. 17 (September 1, 1971): 892; Meselson et al., "Hearing"; J. B. Nielands, "Vietnam: Progress of the Chemical War," *Asian Survey* 10, no. 3 (March 1970): 209; Zierler, *Invention of Ecocide*, 128–30.

66. Anthony Lewis, "Death in the Abstract," *New York Times*, January 4, 1971, 31; Walter Sullivan, "A.A.A.S. Disputes That Were Not on the Agenda," *New York Times*, January 3, 1971, E6.

67. "Science, Ecology and War," *Science News* 92, no. 22 (November 25, 1967): 511.

68. Arthur Galston, Letter to the Editor, *Science*, n.s., 168, no. 3939 (June 26, 1970): 1067.

69. Jean Meyer, "Starvation as a Weapon: Herbicides in Vietnam, I," in special issue "Chemical and Biological Warfare," *Scientist and Citizen* 9, no. 7 (1967): 115–21; Arthur Galston, "Changing the Environment: Herbicides in Vietnam, II," in special issue "Chemical and Biological Warfare," *Scientist and Citizen* 9, no. 7 (1967): 122–29.

Chapter Three

1. Phillip B. Davidson, *Vietnam at War: The History, 1945–1975*, 2nd ed. (New York: Oxford University Press, 1991), 28; Douglas Pike, *Viet Cong: The Organization and Techniques of the National Liberation Front of South Vietnam* (Cambridge, MA: MIT Press, 1966), 385, 388. This work relies upon translated documents. But given the focus on protest, and that many of the Vietnamese and international critiques sought an American audience, I do not think the research suffers irreparably from this dependence. Citations use the language accents found in the original primary and secondary documents but are not replicated in my written text.

2. Edwin A. Martini, *Agent Orange: History, Science, and the Politics of Uncertainty* (Amherst: University of Massachusetts Press, 2012), 53–96; and Bui Thi Phuong-Lan,

"When the Forest Became the Enemy and the Legacy of American Herbicidal Warfare in Vietnam" (Cambridge: Diss. Harvard University, 2003).

3. Pike, *Viet Cong*, 387.

4. Democratic Republic of Vietnam, Commission for Investigation on the American Imperialists' War Crimes in Vietnam, *American Crimes in Vietnam* (unknown location: October 1966), 21, 22, Folder 02, Box 01, Michael Cook Collection, TTU Vietnam Archive and Archive, Texas Tech University, http://www.vietnam.ttu.edu/virtualarchive, hereafter cited as the TTU Vietnam Archive.

5. *American Uses of War Gases and World Public Opinion* (Hanoi: Foreign Languages Publishing House, 1966), 6, emphasis in the original, Folder 10, Box 08, George J. Veith Collection, TTU Vietnam Archive.

6. *American Use of War Gases*, 12.

7. Different pamphlets with the same text, *U.S. Crimes in Vietnam* (Hanoi: Vietnam Courier, 1966), 9, 10; "Crimes of US Imperialists in Vietnam," Folder 01, Box 38, Douglas Pike Collection: Unit 03—War Atrocities, TTU Vietnam Archive; Joseph Dougherty, "The Use of Herbicides in Southeast Asia and Its Criticism," Professional Study (Maxwell Air Force Base, AL: Air War College, April 1, 1972), 27, Folder 13, Box 02, Paul Cecil Collection, TTU Vietnam Archive; David Zierler, *The Invention of Ecocide: Agent Orange, Vietnam, and the Scientists Who Changed the Way We Think about the Environment* (Athens: University of Georgia Press, 2011), 95.

8. Martini, *Agent Orange*, 72–79; Duong Van Mai Elliott, *RAND in Southeast Asia: A History of the Vietnam War Era* (Santa Monica, CA: RAND, 2010), 225; R. R. Betts and Frank Denton, *An Evaluation of Chemical Crop Destruction in Vietnam* (Santa Monica, CA: RAND, 1975), xiii.

9. Don Luce, "A Decade of Atrocity," in Browning and Forman, eds., *The Wasted Nations: Report of the International Commission of Enquiry into United States Crimes in Indochina, June 20–25, 1971* (New York: Harper Colophon Books, 1972), 8.

10. Jay Cravens, *A Well Worn Path* (Huntington, WV: University Publications, 1994), 339.

11. Martini, *Agent Orange*, 75–80; Edgar Lederer, "Report of the Sub-Committee on Chemical Warfare in Vietnam," testimony at the Bertrand Russell War Tribunal, in *Against the Crime of Silence: Proceedings of the Russell International War Crimes Tribunal*, ed. John Duffett (New York: O'Hare Books, 1968), 350; Bui, "When the Forest," 33–35.

12. Fredrik Logevall, "America Isolated: The Western Powers and the Escalation of the War," in *America, the Vietnam War, and the World: Comparative and International Perspectives*, ed. Andreas W. Daum, Lloyd C. Gardner, and Wilfried Mausbach (New York: German Historical Institute and Cambridge University Press, 2003), 177–79, 193.

13. Michael E. Latham, "Knowledge at War: American Social Science and Vietnam," in *A Companion to the Vietnam War*, ed. Marilyn B. Young and Robert Buzzanco (Malden: Blackwell, 2002), 442–44; Anthony J. Russo, *A Statistical Analysis of the U.S. Crop Spraying Program in South Vietnam* (Santa Monica, CA: RAND, 1967), iii.

14. Elliott and Thompson, *RAND in Southeast Asia*, 220–28; Russell Betts and Frank Denton, *An Evaluation of Chemical Crop Destruction in Vietnam* (Santa Monica, CA: RAND, 1975), xii; Russo, *Statistical Analysis*, ix.

15. Russo, *Statistical Analysis*, 32.

16. Betts and Denton, *Evaluation of Chemical Crop Destruction in Vietnam*, iii; Russo, *Statistical Analysis*, iii.

17. Russo later helped his former colleague Daniel Ellsberg publish what became known as the *Pentagon Papers* in 1969. Elliott and Thompson, *RAND in Southeast Asia*; Russo, *Statistical Analysis*, ix, 32; and Anthony J. Russo, "Inside the RAND Corporation and Out: My Story," *Ramparts* 10, no. 10 (April 1972): 52.

18. Arthur W. Blaser, "How to Advance Human Rights without Really Trying: An Analysis of Nongovernmental Tribunals," *Human Rights Quarterly* 14, no. 3 (August 1992): 342–44; and Michitake Aso and Annick Guénel, "The Itinerary of a North Vietnamese Surgeon: Medical Science and Politics during the Cold War," *Science, Technology & Society* 18, no. 3 (2013): 301.

19. Arthur Jay Klinghoffer and Judith Apter Klinghoffer, *International Citizens' Tribunals: Mobilizing Public Opinion to Advance Human Rights* (New York: Palgrave, 2002), 103–62; Anthony D'Amato, "Book Reviews: *Against the Crime of Silence: Proceedings of the Russell International War Crimes Tribunal* by John Duffett; *On Genocide* by Jean-Paul Sartre," *California Law Review* 57, no. 4 (October 1969): 1033–38; Ralph Schoenman, "Foreword," in *Against the Crime of Silence: Proceedings of the Russell International War Crimes Tribunal*, ed. John Duffett (New York: O'Hare Books, 1968), 6; Abraham Behar Testimony, "Incendiary Weapons, Poison Gas, Defoliants Used in Vietnam," Duffett, *Against the Crime of Silence*, 327–31; Alexandre Minkowski Testimony, "On Chemical and Biological Warfare in Vietnam," Duffett, *Against the Crime of Silence*, 331–37; Graeme MacQueen and Joanna Santa-Barbara, "Peace Building through Health Initiatives," *British Medical Journal* 321, no. 7256 (July 29, 2000): 293–96. Describing the herbicides as "weed killers" showed the shared intellectual influence within leftist circles.

20. Behar and Minkowski testimonies, in Duffett, *Against the Crime of Silence*.

21. Edgar Lederer, "Report," in Duffett, *Against the Crime of Silence*, 363.

22. "Extracts from a Report on Agricultural Chemicals Used in Vietnam," testimony by the Japanese Scientific Committee, in Duffett, *Against the Crime of Silence*, 373; Klinghoffer and Klinghoffer, *International Citizens' Tribunals*, 156; and Richard A. Falk, "Law and Responsibility in Warfare: The Vietnam Experience," *Instant Research on Peace and Violence* 4, no. 1 (1974): 2.

23. Carl Oglesby, "Introduction," *Ramparts* 6, no. 7 (February 1968): 35–42, https://www.unz.org. Michael Freeman argues that the tribunal and especially Sartre's subsequent reflection, *On Genocide* (Boston: Beacon Hill Books, 1968), represent the first major use of genocide as a concept after the Holocaust and World War II. See Michael Freeman, "Speaking about the Unspeakable: Genocide and Philosophy," *Journal of Applied Philosophy* 8, no. 1 (1991): 4.

24. Pierre Biquard, "Proceedings of the WFSW Conference on Chemical Warfare in Vietnam," *Scientific World* 14, 4.

25. Lars Gogman, "Vietnam in the Collections," Labor Movement and Archives, http://www.arbark.se/wib/vietnam-in-the-collections.pdf.

26. Susan Sontag, "A Letter from Sweden, 1970," Folder 05, Box 08, Douglas Pike Collection: Unit 03—Antiwar Activities, TTU Vietnam Archive; Richard Falk, "Introduction," in Frank Browning and Dorothy Forman, eds., *The Wasted Nations: Report of the International Commission of Enquiry into United States Crimes in Indochina, June 20–25, 1971* (New York: Harper Colophon Books, 1972), xv.

27. Bert Pfeiffer, quoted in Browning and Forman, *Wasted Nations*, 117.

28. Browning and Forman, *Wasted Nations*, 127.

29. Arthur Westing, "The Environmental Disruption of Indochina," in *The Effects of Modern Weapons on the Human Environment in Indochina: Documents presented at a hearing organized by International Commission in cooperation with the Stockholm Conference on Vietnam and the Swedish Committee for Vietnam, Stockholm June 2–4, 1972*, 1:2, Folder 15, Box 10, Douglas Pike Collection: Unit 11—Monographs, TTU Vietnam Archive. The People's Forum, a Swedish coalition group, sponsored the panel presentations on the Indochina War. See "Environment Conference Will Offer Some Sideshows," *New York Times*, June 5, 1972, 24; and Lars Emmelin, "The Stockholm Conferences," *Ambio* 1, no. 4 (1972): 140.

30. Other alternative groups that met in conjunction with the United Nations event included the Environment Forum and the international peace group Dai Dong, which addressed Agent Orange with a day of panels. See Tord Björk, "Challenging Western Environmentalism at the United Nations Conference on Human Environment at Stockholm 1972," *RIO + 20 / STH + 40*, Paper II, June 2012, 2:21, http://www.aktivism.info/rapporter/ChallengingUN72.pdf.

31. John H. E. Fried, "War by Ecocide: Some Legal Observations," in *The Effects of Modern Weapons on the Human Environment in Indochina: Documents presented at a hearing organized by International Commission in cooperation with the Stockholm Conference on Vietnam and the Swedish Committee for Vietnam, Stockholm June 2–4, 1972*, 2:2, Folder 15, Box 10, Douglas Pike Collection: Unit 11—Monographs, TTU Vietnam Archive, emphasis in the original.

32. Pike, *Viet Cong*, 309.

33. Pike, *Viet Cong*, 441, 459, 462, and 465.

34. "Dr. Gerhard Grümmer Describes Trip to DRV," December 8, 1970, Folder 20, Box 03, Douglas Pike Collection: Unit 06—Democratic Republic of Vietnam, TTU Vietnam Archive.

35. "Dr. Gerhard Grümmer Describes Trip to DRV," TTU Vietnam Archive.

36. Gerhard Grümmer, *Herbicides in Vietnam* (Berlin: Vietnam Commission of the Afro-Asian Solidarity Committee of the GDR, 1969).

37. Grümmer, *Herbicides in Vietnam*.

38. Gerhard Grümmer, *Genocide with Herbicides: Report—Analysis—Evidence* (Berlin: Afro-Asian Solidarity Committee of the German Democratic Republic, 1971), 10.

39. Grümmer, *Genocide with Herbicides*, 11–18.

40. Grümmer, *Genocide with Herbicides*, 104–8.

41. Gerhard Grümmer, *Accusation from the Jungle* (Berlin: Vietnam Commission of the Afro-Asian Solidarity Committee of the German Democratic Republic, 1972), 1.

42. Grümmer, *Accusation from the Jungle*, 20.

43. Grümmer, *Accusation from the Jungle*, 20.

44. Grümmer, *Accusation from the Jungle*, 43.

45. Grümmer, *Accusation from the Jungle*, 46.

46. Grümmer, *Accusation from the Jungle*, 69.

47. Grümmer, *Accusation from the Jungle*, 46–68; 68, 69.

48. Gerhard Grümmer, *Giftküchen des Teufels* (Berlin: Militärverl. d. Dt. Demokrat. Republik, 1988).

49. Henry Kamm, "Nutrition Is Called Major Hanoi Worry: Leading Vietnamese Doctor Says Cuts in Aid by China and West Have Increased the Problem," *New York Times*, August 20, 1979, A9; Thomas S. Helling and Daniel Azoulay, "Ton That Tung's Livers," *Annals of Surgery* 259, no. 6 (June 2014): 1245–52, doi:10.1097/SLA.0000000000000370; "How Did Hanoians Fight and Win?" *Vietnam Courier*, November 6, 1967, Folder 15, Box 09, Douglas Pike Collection: Unit 02—Military Operations, TTU Vietnam Archive.

50. VWP-DRV Leadership, 1960 to 1973, Part II—The Government, July 1973, Folder 07, Box 04, United States Department of State Collection, TTU Vietnam Archive; *Vietnamese Intellectuals against U.S. Aggression* (Hanoi: Foreign Languages Publishing House, 1966); Aso and Guénel, "Itinerary of a North Vietnamese Surgeon," 292–301.

51. Quoted in Robert Dreyfuss, "Apocalypse Still," *Mother Jones*, January 1, 2000, http://www.motherjones.com.

52. Jane Fonda, *My Life So Far* (New York: Random House, 2005), 302.

53. "Hanoi Scientist Links Defoliants to Cancer," *Washington Post*, October 14, 1972, A13.

54. Aso and Guénel, "North Vietnamese Surgeon," 302; Typescript: Midwest Itinerary for Professor Ton That Tùng, and Advertisement for Special Lecture: Incidence of Primary Liver Cancer as Related to Dioxin—Agent Orange, given by Ton-That-Tung, M.D., April 18, 1979, Item No. 1288, Box 49, "Vietnam," Series III Subseries II, Young Collection, http://specialcollections.nal.usda.gov/agentorange/agent-orange-item-id-1288; Steven Smith, "Vietnamese Doctor Links Spray, Cancer," [Eugene] *Register-Guard*, April 28, 1979, 4A; and "U.S. Defoliant Is Linked to Liver Cancer in Vietnam," *New York Times*, May 6, 1979, 25.

55. Ton That Tùng, "U.S. Uses Science to Destroy Vietnamese Land, People," translated article, *Nhân Dân*, August 30, 1972, TTU Vietnam Archive.

56. Ton That Tùng, "U.S. Uses Science to Destroy Vietnamese Land, People."

57. Biochemical Warfare and Ecological Disruption in Vietnam, December 1975, Folder 19, Box 03, Douglas Pike Collection: Unit 11—Monographs, TTU Vietnam Archive.

58. *Herbicides and Defoliants in War: The Long-Term Effects on Man and Nature* (Honolulu: University Press of the Pacific, 2003), 11–20. The symposium papers were compiled by the staff of the *Vietnam Courier.*

59. *Herbicides and Defoliants,* 15.

60. Frank Egler, "Chemical Warfare in Southeast Asia," Review of *Harvest of Death,* by J. B. Neilands, G. H. Orians, E. W. Pfeiffer, Alje Vennema, and Arthur Westing, *Ecology* 53, no. 6 (1972): 1207.

Chapter Four

1. Billee Shoecraft, *Sue the Bastards!* (Phoenix: Franklin Press, 1971), 1–6; Cathy Trost, *Elements of Risk: The Chemical Industry and Its Threat to America* (New York: Times Books, 1984), 95, 96.

2. Rachel Carson, *Silent Spring* (New York: Houghton Mifflin, 2002 [Anniversary Edition]), 64.

3. Carson, *Silent Spring,* 72.

4. Shoecraft, *Sue the Bastards!,* 6.

5. Joshua Blu Buhs, *The Fire Ant Wars: Nature, Science, and Public Policy in Twentieth-Century America* (Chicago: University of Chicago Press, 2004), 68.

6. Jack Doyle, *Trespass against Us: Dow Chemical & the Toxic Century* (Monroe, ME: Common Courage Press, 2004), 68, 69. For the various DDT campaigns and history of its regulation, see Thomas Dunlap, *DDT: Scientists, Citizens, and Public Policy* (Princeton: Princeton University Press, 1981); Pete Daniel, *Toxic Drift: Pesticides and Health in the Post–World War II South* (Baton Rouge: Louisiana State Press and Smithsonian National Museum of American History, 2005); Edmund Russell, *War and Nature: Fighting Humans and Insects with Chemicals from World War I to "Silent Spring"* (New York: Cambridge University Press, 2001); Joshua Blu Buhs, *The Fire Ant Wars;* and David Kinkela, *DDT and the American Century: Global Health, Environmental Politics, and the Pesticide That Changed the World* (Chapel Hill: University of North Carolina Press, 2011). Alan Marcus examined the agricultural uses of DES in *Cancer from Beef: DES, Federal Food Regulation, and Consumer Confidence* (Baltimore: Johns Hopkins University Press, 1994).

7. Shepard Krech III, *The Ecological Indian: Myth and History* (New York: W. W. Norton, 2000), 45–72; and Karen L. Smith, *The Magnificent Experiment: Building the Salt River Reclamation Project, 1890–1917* (Tucson: University of Arizona Press, 1987), 58–59.

8. Cody Ferguson, *This Is Our Land: Grassroots Environmentalism in the Late Twentieth Century* (New Brunswick: Rutgers University Press, 2015), 75; and Karen Smith, "The Campaign for Water in Central Arizona, 1890–1903," *Arizona and the West* 23, no. 2 (Summer 1981): 128–30.

9. Ferguson, *This Is Our Land*, 71–90; Wendy Espeland, *The Struggle for Water: Politics, Rationality, and Identity in the American Southwest* (Chicago: University of Chicago Press, 1998), 183–282.

10. Aldo Leopold, "Grass, Brush, Timber, and Fire in Southern Arizona," *Journal of Forestry* 22, no. 6 (October 1, 1924): 1–10, http://www.nps.gov/seki/learn/nature /upload/leopold24.pdf; Aldo Leopold, "The Virgin Southwest," in *The River of the Mother of God: and Other Essays by Aldo Leopold*, ed. Susan L. Flader and J. Baird Callicott (Madison: University of Wisconsin Press; reprint ed. 1992); Adam Sowards, "Reclamation, Ranching, and Reservation: Environmental, Cultural, and Governmental Rivalries in Transitional Arizona," *Journal of the Southwest* 22, no. 3 (1998): 338; Adam Sowards, "Administrative Trials, Environmental Consequences, and the Use of History in Arizona's Tonto National Forest, 1926–1996," *Western Historical Quarterly* 31, no. 4 (2000): 195–98.

11. Ira Judd, "Range Reseeding Success on the Tonto National Forest, Arizona," *Journal of Range Management* 19, no. 5 (September 1966): 296, 300; Espeland, 135–82; Sowards, "Administrative Trials," 192, 199, 200–202.

12. Stephen Pyne, *The Southwest: A Fire Survey* (Tucson: University of Arizona Press, 2016), 126–28; Stephen Pyne, *Fire in America: A Cultural History of Wildland and Rural Fire* (Seattle: University of Washington Press, 1982), 95; Peter F. Ffolliott, Leonard F. DeBano, and Malchus B. Baker Jr., "Arizona Watershed Management Program," *Journal of the Arizona-Nevada Academy of Science* 35 (2003): 5–10.

13. Jay Cravens, *A Well Worn Path* (Huntington, WV: University Editions, 1994), 74 and all subsequent quotes.

14. "Forest Service Launches Chaparral Control Tests," *Tucson Daily Citizen*, February 13, 1958, 18.

15. "Forest Service Launches Chaparral Control Tests."

16. Alden R. Hibbert, Edwin A. Davis, and David G. Scholl, *Chaparral Conversion Potential in Arizona, Part I: Water Yield Response and Effects on Other Resources*, Series: USDA Forest Service Research Paper RM-126 (Fort Collins, CO: Rocky Mountain Forest and Range Experiment Station, Forest Service, US Dept. of Agriculture, 1974), 12.

17. Hibbert et al., *Chaparral Conversion*, 13.

18. Sowards, "Administrative Trials," 202–7, quote on 205.

19. Shoecraft, *Sue the Bastards!*, 17.

20. "Beauty Is Expressed through Poetry," *Arizona Republic*, September 11, 1960, 123; Doyle, *Trespass against Us*, 68; Wilbur Haak, Lynn F. Haak, and Gila County Historical Museum Archive, *Globe* (Charleston, SC: Arcadia, 2008), 21; Shoecraft, *Sue the Bastards!*, 15–18.

21. "Beauty Is Expressed through Poetry"; Wade Cavanaugh, "A House for Ice Canyon," *Arizona Republic: Home and Garden Section*, January 12, 1969, 142, 150.

22. Steven V. Roberts, "Deformities and Hemorrhaging Laid to Forest Spray in Arizona," *New York Times*, February 8, 1970, 60.

23. Shoecraft, *Sue the Bastards!*, 5, 6.

24. Quote in Trost, *Elements of Risk*, 98.

25. Trost, *Elements of Risk*, 98–100; Shoecraft, *Sue the Bastards!*, 22.

26. Shoecraft, *Sue the Bastards!*, 22, 23; Trost, *Elements of Risk*, 100.

27. Shoecraft, *Sue the Bastards!*, 20–40.

28. "Burning Suspended on Tonto Forest Lands," *Arizona Republic*, October 18, 1969, 74.

29. "Burning Suspended on Tonto Forest Lands," 74.

30. Trost, *Elements of Risk*, 105.

31. Trost, *Elements of Risk*, 106.

32. Nyla Crone, "Damage From Herbicides near Globe Called Slight," *Tucson Daily Citizen*, December 13, 1969, 8.

33. Trost, *Elements of Risk*, 104–15; Shoecraft, *Sue the Bastards!*, 75–90.

34. Wayne Binns, Typescript "Investigation of Spray Project near Globe, Arizona," 1970, 21, Young Collection, https://specialcollections.nal.usda.gov/agentorange/agent-orange-item-id-2908.

35. Range Seeding Equipment Committee, *Chemical Control of Range Weeds: Handbook* (Washington, DC: U.S. Department of Agriculture, U.S. Department of the Interior, Range Seeding Equipment Committee, rev. January 1969), iii; Binns, Typescript "Investigation of Spray Project," 19–24, 28, 29. The Tshirley report remains a "definitive" assessment of the lack of herbicide damage at Globe; the incident is cited as a case study demonstrating the phenoxy herbicides' safety in Rodney W. Bovey, *Woody Plants and Woody Plant Management: Ecology: Safety, and Environmental Impact* (New York: Marcel Dekker, 2001).

36. Richard D. McCarthy, *The Ultimate Folly: War by Pestilence, Asphyxiation and Defoliation* (New York: Alfred A. Knopf, 1969), 87.

37. "McCarthy Will Meet with Pinal Herbicide Foes," *Tucson Daily Citizen*, February 5, 1970, 19.

38. "McCarthy Will Meet with Pinal Herbicide Foes"; "Globe Spray Report Expected by Next Week," *Tucson Daily Citizen*, February 10, 1970, 6; newspapers from Connecticut, Iowa, Nebraska, Texas, and Colorado, and the *New York Times*, covered McCarthy's visit.

39. "Rep. McCarthy Opens Hearings on Herbicide," *Lincoln Evening Journal*, February 12, 1970, 17; "McCarthy Says White House Easing Restrictions on Spray," *Tucson Daily Citizen*, February 13, 1970, 27; Vince Taylor, "Herbicide Damage Suits Due," *Arizona Republic* [Phoenix], February 13, 1970, 27; "Major Congressional Action: Effects of Herbicides," *1970 CQ Almanac* 26 (February 1970): 495.

40. "Rep. McCarthy Opens Hearings on Herbicide," *Lincoln Evening Journal*, February 12, 1970, 17; "McCarthy Says White House Easing Restrictions on Spray"; Paul Dean, "Sniffing Something besides Herbicides," *Arizona Republic*, February 19, 1970, 39; "Globe's Mystery," *Time* 95, no. 8 (February 23, 1970): 42.

41. Steven V. Roberts, "Deformities and Hemorrhaging Laid to Forest Service Spray in Arizona," *New York Times*, February 8, 1970, 60; Trost, *Elements of Risk*, 148, 149.

42. Quote from "Major Congressional Action: Effects of Herbicides"; information on the Bionetics study and responses to it from Peter Sills, *Toxic War: The Story of Agent Orange* (Nashville, TN: Vanderbilt University Press, 2014), 109–20; Douglas Martin, "Thomas Whiteside, 79, Dies; Writer Exposed Agent Orange," *New York Times*, October 12, 1997, 44.

43. Thomas Whiteside, *Defoliation* (New York: Ballantine/Friends of the Earth Book, 1970), 44. Whiteside's extensive *New Yorker* series was published later that year with several appendices, including military materials, statements by Nixon and his science adviser, and information from the AAAS.

44. Whiteside, "Defoliation," *New Yorker*, February 7, 1970, 32, 33; Martin, "Thomas Whiteside, 79, Dies"; Sills, *Toxic War*, 109–20.

45. Harrison Wellford, "Statement of Harrison Wellford," Center for the Study of Responsive Law, Washington, DC, *Hearings before the Subcommittee on Energy, Natural Resources, and the Environment of the Committee on Commerce; United States Senate, Ninety-First Congress, Second Session on Effects of 2,4,5-T on Man and the Environment, April 7 and 15, 1970* (Washington, DC, 1970), 10–20; hereafter cited as Hart Hearing.

46. Dr. Ned D. Bayley, "Statement of Dr. Ned Bayley," Hart Hearing, 32–44; Trost, *Elements of Risk*, 156, 157; Dr. Julius E. Johnson, "Statement of Dr. Julius E. Johnson," Vice President and Director of Research, The Dow Chemical Co., 376, Hart Hearing.

47. Dr. Samuel Epstein, "Statement of Dr. Samuel Epstein," Professor at Harvard Medical School and cochairman of the advisory panel on teratogenicity of pesticides, April 7 and 15, 1970, 414, Hart Hearing. Emphasis in the original.

48. Johnson testimony, Hart Hearing.

49. Trost, *Elements of Risk*, 172.

50. Shoecraft, *Sue the Bastards!*, 81.

51. Shoecraft, *Sue the Bastards!*, 151.

52. "Six Families Sue for $4.5 Million in Spray Damage," *Arizona Republic*, June 2, 1970, 40.

53. "U.S. Sued in Forest Defoliation," *Tucson Daily Citizen*, June 7, 1971, 3.

54. Trost, *Elements of Risk*, 95–110.

55. Memo from Arvin L. White, Director of Fiscal and Accounting Management, Region 3, to Fiscal and Accounting Management Files and Others, April 22, 1985, Subject File, "Insects and Diseases, Pesticides and Herbicides," Forest History Society, Durham, North Carolina, hereafter FHS.

56. Shoecraft, *Sue the Bastards!*, "Foreword," xiii.

57. "Recruits Sought by Herbicide Foe," *Tucson Daily Citizen*, December 1, 1971, 13.

58. Ferguson, *Our Land*, 76–80.

59. Bill Morrell, "All Outdoors," *Arizona Daily Sun*, October 1, 1974, 8.

60. Morrell, "All Outdoors," 8.

61. Cromwell Warner, "Is Water Worth the Price?" *Arizona Daily Sun*, August 7, 1975, 1; "Interior Rules against BLM Herbicide Spraying," Sierra Club, Grand Canyon Chapter, *Canyon Echo* 20, no. 7 (September 1984), 1.

62. "Dangerous Herbicide Being Used in State, Critics Say," *San Bernardino Sun*, January 27, 1971.

Chapter Five

1. Ida Honorof, "Vigilance or Villainy," *A Report to the Consumer* 6, no. 125 (May 1976): 3. All quotes in this paragraph come from this newsletter issue.

2. Mike Davis, *Ecology of Fear: Los Angeles and the Imagination of Disaster* (New York: Vintage Books, 1998).

3. See John C. McMillian, *Smoking Typewriters: The Sixties Underground Press and the Rise of Alternative Media in America* (New York: Oxford University Press, 2011). McMillian points out that new technologies like the mimeograph empowered new voices, chapters 1 and 7.

4. Elaine Woo, "Obituary: Ida Honorof, 93; Radio Host Crusaded for the Environment," *Los Angeles Times*, March 18, 2007, https://www.latimes.com; "Health Federation to Honor Valley Radio Personality," *Valley News*, December 25, 1970, 13; "Action Line," *Pasadena Independent Topics*, April 21, 1971, 1; "Drugs in Meat Radio Topic," *San Bernardino Sun*, April 10, 1972; "Conference on Nutrition and Ecology," *Los Angeles Times*, January 19, 1973, part II, 4. Alan Marcus questions Honorof's identification as a "consumer advocate" in *Cancer from Beef: DES, Federal Food Regulation, and Consumer Confidence* (Baltimore: Johns Hopkins University Press, 1994), 112. Nancy Langston examines the history of DES in *Toxic Bodies: Hormone Disruptors and the Legacy of DES* (New Haven: Yale University Press, 2010).

5. M. Kat Anderson, Michael G. Barbour, and Valerie Whitworth, "A World of Balance and Plenty: Land, Plants, Animals, and Humans in a Pre-European California," *California History* 76, no. 2/3 (July 1, 1997): 19.

6. Anderson et al., "World of Balance," 19.

7. Anderson et al., "World of Balance," 20.

8. Davis, *Ecology of Fear*, 11.

9. Stephen Pyne, *California: A Fire Survey* (Tucson: University of Arizona Press, 2016), 7–8.

10. Thomas Sanchez and Lawrence Clark Powell, *Angels Burning: Native Notes from the Land of Earthquake and Fire* (Santa Barbara, CA: Capra Press, 1987), 22, 23. All subsequent quotes in this paragraph come from pages 23–30.

11. Sanchez and Powell, *Angels Burning*, 23.

12. Sanchez and Powell, *Angels Burning*, 27.

13. Sanchez and Powell, *Angels Burning*, 27.

14. Sanchez and Powell, *Angels Burning*, 29.

15. Davis, *Ecology of Fear*, 75.

16. Pyne, *Fire in America*, 388.

17. For the history of napalm use in WWII, see Robert M. Neer, *Napalm: An American Biography* (Cambridge, MA: Belknap Press, 2013), 62–86; David Carle, *Burning Questions: America's Fight with Nature's Fire* (Westport, CT: Praeger, 2002), 71–73; Pyne, *Fire in America*, 394–400.

18. Davis, *Ecology of Fear*, 98.

19. Bob Geggie, "Training Fire Fighters on Career Basis Advocated by Veteran Forest Official," *San Bernardino Sun*, January 4, 1957, 17.

20. Quoted in Ida Honorof, "Herbicides," *Report to the Consumer* 1, no. 5 (March 1971): 8.

21. Ida Honorof, "Defoliation: Los Angeles County," *Report to the Consumer* 6, unknown no. and date, 1.

22. Honorof, "Defoliation," 3, emphasis in the original.

23. Honorof, "Defoliation," 4.

24. Herbicide Hearing Committee, "Report on the Use and Control of Simazin [*sic*] and Other Herbicides by the County of Los Angeles," March 7, 1972, County of Los Angeles Health Department, http://www.cdpr.ca.gov.

25. Steven Stoll, *The Fruits of Natural Advantage: Making the Industrial Countryside in California* (Berkeley: University of California Press, 1998), 4.

26. Stoll, *Fruits of Natural Advantage*, 28–30.

27. Carey McWilliams, "The Farmers Get Tough," *American Mercury* 33 (October 1934): 245, http://www.unz.org.

28. Kathryn S. Olmsted, *Right out of California: The 1930s and the Big Business Roots of Modern Conservatism* (New York: New Press, 2015), 108; Linda Nash, *Inescapable Ecologies: A History of Environment, Disease, and Knowledge* (Berkeley: University of California Press, 2006), 82–87, 130.

29. Stoll, *Fruits of Natural Advantage*, 107.

30. David Vaught, *Cultivating California: Growers, Specialty Crops, and Labor, 1875–1920* (Baltimore: Johns Hopkins University Press, 1999), 49; Stoll, *Fruits of Natural Advantage*, 107–15.

31. Chester A. Perry, Cyrus M McKell, Joe R. Goodin, and Thomas M. Litter, "Chemical Control of an Old Stand of Chaparral to Increase Range Productivity," *Journal of Range Management* 20, no. 3 (1967): 166.

32. Cyrus M. McKell, J. R. Goodin, and Cameron Duncan, "Chaparral Manipulation Affects Soil Moisture Depletion Patterns and Seedling Establishment," *Journal of Range Management* 22, no. 3 (May 1969): 159.

33. Perry et al., 166, 168, 169; McKell et al., "Chaparral Manipulation," 160–64.

34. Stoll, *Fruits of Natural Advantage*, 116.

35. Nash, *Inescapable Ecologies*, 132–35; "State Clamps Tighter Rule on Chemical Weed Controls," *Bakersfield Californian*, April 1, 1950, 7; Herbicide Hearing Committee, "Report."

36. Robert P. Laurence, "'Birth Defect' Defoliant Sold in California," *Independent Press-Telegram*, February 14, 1970, 15.

37. Honorof, "Herbicides," 1.

38. "State Adopts Strict Procedures for Spraying Herbicides," *Red Bluff Daily News*, February 20, 1970, 7; "Spray Rules for Copters," [Eureka] *Times Standard*, February 21, 1970, 10.

39. "Consumer Advocates Claim Viet Herbicides Used in U.S.," *Independent*, January 27, 1971, 6. All quotes in this paragraph appear in the article.

40. "Ida Honorof Set to Speak on Food Pollution," *Van Nuys News*, May 23, 1971, 36; "Pollution Film," *Independent Press-Telegraph*, September 18, 1971, 8.

41. "Chavez Requests Probe of Lettuce," *Arizona Daily Sun*, February 26, 1973, 1; "Chavez Calls for Congressional Probe into Lettuce Contamination," *Reno* [Nevada] *Evening Gazette*, February 26, 1973, 12; Jay Perkins, "Probe of Lettuce Pesticide Asked," *San Bernardino Sun*, February 27, 1973, 4; Randy Shaw, *Beyond the Fields: Cesar Chavez, the UFW, and the Struggle for Justice in the 21st Century* (Berkeley: University of California Press, 2008), 134.

42. Ida Honorof, "Lethal Lettuce: U.S. Head Lettuce Halted at the Canadian Borders," *Report to the Consumer* 3, no. 48 no. 3/3 (1973): 3.

43. Honorof, "Lethal Lettuce," emphasis in the original.

44. Elaine Woo, "Obituary: Ida Honorof, 93; Radio Host Crusaded for the Environment," *Los Angeles Times*, March 18, 2007, https://www.latimes.com.

45. Laura Pulido, *Environmentalism and Economic Justice: Two Chicano Struggles in the Southwest* (Tucson: University of Arizona Press, 1996); Nash, *Inescapable Ecologies*; Adam Tompkins, *Ghostworkers and Greens: The Cooperative Campaigns of Farmworkers and Environmentalists for Pesticide Reform* (Ithaca: IRL Press-Cornell University Press, 2016). Chapter 9 discusses the UFW and its campaign against pesticides in greater detail.

46. "Shouts Punctuate Hearing on Effects of Pesticide," *Van Nuys News*, March 11, 1973, 7.

47. Mark Oberle, "Forest Fire Suppression Policy Has Its Ecological Drawbacks," *Science*, n.s., 165, no. 3893 (August 8, 1969): 571.

48. Lawrence W. Hill and Raymond M. Rice, "Converting from Brush to Grass Increases Water Yield in Southern California," *Journal of Range Management* 16, no. 6 (November 1963): 300, 305.

49. August L. Hormay, Fred J. Alberico, and P. B. Lord, "Experience with 2,4-D Spraying on the Lassen National Forest," *Journal of Range Management* 15, no. 6 (November 1962): 325.

50. Hormay et al., "Experiences with 2,4-D Spraying," 325–28; Edward C. Stone, "Preserving Vegetation in Parks and Wilderness," *Science*, n.s., 150, no. 3701 (December 3, 1965): 1262–67; Perry et al., 166–69.

51. Perry et al., "Chemical Control of an Old Stand of Chaparral," 169; McKell,

Goodin, and Duncan, "Chaparral Manipulation Affects Soil Moisture Depletion Patterns and Seedling Establishment," 160–64; Harold H. Biswell, "Water Control by Rangeland Management," *Journal of Range Management* 22, no. 4 (July 1969): 229; Oberle, "Forest Fires," 572, 573; "Forest Service Planning Controlled Burn Tuesday," *Redlands Daily Facts*, April 20, 1970, 4; "Brushland Conversion Subject of Workshop at Cahto Peak," *Ukiah Daily Journal*, June 1, 1972, 7.

52. Mrs. Teddy Ann Wilson, "Letter to the Editor," *County Sun*, May 22, 1971, 23.

53. Burton N. Brin, "Letter to the Editor," *County Sun*, May 22, 1971, 23.

54. "Dangerous Herbicide Being Used in State, Critics Say," *County Sun*, January 27, 1971, 3; "Forest Will Keep Spray," *Times Standard*, June 25, 1971, 13.

55. Everett R. Holles, "Use of Defoliant as Fire Guard Called Off," *County Sun*, July 12, 1971, A-2.

56. "Herbicide Spraying Set," *Sentinel*, May 9, 1973, 4.

57. Molly Burrell, "Group Sues over Herbicide Toxins in National Forests," *Independent*, July 30, 1973, 3.

58. Ruth Croft, "Young Ecologists Will Challenge Herbicide Use in National Forests: Suit Seeks to Restrain U.S. Chemical Spraying," *Van Nuys News*, August 12, 1973, 1.

59. Ida Honorof, "Vigilance or Villainy," 1; Burrell, "Group Sues"; Croft, "Young Ecologists."

60. Ruth Croft, "Two Scientists Will Testify against Use of Herbicides," *Van Nuys News*, August 26, 1973, 1.

61. Croft, "Two Scientists," 1.

62. "Ask Ban on Herbicide in Forests," *Van Nuys News*, August 26, 1973, 1; Ruth Croft, "Ex-Valley Assemblyman Defends Herbicide Use," *Van Nuys News*, September 2, 1973, 6.

63. "Reader Hits Conrad over Statements," Ida Honorof, Letter to the Editor, *News*, September 14, 1973, 2; "Herbicide Issue Explained by Reader," Charles Conrad, Letter to the Editor, *News*, September, 16 1973, 1; Ida Honorof, "Genocide—USDA Forest Service Style," *Report to the Consumer* 3, no. 56 (7/3): 1–4.

64. "Group Seeks Pesticide Ban," *Van Nuys News*, September 25, 1973, 5; "Cranston Aide to Attend Rally against Pesticides," *Van Nuys News*, September 28, 1973, 3.

65. Ruth Croft, "Schedule 2nd Rally to Protest Pesticides," *Van Nuys News*, September 27, 1973, 1; "Cranston Aide," 3.

66. Ida Honorof, "A Moment's Reprieve from Death!" *Report to the Consumer* 6, no. 136 (October 1976): 3.

67. Honorof, "Moment's Reprieve," 2–4.

Chapter Six

1. Carol Van Strum, *A Bitter Fog: Herbicides and Human Rights* (San Francisco: Sierra Book Club, 1983), 1.

2. Van Strum, *Bitter Fog*, 1–5.

3. Linda Lear, "Bombshell in Beltsville: The USDA and the Challenge of 'Silent Spring,'" in special issue "History of Agriculture and the Environment," *Agricultural History* 66, no. 2, *History of Agriculture and the Environment* (Spring 1992): 151–56; Emily Brock, *Money Trees: The Douglas Fir and American Forestry* (Corvallis: Oregon State University Press, 2015), 16; Joshua Blu Buhs, *The Fire Ant Wars: Nature, Science, and Public Policy in 20th-Century America* (Chicago: University of Chicago Press, 2004), 9–39; and Pete Daniel, *Toxic Drift: Pesticides and Health in the Post–World War II South* (Baton Rouge: Louisiana State University Press, 2005), 5.

4. For example, "To Fight Fires: Forest Service to Make Effort to Stop Destroying Agent," *Weekly Oregon Statesman*, May 12, 1905.

5. Nancy Langston, *Forest Dreams, Forest Nightmares: The Paradox of Old Growth in the Inland West* (Seattle: University of Washington Press, 1996), 86–89; William G. Robbins, *Lumberjacks and Legislators: Political Economy of the U.S. Lumber Industry, 1890–1941* (College Station: Texas A&M Press, 1982), 8, 9.

6. William Robbins, *Lumberjacks and Legislators*, 14.

7. Brock, *Money Trees*; Langston, *Forest Dreams*, 86, 87; William G. Robbins, *Landscapes of Conflict: The Oregon Story, 1940–2000* (Seattle: University of Washington Press, 2004), 80–82.

8. Langston, *Forest Dreams*, 267.

9. Ronald McCormick, *Plain Green Wrapper: A Forester's Story* (LaVergne, TN: Ingram, 2009), 81.

10. McCormick, *Plain Green Wrapper*, 78–81; Dennis Tavares, *How Mendocino County Went to Pot: Memories of Life in Mendocino Redwood Country in the Last Half of the 1900s*, ch. "The 1960s in the Forest Industry: Union Lumber's Empire" (n.p.: Trafford, 2014).

11. Beverly Brown, *In Timber Country: Working People's Stories of Environmental Conflict and Urban Flight* (Philadelphia: Temple University Press, 1995), 24.

12. William G. Robbins, "The Social Context of Forestry: The Pacific Northwest in the Twentieth Century," *Western Historical Quarterly* 16, no. 4 (October 1985): 414, 415, 423; Stephen Pyne, "Fire Policy and Fire Research in the U.S. Forest Service," *Journal of Forest History* 25, no. 2 (April 1981): 74; Brown, *In Timber Country*, 24; Robbins, *Landscapes of Conflict*, 114–24, 130, 169, 170.

13. "Spraying of Herbicide Helps Fir," [Eureka] *Humboldt Standard*, April 10, 1965, 53.

14. McCormick, *Plain Green Wrapper*, 78–82; Tavares, *How Mendocino County*, 77; Robbins, *Landscapes of Conflict*, 115–35; Langston, *Forest Dreams*, 253.

15. Brock, *Money Trees*, 132, 133; Langston, *Forest Dreams*, 110–14; Darren Speece, *Defending Giants: The Redwood Wars and the Transformation of American Environmental Politics* (Seattle: University of Washington Press, 2017), 90.

16. Joe Mosley, "Oregon Woman Wages War against 'Bitter Fog,'" *Seattle Times*, July 7, 1991.

17. William G. Robbins, "Cornucopian Dreams: Remaking Nature in Postwar Oregon," *Agricultural History* 76, no. 2 (Spring 2002): 215; James Kopp, *Eden within Eden: Oregon's Utopian Heritage* (Corvallis: Oregon State University Press, 2009), chs. 6 and 7, for a history of Oregon's twentieth-century utopian communal experiments; Van Strum, *Bitter Fog*, 24.

18. Van Strum, *Bitter Fog*, 1–12; and Thomas Whiteside, *The Withering Rain: America's Herbicidal Folly* (New York: Dutton, 1971).

19. Michael Newton and L. T. Burcham, "Defoliation Effects on Forest Ecology," Letters to the Editor, *Science*, n.s., 161, no. 3837 (July 12, 1968): 109; Michael Newton and Logan A. Norris, and Arthur W. Galston, "Herbicide Usage," Letters to the Editor, *Science*, n.s. 168, no. 3939 (June 26, 1970): 1606; Van Strum, *Bitter Fog*, 20–26. Newton had extensive involvement with the phenoxy herbicides, having visited Vietnam as a member of the National Academy of Science report team. He is also the scientist Ida Honorof thought should be thrown in jail.

20. Van Strum, *Bitter Fog*, 24.

21. Van Strum, *Bitter Fog*, 80.

22. Van Strum, *Bitter Fog*, 82.

23. Van Strum, *Bitter Fog*, 82.

24. Van Strum, *Bitter Fog*, 86.

25. Sylvia Noble Tesh, *Uncertain Hazards: Environmental Activists and Scientific Proof* (Ithaca: Cornell University Press, 2000).

26. Van Strum, *Bitter Fog*, 92; Robbins, *Landscapes of Conflict*, 197.

27. Van Strum, *Bitter Fog*, 96.

28. Van Strum, *Bitter Fog*, 95–100; Ida Honorof, "A Moment's Reprieve from Death!" *A Report to the Consumer* 6 no. 136 (October 1976): 3. Tshirley was heavily involved in evaluating the effects of the herbicides in the Globe episode.

29. Ida Honorof, "EPA Must Protect the Public's Health and the Environment!" *A Report to the Consumer* 8, no. 145 (March 1977): 1.

30. "'Herbicides' Topic of Forum Tonight," *Ukiah Daily Journal*, January 14, 1977, 9; "Herbicide Foes Plan Meeting," *Times Standard*, February 2, 1977, 2; "Herbicide Foes Argue for Alternative," *Times Standard*, February 10, 1977, 5; "Milk Samples Contain Spray," *Daily Chronicle*, February 18, 1977, 4; "Poison Found in Texas Tests of Breast Milk," *San Antonio Express*, April 9, 1977, 17. Jessica Martucci discusses the La Leche League's role in identifying toxic contamination in breast milk across the country. in *Back to the Breast: Natural Motherhood and Breastfeeding in America* (Chicago: University of Chicago Press, 2015), 144–57.

31. Van Strum, *Bitter Fog*, 105, 106.

32. Michael Newton, "Testimony on House Bill 3230 for the House State Government Operations Committee," May 31, 1977, Subject Files, "Insects and Diseases: Pesticides and Herbicides," Forest History Society, Durham, North Carolina; hereafter, FHS.

33. Newton and Burcham, "Defoliation Effects," 3.

34. Newton and Burcham, "Defoliation Effects," 3.

35. Mosley, "Oregon Woman Wages War."

36. Van Strum, *Bitter Fog*, 219–25; Robbins, *Landscapes of Conflict*, 197–99.

37. Honorof, "'Agent Orange' Is Alive and Thriving in the U.S.A.," *Report to the Consumer* 7, no. 150 (May 1977): 3.

38. R. J. Williams, "Attitudes toward the War, The Use of Chemical Weapons, and the Dow Chemical Company," March 1968, 4, Folder 00103, Box 1 Napalm, Dow Chemical Company, Dow Collection, Chemical Heritage Foundation, Philadelphia, Pennsylvania; hereafter cited as Dow Collection; Peter Sills, *Toxic War: The Story of Agent Orange* (Nashville: Vanderbilt University Press, 2014), 109–12; Dr. Julius E. Johnson, "Statement of Dr. Julius E. Johnson," Vice President and Director of Research, The Dow Chemical Co., 376, Hart Hearing.

39. "State Says Fish May Be Tainted," *News Palladium*, June 30, 1978, 19; R. Jeffrey Smith, "Dioxins Have Been Present since the Advent of Fire, Says Dow," *Science*, n.s., 202, no. 4373 (December 15, 1978): 1166.

40. R. Jeffrey Smith, "Dioxins," 1166.

41. R. Jeffrey Smith, "EPA Halts Most Use of Herbicide 2,4,5-T," *Science*, n.s., 203, no. 4385 (March 16, 1979): 166.

42. The Chlorinated Dioxin Task Force, the Michigan Division, Dow Chemical U.S.A. "Typescript: The Trace Chemistries of Fire—A Source of and Routes for the Entry of Chlorinated Dioxins into the Environment 1978" (Midland: Dow Chemical Company, 1978), 2–5, Box 120, Folder 3552, Young Collection.

43. Warren B. Crummett, *Decades of Dioxin: Limelight on a Molecule* (Philadelphia: Xlibris, 2002), 223.

44. There are several different stories about when Hill became conscious of the potential link between phenoxy herbicide exposure and miscarriages. Tesh suggests she became aware when she found CATS literature. Another 1979 source dates Hill's awareness as early as 1975, just a few months after her miscarriage. In a 1981 *Mother Earth News* interview, Hill herself identifies her interest as happening in 1977 when she found Allen's research while attending classes at the University of Oregon. In addition to Tesh, see Roy Murphy, "Chemicals Can Kill," originally published in *Northwestern Magazine*, Portland, Oregon, 1979, and now available on Roy Murphy, Phoenix Story Productions, http://www.murphyroy.com/adventure/chemicals.html.

45. Sara Pacher, "Bonnie Hill: Oregon Environmental Activist," the Plowboy Interview, *Mother Earth News*, November/December 1981, http://www.motherearthnews.com.

46. Bonaccorsi, Aurora, Roberto Fanelli, and Gianni Tognoni, "In the Wake of Seveso," *Ambio* 7, no. 5/6 (1978): 234–39; Honorof, "Agent Orange Is Alive and Thriving," 2.

47. Tesh, "Uncertain Hazards," 15, 16; Van Strum, *Bitter Fog*, 161–68; Linda Layne, "In Search of Community: Tales of Pregnancy Loss in Three Toxically Assaulted U.S. Communities," in special issue, "Earthwork: Women and Environments," *Women's Studies*

Quarterly 29, no. 1/2 (Spring–Summer 2001): 35–40; Robert D. McFadden, "E.P.A., Citing Miscarriages, Restricts 2 Herbicides," *New York Times*, March 2, 1979, A10.

48. Layne, "Community," 35–40.

49. Smith, "EPA Halts," 1090–92; "Dow Attacks Study Used to Ban 2,4,5-T," *Science News* 115, no. 11 (March 17, 1979): 166; Van Strum, *Bitter Fog*, 163.

50. E. N. Brandt, *Growth Company: Dow Chemical's First Century* (East Lansing: Michigan State University Press, 1997), 367. Brandt notes that this was the "first and only time Dow used [this] argument."

51. The Chlorinated Dioxin Task Force, the Michigan Division, "Trace Chemistries of Fire," Young Collection; Crummett, *Decades of Dioxin*, 187.

52. Robbins, *Landscapes of Conflict*, 200–203.

53. Robin Marantz Henig, "Agriculture's Strange Bedfellows: CAST-Industry Tie Raises Credibility Concerns," *BioScience* 29, no. 1 (January 1979): 9. The May 1979 issue contained several letters to the editor regarding the CAST article, among them one from John E. Donalds, Agricultural Products Department, Dow Chemical U.S.A. Donalds questioned whether "highly respected and dedicated scientists" could be bought for $5,000 annual dues. "CAST Profile Evokes Avid Response," *BioScience* 29, no. 5 (May 1979): 279. CAST issued a report in 1975 questioning concerns about the defoliation program in South Vietnam.

54. Letter from John Davidson, Dow Technical Advisor, to Western Timber Association, June 15, 1978, 1 Insect and Diseases, FHS.

55. Davidson letter, FHS.

56. Bob Pfohman, "Herbicide Study: Livelihoods in State of Limbo," *Statesman Journal*, September 4, 1977, 44.

57. Letter from Erik Jansson, Friends of the Earth, to Stuart Eizenstat, Domestic Advisor to the President, August 9, 1978, FHS; Press Release, "Additional Criteria Announced for 2,4,5-T Use in National Forests," August 15, 1978, *Forest Service News* Western Timber Association, Insect and Diseases, FHS; Larry Kramer, "8 Oregon Women Seek Link of Herbicide to Miscarriages," *Washington Post*, August 15, 1978, https://www.washingtonpost.com; "Herbicide Spray Probed in Miscarriages," Energy, Environment, *News in Brief*, August 15, 1978, Insect and Diseases, FHS; Larry Kramer, "Herbicides under Study as Cause of Miscarriages," August 20, 1978, 44, *The Pantagraph* (and reprinted story with new headline; "Dangerous Herbicide Sprayed across Md.," August 18, 1978, 26); "Oregon Women Suggest Link between Herbicide and High Incidence of Miscarriages in Alsea," August 21, 1978, *NBC News*, NBCUniversal Archives, www.nbcuniversalarchives.com.

58. Letter from James Plumb, Vice President Communications, American Forest Institute, to Richard G. Reid, Information Forester for Western Timber Association, August 28, 1978, Insect and Diseases, FHS.

59. Jane See White, "'Agent Orange' Still Used in U.S.," *Paris News*, October 22, 1978, 1.

60. The White articles appeared in the October 22, 1978, edition of the *Paris News*, and with different headlines in papers across the country, including: "Chemical Banned in Vietnam Is Still Being Used in U.S.," *Kokomo Tribune* [Indiana], 7; "Agent Orange's 2,4,5-T: Safe Herbicide or Crippling Spray?" *Southern Illinoisan* [Carbondale], 24; and "Agent Orange Is Getting Closer to Home," *Sentinel* [Santa Cruz, California], 17.

61. AIM, "A Bad Case of Bumblefoot," *AIM Report* 7, no. 20 (October 1978, Part 2): 1, Insect and Diseases, FHS.

62. AIM, "Bad Case of Bumblefoot," 2.

63. AIM, "Bad Case of Bumblefoot," 3.

64. "Dow Attacks Study Used to Ban 2,4,5-T," *Science News* 115, no. 11 (March 17, 1979): 166.

65. Jay Heinrichs, "'T' on Trial, Part 2," *American Forester*, March 1979, 19, Insect and Diseases, FHS.

66. Heinrichs, "'T' on Trial, Part 2," 55.

67. George A. Craig [Executive Vice President, Western Timber Association], "EPA Defends 2,4,5-T Ban," Timber Thoughts, *Wood Review*, April 6, 1979 [March 29, 1979], Insect and Diseases, FHS.

68. "'T' Memo," August 31, 1979, Western Timber Association Files, FHS.

69. Gary Jones, "Dow News Release," April 23, 1979, and Notice of the Dow Chemical Company's Intent to Proceed Immediately with Cancellation Hearings and the Withdraw from Suspension Hearings in re: Emergency Suspension Orders for 2,4,5-T and Silvex, FIFRA Docket Nos. 409, 410, April 23, 1979, Young Collection.

70. Jones, "Dow News Release"; Fred Wilcox, *Waiting for an Army to Die: The Tragedy of Agent Orange* (Cabin John, MD: Seven Locks Press, 1989), 160–70.

71. There is a vast literature on responses to *Silent Spring*. See Michael B. Smith, "'Silence, Miss Carson!' Science, Gender, and the Reception of 'Silent Spring,'" *Feminist Studies* 27, no. 3 (Autumn 2001); Linda Lear, *Rachel Carson: Witness for Nature* (New York: Penguin, 2009 [1997]), 430–40; H. Patricia Hynes, *The Recurring Silent Spring* (New York: Pergamon Press, 1989); Maril Hazlett, "'Woman vs. Man vs. Bugs': Gender and Popular Ecology in Early Reactions to Silent Spring," *Environmental History* 9, no. 4 (2004): 701–29.

72. United States Environmental Protection Agency before the Administrator, in re: The Dow Chemical Company et al., FIFRA Docket Nos. 415 et al., Respondent's Comments on Recent Dow Memoranda in Support of Compulsory Document Discovery against Dr. Allen, January 30, 1980, Young Collection.

73. In the United States District Court for the Western District of Wisconsin, *United States of America v. James R. Allen, Defendant*, Information No. 790R71, 18 USCC 641, with transmittal slip from M. Breinholt, U.S. Department of Agriculture (USDA) to Alvin L. Young, dated February 21, 1980, Young Collection.

74. Memo with attachments (newsletter, newspaper clippings) from John E. Benneth, American Forest Institute, to Larry Burke, Al Wilson, Mike Sullivan, Ennis

Cooper, Tony McCann, Dick Reid, Western Timber Association, February 14, 1980, FHS. The attachments included a CATS newsletter discussing Allen's situation, and two newspaper clippings detailing the Allen fraud case.

75. Letter from Forest Chemicals Communication Task Force Group to Forest Industry Chemicals Committee, January 8, 1980, Subject: Press Stories on Forest Chemical issues, FHS; newspaper clipping, "Swan-Area Miscarriages Prompt Suit," *Missoulian*, February 16, 1980, 11, FHS.

76. Kass Green, *Forests, Herbicides and People* (New York: Council on Economic Priorities, 1982), 14.

77. Green, *Forests*, 114–50.

78. "Private Dioxin Study Unleashes Public Storm," *County Sun*, June 3, 1983, 1; and "Dioxin Documents: Court Papers Show Companies Discussed Defoliant's Dangers," *Daily News*, July 6, 1983, 28.

79. "Dow's Bad Chemistry," *Newsweek*, July 18, 1983, FHS; Press release, "Dow Chemical Withdraws from 2,4,5-T Business in the U.S.; Exits EPA Policy Proceedings on the Herbicide," October 14, 1983, Young Collection.

Chapter Seven

1. Carol Van Strum, "Melyce," *Synthesis/Regeneration* 7–8, Summer 1995, http://www.greens.org/s-r/078/07-22.html.

2. Quoted in Linda Layne, "In Search of Community: Tales of Pregnancy Loss in Three Toxically Assaulted U.S. Communities," in special issue "Earthwork: Women and Environments," *Women's Studies Quarterly* 29, no. 1/2 (Spring–Summer 2001): 39.

3. E. N. Brandt, *Growth Company: Dow Chemical's First Century* (East Lansing: Michigan State University Press, 1997); "The Flower Children" is the title of chapter 13.

4. Quoted in Gerard J. De Groot, "Ronald Reagan and Student Unrest in California, 1966–1970," *Pacific Historical Review* 65, no. 1 (February 1996), 107.

5. De Groot, "Ronald Reagan," 107.

6. Dan Baum, "Legalize It All: How to Win the War on Drugs," *Harper's Magazine*, March 22, 2016, https://harpers.org. Former Nixon aides questioned the veracity of the Ehrlichman quote when interviewed, although they focused on the charges that the administration's initiative was racially motivated. See Hilary Hanson, "Nixon Aides Suggest Colleague Was Kidding about Drug War Being Designed to Target Black People," *Huffington Post*, http://www.huffingtonpost.com

7. Michelle Alexander, "The War on Drugs and the New Jim Crow," *Race, Poverty & the Environment* 17, no. 1 (Spring 2010), 75–77.

8. For works charting these societal transformations, see Peter Braunstein and Michael William Doyle, eds., *Imagine Nation: The American Counterculture of the 1960's and 70's* (New York: Routledge, 2001); Beth Bailey, *Sex in the Heartland* (Cambridge: Harvard University Press, 1999); Bruce Schulman, *The Seventies: The Great Shift in American Culture, Society, and Politics* (New York: Free Press, 2001); Beth Bailey and

David Farber, eds., *America in the Seventies* (Lawrence: University of Kansas Press, 2004).

9. Quoted in David Musto, "The History of the Marihuana Tax Act of 1937," Schaffer Library of Drug Policy, https://druglibrary.net. The article first appeared in the *Archives of General Psychiatry* 26 (February 1972).

10. Musto, "The History of the Marihuana Tax Act"; Isaac Campos, *Home Grown: Marijuana and the Origins of Mexico's War on Drugs* (Chapel Hill: University of North Carolina Press, 2012), chs. 7 and 8 (155–202).

11. Kathleen Frydl, *The Drug Wars in America, 1940–1973* (Cambridge: Cambridge University Press, 2013), 125; Daniel Weimer, *Seeing Drugs: Modernization, Counterinsurgency, and U.S. Narcotics Control in the Third World, 1969–1976* (Kent: Kent State University Press, 2011), 39; Matthew R. Pembleton, "The Voice of the Bureau: How Frederic Sondern and the Bureau of Narcotics Crafted a Drug War and Shaped Popular Understanding of Drugs, Addiction, and Organized Crime in the 1950s," *Journal of American Culture* 38, no. 2 (June 2015), 113–16; and John C. McWilliams, "Unsung Partner against Crime: Harry J. Anslinger and the Federal Bureau of Narcotics, 1930–1962," *Pennsylvania Magazine of History and Biography* 113, no. 2 (April 1989), 210.

12. Frank Browning and Banning Garrett, "The New Opium War," *Ramparts*, May 1971, 32, https://www.unz.org.

13. Erika Dyck discusses LSD's journey from military testing to campus recreational drug in *Psychedelic Psychiatry: LSD from Clinic to Campus* (Baltimore: Johns Hopkins University Press, 2008).

14. Doug Rossinow, "The New Left in the Counterculture: Hypotheses and Evidence," *Radical History Review*, no. 67 (Winter 1997), 80.

15. Joshua Clark Davis, *From Head Shops to Whole Foods: The Rise and Fall of Activist Entrepreneurs* (New York: Columbia University Press, 2017), 82–87.

16. Kevin Stewart, *Tales of the Emerald Triangle: Memoirs of a Marijuana Grower* (self-pub., 2002), chs. 1–9.

17. Emily Brady, *Humboldt: Life on America's Marijuana Frontier* (New York: Grand Central, 2013), 58–73.

18. "Nixon Asks for Funds to War on Drug Pushers," *Lowell Sun*, June 17, 1971, 57.

19. Jeremy Kuzmarov, *The Myth of the Addicted Army: Vietnam and the Modern War on Drugs* (Amherst: University of Massachusetts Press, 2009), 78–91.

20. Editorial, "All-Out War Needed to Fight Drug Abuse," *Oneonta Star*, June 4, 1971, 4.

21. Editorial, "All-Out War."

22. James Reston, "Want to Duck Combat? Get Hooked on Drugs," *Oneonta Star*, June 4, 1971, 4.

23. Reston, "Want to Duck Combat?" 4.

24. "Nixon Should Keep Open Mind on Taut Issue of Drug Usage," *Salt Lake Tribune*, June 7, 1971, 28.

25. "Nixon Should Keep Open Mind," 28.

26. "U.S. Wages Big Border War to Halt Marijuana Smuggling," *Sentinel*, September 14, 1969, 2.

27. "Our 'Grass' Shortage," *The Argus*, September 23, 1969, 8.

28. "Operation Intercept Nets Inconveniences," *The Times*, September 29, 1969, 3; "Marijuana Hunt Hurts Innocents: Intercept to Continue Indefinitely," *Progress Bulletin*, 33.

29. Quote in Frydl, 380; Richard R. Craig, "La campaña permanente: Mexico's Antidrug Campaign," *Journal of Interamerican Studies and World Affairs* 20, no. 2 (May 1978): 109.

30. "Here's a Backward Glance at Passing 1969," *Chula Vista Star-News*, January 1, 1970, A6; "U.S. Attack Aims at Cutting Drug Traffic: Planes Patrol Mexican Border," *Progress Bulletin*, September 14, 1969, 1.

31. Craig, "La campaña permanente," 115–20; Report of the Comptroller General of the United States, "Opium Eradication Efforts in Mexico: Cautious Optimism Advised," February 18, 1977, 21, https://www.archives.gov; Weimer, *Seeing Drugs*, 187.

32. "Battle against Heroin, Marijuana," *Sun*, March 10, 1970, 2; Craig, "La campaña permanente," 109; Weimer, *Seeing Drugs*, 178–81; Isaac Campos, *Home Grown*, 225–31; "'Operation Cooperation': U.S., Mexico Join Forces in the War on Drugs," *Register-Guard*, June 29, 1970, 4A.

33. Weimer, *Seeing Drugs*, 181.

34. Craig, "La campaña permanente," 128.

35. Richard Craig, "Operation Condor: Mexico's Antidrug Campaign Enters a New Era," *Journal of Interamerican Studies and World Affairs* 22, no. 3 (August 1980), 349.

36. Weimer, *Seeing Drugs*, 193, 196.

37. Ida Honorof, "Halt Home-Front Chemical Warfare," *Report to the Consumer* 4, no. 128 (June 1976), 1. All subsequent quotes in this paragraph are from the newsletter unless otherwise noted.

38. Honorof, "Halt Home-Front Chemical Warfare," 1.

39. From *Newsworks*, March 1, 1976, quoted in Ida Honorof, "The Chemical Companies: Public Enemy Number One!" *Report to the Consumer* 6, no. 121 (March 1976): 4.

40. Honorof, "Halt Home-Front Chemical Warfare," 1.

41. Craig, "Operation Condor," 351.

42. Philip J. Landrigan, MD, MSc, Kenneth E. Powell, MD, MPH, Levy M. James, and Philip R. Taylor, MD, "Paraquat and Marijuana: Epidemiologic Risk Assessment," *American Journal of Public Health* 73, no. 7 (1983), 784–88; L. Garmon, "Pot-Smokers May Be Imperiled by Paraquat-Spraying Program," *Science News* 124, no. 4 (July 23, 1983): 55; "Paraquat Contamination of Marijuana—United States," *Morbidity and Mortality Weekly Report* 28, no. 8 (March 2, 1979): 93–94.

43. Brady, *Humboldt*, 71.

44. Brady, *Humboldt*, 69–72; Bruce Anderson, *The Mendocino Papers, Volume I*

(Mendocino, CA: Bruce Anderson, 2008), 159; Samantha Kaye, "Crop Report: Timber Down, Ag Crops Up," *Daily Journal*, April 27, 1980, 1; Evelyn Simpson, "Marijuana Not a Crop, Barbero Tells Board," July 9, 1980, *Daily Journal*, 1; "Pot Crop Flap: Eriksen [*sic*] Says He Will Leave Job," *Daily Journal*, February 24, 1981, 1.

45. John Diehl, "Forest Herbicide Use Controversy Boils Over," *Times Standard*, October 6, 1976, 3.

46. Carol Pogash, "A Hard Look / Mendocino's Fierce Debate over Herbicide Safety," *San Francisco Examiner*, April 12, 1979, 22.

47. T. M. Lucas, "Letters to the Editor," *Daily Journal*, October 24, 1974, 19.

48. News Center 4, "Initiative to Ban Spraying in Mendocino County," interview of Betty Lou Whaley by Thayer Walker, April 24, 1979.

49. News Center 4, "Initiative."

50. Dow Chemical Company, ". . . Let's Consider Our Critics for a Moment," *Bottom Line*, March 1979, 4.

51. Quote taken from James Harris, "Newcomers vs. Oldtimers . . . Battle over Herbicide on the Ballot," *Sacramento Bee*, May 28, 1979, Forest Industry news clippings, Forest History Society; hereafter, FHS.

52. Quoted in Harris, "Newcomers vs. Oldtimers."

53. Harris, "Newcomers vs. Oldtimers."

54. Harris, "Newcomers vs. Oldtimers."

55. Darren Frederick Speece, *Defending Giants: The Redwood Wars and the Transformation of American Environmental Politics* (Seattle: University of Washington Press, 2017), 87, 88, 176–78.

56. *Daily Journal*, September 15, 1978, 1; *Daily Journal*, December 31, 1978, 2; *Daily Journal*, February 4, 1979, 1; Pogash, "Hard Look," 22. Herbicides would remain linked to clearcutting activism environmentalist Julia Butterfly Hill, who lived in a redwood tree for two years to protest logging practices, which included spraying herbicides to allow replanting of tree stock.

57. "The Point Is . . . : A Summary of Public Issues Important to the Dow Chemical Company," no. 8, December 21, 1979, Folder 6471, Box 215, Series IX "Media Response," Unprocessed, Young Collection.

58. Dow Chemical Company, "A Familiar Scenario with a New Cast . . . ," *Bottom Line*, January 1981, Folder 6168, Box 212, Young Collection.

59. Dow Chemical Company, "A Familiar Scenario with a New Cast . . . ," *Bottom Line*, January 1981; Forest Industry Chemical Committee, National Forest Products Association, "'T' Memo," June 11, 1979, Forest Industry news clippings, FHS.

60. John E. Benneth, "Memorandum: Serpent Fruits," American Forest Industry, July 19, 1979; and letter from John E. Benneth to Senator Bob Packard, April 25, 1980, Forest Industry news clippings, FHS.

61. Dow Chemical Company, "A Familiar Scenario with a New Cast . . . ," *Bottom Line*, January 1981.

62. Dow Chemical Company, "A Familiar Scenario with a New Cast . . . ," *Bottom Line*, January 1981, 3.

Chapter Eight

1. Quoted in Fred Wilcox, *Waiting for an Army to Die: The Tragedy of Agent Orange* (Cabin John, MD: Seven Locks Press, 1989), xi.

2. Associated Press, New York, July 25, 1978, LexisNexis Academic; Associated Press, January 9, 1979, LexisNexis Academic.

3. Two important works that discuss the physical and psychological effects of the war on US soldiers are Fred A. Wilcox, *Waiting for an Army to Die: The Tragedy of Agent Orange* (New York: Vintage Books, 1983); and Wilbur Scott, *Vietnam Veterans since the War: The Politics of PTSD, Agent Orange, and the National Memorial* (Norman: University of Oklahoma Press, 2004).

4. See Michael Gough, *Dioxin, Agent Orange: The Facts* (New York: Springer, 1986); Edwin Martini, *Agent Orange: History, Science, and the Politics of Uncertainty* (Amherst: University of Massachusetts Press, 2012).

5. Psychiatrist Robert Jay Lifton developed the idea of "survivor" in three influential books he wrote: *Death in Life: Survivors of Hiroshima* (1967), *Home from the War: Vietnam Veterans—Neither Victims nor Executioners* (1973), and *The Nazi Doctors: Medical Killing and the Psychology of Genocide* (1986). These works led to the recognition of a new mental health concept, Post-Traumatic Stress Disorder, wherein individuals continue to experience the negative emotional and mental effects after a traumatic experience. Kirsten Fermaglich examines the link between Lifton's theory and the influence of the Holocaust concentration camps on the Jewish Lifton. See Kirsten Fermaglich, *American Dreams and Nazi Nightmares: Early Holocaust Consciousness and Liberal America, 1957–1965* (New York: Brandeis University Press, 2007).

6. John Langston Gwaltney, "Maude DeVictor," in *The Dissenters: Voices from Contemporary America* (New York: Random House, 1986), 120.

7. Gwaltney, *Dissenters*, 106–12.

8. Gwaltney, *The Dissenters*, 106–11, quotes from 111.

9. "Maude DeVictor: Fired for Helping Agent Orange Victims," *The Veteran* 15, no. 1 (Spring 1985): 3.

10. Wilbur Scott, "Competing Paradigms in the Assessment of Latent Disorders: The Case of Agent Orange," *Social Problems* 35, no. 2 (1988), 149. Scott details the efforts Kurtis took to attract attention to the documentary, which included sending copies of the documentary to local and state elected officials.

11. This rough survey of Agent Orange stories was done by searching LexisNexis Academic News Wire sources for the dates listed.

12. Richard Severo, "Two Crippled Lives Mirror Disputes on Herbicides, Agent Orange: A Legacy of Suspicion," May 27, 1979, *New York Times*, 1. The other articles in the series, all written by Richard Severo, were "U.S., Despite Claims of Veterans Says

None Are Herbicide Victims," May 28, 1979, and "Herbicides Pose a Bitter Mystery in U.S. Decades after Discovery," May 29, 1979.

13. The information on Agent Orange stories was compiled through a survey of Lexis Nexis database of news sources. Despite extensive research, I have not been able to determine the exact date Reutershan appeared on *The Today Show*. Most sources simply state "early in 1978." There is documentation that Reutershan had heard about Maude DeVictor's research with veterans exposed to Agent Orange before his appearance. Because of this, I do not make causal claims with respect to his appearance and the Kurtis documentary.

14. Peter Sills, *Toxic War: The Story of Agent Orange* (Nashville: Vanderbilt University Press, 2014), 130.

15. Associated Press, New York, July 25, 1978, LexisNexis Academic; and Associated Press, January 9, 1979, LexisNexis Academic.

16. News Wire, Associated Press, March 24, 1978, PM cycle, LexisNexis Academic.

17. Mike Shanahan, News Wire, Associated Press, October 11, 1978, AM cycle, LexisNexis Academic.

18. P. Gunby, "Dispute over Some Herbicides Rages in Wake of Agent Orange," *Journal of the American Medical Association*, April 6, 1979 241, no. 14: 1443–44; "Letter from Gilbert Boger, M.D., to the Editor of the *Journal of the American Medical Association* 242, no. 22 (November 30, 1979)," cited in Wilcox, Appendix, 195; Sills, 131, 132.

19. Sills, *Toxic War*, 134–36.

20. Sills, *Toxic War*, 142–44; Peter H. Schuck, *Agent Orange on Trial: Mass Toxic Disasters in the Courts* (Cambridge: Belknap Press of Harvard University Press, 1987).

21. Sills, *Toxic War*, 144–50.

22. Mike Sutton, "Agent Orange Hearings: Studies While Nam Vets Suffer," *The Veteran*, February/March 1982 (vol. 12, no. 1), 3.

23. Tim Jones, "Agent Orange: Lethal Legacy," *Chicago Tribune*, December 6, 2009; Sills, 150–55.

24. Sills, *Toxic War*, 150–54; E. N. Brandt, *Growth Company: Dow Chemical's First Century* (East Lansing: Michigan State University Press, 1997), 163–66.

25. Sills, *Toxic War*, 162.

26. Sills, *Toxic War*, 162–65.

27. Schuck, *Agent Orange on Trial*, 149, 150; Sills, *Toxic War*, 165, 166.

28. Ralph Blumenthal, "Vietnam Agent Orange Suit by Veterans Is Going to Trial," *New York Times*, May 6, 1984, 2.

29. Andrew E. Hunt, *The Turning: The History of Vietnam Veterans Against the War* (New York: New York University Press, 1999), 183.

30. "Chemical Time Bomb in Vietnam Veterans: Defoliant Agent Orange," *The Veteran* 8, no. 2 (Summer 1978), 1.

31. "Chemical Time Bomb in Vietnam Veterans," 7.

32. "Chemical Time Bomb in Vietnam Veterans," 7.

33. "Defoliant Agent Orange Exposed," 11.

34. "Defoliant Agent Orange: Chemical Time Bomb in Vietnam Veterans," *The Veteran* 8, no. 3 (Fall 1978), 7.

35. "Defoliant Agent Orange," *The Veteran* 8, no. 3 (Fall 1978), 7.

36. "Treat Effects of Agent Orange: Demand VA Action," *The Veteran* 8, o. 3 (Fall 1978), 7.

37. "Defoliant Agent Orange: Chemical Time Bomb in Vietnam Veterans," *The Veteran* 9, no. 2 (Spring 1979), 3, 18.

38. "Defoliant Agent Orange: Chemical Time Bomb in Vietnam Veterans," *The Veteran*, 9, no. 2 (Spring 1979), 18; Wilcox, *Waiting for an Army*, ch. 10, "The Vietnamization of America." Wilcox interviews a cast of familiar characters in the chapter, including Bob McKusick (Globe), Bonnie Hill (Alsea), and various Vietnamese people.

39. "National Conference and Actions on Agent Orange," *The Veteran* 11, no. 2 (Spring 1981), 1.

40. "Agent Orange Shorts," *The Veteran* 11, no. 4 (November/December 1981), 13.

41. Mike Sutton, "The Battle Goes Forward: Agent Orange," *The Veteran* 13, no. 1 (February/March 1983): 3, 13.

42. "Hot off the Press: Agent Orange Booklets," *The Veteran* 13, no. 3 (June/July 1983): 4.

43. Bill Davis, "Agent Orange Studies: One & One," *The Veteran* 17, no. 3 (Fall 1987): 19.

44. Davis, "Agent Orange Studies," 19.

45. John Lindquist, "Vietnam Veterans Are Still Dying from Agent Orange," *The Veteran* 20, no. 2 (Summer 1990): 21.

46. "Vets Notes: Cutting through Red Tape, Making Sense of the Regs," *The Veteran* 10, no. 3 (Summer 1980): 2, 10.

47. "Vets Notes," 2.

48. John Lindquist, "Agent Orange, Supreme Court: Chemical Companies Get Over," *The Veteran* 12, no. 1 (February/March 1982): 2.

49. "Agent Orange Civil Suit," *The Veteran* 10, no. 3 (Summer 1980): 3.

50. "Agent Orange 'Settled': Everyone Wins! Except the Victims," *The Veteran* 14, no. 2 (Spring 1984): 4.

51. "Agent Orange 'Settled,'" 4.

52. "Settlement or Sellout? Agent Orange Trial," *The Veteran* 14, no. 3 (July 1984): 4.

53. John Lindquist, "Agent Orange Lawsuit: This Settlement Wasn't Made for Me and You," *The Veteran* 19, no. 1 (Annual 1989): 6.

54. "Final Agent Orange Settlement Approved," *The Veteran* 15, no. 3 (Summer 1985): 1; Barry Romo, "Agent Orange: In Another 'Vietnam,'" *The Veteran* 19, no. 1 (Annual 1989): 5; Lindquist, "Agent Orange Lawsuit," 6; Lindquist, "Vietnam Veterans Are Still Dying from Agent Orange," 21.

55. "Treat Effects of Agent Orange: Demand VA Action," *The Veteran* 8, no. 3 (Fall 1978): 7.

56. Mike Sutton, "Agent Orange Hearings: Going Secret with the Truth," *The Veteran*

12, no. 3 (Summer 1982): 7. Nimmo would resign in part over charges he misused taxpayer funds but also because of his mishandling of the Agent Orange issue. Suzanne F. Green, "Robert Nimmo Quit Monday as Veterans Administration Chief Citing . . . ," October 4, 1982, UPI Archives, emphasis in the original.

57. Bill Davis, "Agent Orange Studies: One & One," *The Veteran* 17, no. 3 (Fall 1987): 19.

58. John VVAW, "Vets Retested: Original Agent Tests a Phony," *The Veteran* 9, no. 1 (Winter 1978/1979): 15.

59. "Defoliant Agent Orange: Chemical Time Bomb in Vietnam Veterans," *The Veteran*, 9, no. 2 (Spring 1979): 18.

60. Kathy Gauthier, "Listen My People," *The Veteran* 11, no. 4 (November/December 1981): 15.

61. "Chemical Company Hit: 'Ranch Hands' Revenge,'" *The Veteran* 14, no. 2 (Spring 1984): 5.

62. Edwin Martini, *Agent Orange: History, Science, and the Politics of Uncertainty* (Amherst: University of Massachusetts Press, 2012), 179–92; Bill Davis, "Agent Orange & Australian Vets," *The Veteran*, no. 4 (November/December 1981); "Australians Investigate Use of U.S. Defoliants in Vietnam," April 2, 1980, Folder 06, Box 03, Douglas Pike Collection: Unit 03—Technology, TTU Vietnam Archive and Archive, Texas Tech University, http://www.vietnam.ttu.edu/virtualarchive, hereafter cited as the TTU Vietnam Archive.

63. Michael Uhl and Tod Ensign, *GI Guinea Pigs: How the Pentagon Exposed Our Troops to Dangers More Deadly than War* (New York: Playboy Press, 1980), 201, 208, 209.

64. Michael Uhl, *The War I Survived Was Vietnam: Collected Writings of a Veteran and Antiwar Activist* (Jefferson, NC: McFarland, 2016), 238–43, 247–64.

65. Uhl and Ensign, *GI Guinea Pigs*, 224.

66. Uhl and Ensign, *GI Guinea Pigs*, 219–22; Erika Dyck, *Psychedelic Psychiatry: LSD from Clinic to Campus* (Baltimore: Johns Hopkins University Press, 2010); Susan L. Smith, *Toxic Exposures: Mustard Gas and the Health Consequences of World War II in the United States* (New Brunswick, NJ: Rutgers University Press, 2017), 15–41; Barry Commoner, "Citizen Soldier," 4, US Defoliation Program, January 1980, Folder 05, Box 03, Douglas Pike Collection: Unit 03—Technology, TTU Vietnam Archive.

67. Jacki Ochs and Betsy Sussler, "Secret Agent," *BOMB* 1, no. 2 (1982): 44.

68. Ochs and Sussler, "Secret Agent," 45; *Vietnam: The Secret Agent*, dir. Jacki Ochs (1983; New York: Green Mountain Post Films; Human Arts Association; First-Run Features, 1983). I have omitted discussing the congressional hearings in the interest of space and presenting perspectives not offered in other coverage of the veterans' protests. See Leslie Reagan, "'My Daughter Was Genetically Drafted with Me': US-Vietnam War Veterans, Disabilities and Gender," *Gender & History* 28, no. 3 (2016): 833–53, for a more in-depth discussion of the congressional hearings and their optics.

69. Michael E. Hill, "Maude DeVictor," *Washington Post*, November 9, 1986, A6.

70. *Rambo: First Blood*, dir. Ted Kotcheff (Hollywood, CA: Orion Pictures [MGM], 1982); Bobbie Ann Mason, *In Country* (New York: Harper & Row, 1985).

71. Anthony Brice, "Six Degrees Interviews Maude DeVictor—African American Heroine to Military Veterans," *Blog Talk Radio*, http://www.blogtalkradio.com.

Chapter Nine

1. Leslie Patten Wolff, "Vets Face New War at Home," *Buffalo Veteran* 2, no. 2 (July/August 1980): 3.

2. Wolff, "Vets Face New War."

3. "Chemical Time Bomb in Vietnam Veterans: Defoliant Agent Orange Exposed," *The Veteran* 8, no. 2 (Summer 1978): 7.

4. "Chemical Time Bomb," 7.

5. "Chemical Time Bomb," 7; "What Is Agent Orange?" *The Veteran* 8, no. 3 (Fall 1978): 8, 9; "Defoliant Agent Orange: Chemical Time Bomb in Vietnam Veterans," *The Veteran* 8, no. 3 (Fall 1978): 8, 9.

6. "Returning to Vietnam: Vet Attends Agent Orange Conference," *The Veteran* 13, no. 4 (Fall 1983): 3.

7. "Agent Orange Shorts," *The Veteran* 13, no. 2 (April/May 1983): 5.

8. Barry Romo, "Agent Orange: In Another 'Vietnam,'" *The Veteran* 19, no. 1 (Annual 1989): 6.

9. John Zutz, "Agent Orange Still Kills," *The Veteran* 19, no. 1 (Annual 1989): 7.

10. Leslie J. Reagan, "'My Daughter Was Genetically Drafted with Me,'" 837–40; *Vietnam: The Secret Agent*, dir. Jacki Ochs (New York: Green Mountain Post Films; Human Arts Association; First-Run Features, 1983).

11. Clifford Linedecker with Michael and Maureen Ryan, *Kerry: Agent Orange and an American Family* (New York: Dell, 1982).

12. Elmo Zumwalt Jr., Elmo Zumwalt III, and John Pekkanen, *My Father, My Son* (New York: Macmillan, 1986), 163.

13. Peter Sills, *Toxic War: The Story of Agent Orange* (Nashville, TN: Vanderbilt University Press, 2014), 85; Edwin Martini, *Agent Orange: History, Science, and the Politics of Uncertainty* (Amherst: University of Massachusetts Press, 2012), 18.

14. Zumwalt, Zumwalt, and Pekkanen, *My Father, My Son*, 159.

15. Stanford Biology Study Group [Bruce Bartholomew, Matthew Bradley, Patricia Caldarola, Peter Cohen, Howard Edinberg, Lawrence Gilbert, Paul Grobstein, Donald Kennedy, Edward Merrell, Patrice Morrow, and Colin Pittendrigh], *The Destruction of Indochina: A Legacy of Our Presence* (Stanford: Stanford Biology Study Group, 1970), 3.

16. Stanford Biology Study Group, 3; Biochemical Warfare and Ecological Disruption in Vietnam, December 1, 1975, Folder 19, Box 03, Douglas Pike Collection: Unit 11—Monographs, TTU Vietnam Archive and Archive, Texas Tech University, http://www.vietnam.ttu.edu/virtualarchive, hereafter cited as the TTU Vietnam Archive; Joseph M. Dougherty, "The Use of Herbicides in Southeast Asia and Its Criticism" (Maxwell

AFB, AL: Air War College, 1972), The Use of Herbicides in Southeast Asia and Its Criticism, 01 April 1972, Folder 13, Box 02, Paul Cecil Collection, TTU Vietnam Archive; Nguyen Khac Vien, "Lasting Consequences of Chemical Warfare," May 1, 1984, Folder 04, Box 03, Douglas Pike Collection: Unit 03—Technology, TTU Vietnam Archive; "Evidence on Agent Orange Deadly Dioxin," June 1, 1980, Folder 06, Box 03, Douglas Pike Collection: Unit 03—Technology, TTU Vietnam Archive.

17. Gerald Hickey, "Perceived Effects of Herbicides Used in the Highlands of South Vietnam," *The Effects of Herbicides in South Vietnam, Part B Working Papers* (Washington DC: National Academy of Sciences, 1974), 13, Folder 8, Box Two "Technology—Chemical Biological Warfare—(General Literature)," Douglas Pike Collection: Unit Three—Technology, TTU Vietnam Archive.

18. Hickey, "Perceived Effects," 1–14. Hickey expressed his shock and dismay when the cover letter of the 1974 NAS report tried to discredit Vietnamese villagers' accounts of illnesses resulting from herbicide exposure. See Gerald Hickey, *Window on a War: An Anthropologist in the Vietnam Conflict* (Lubbock: Texas Tech University Press, 2002), 355.

19. Hilary A. Rose and Steven P. R. Rose, "Chemical Spraying as Reported by Refugees from South Vietnam," *Science*, n.s., 177, no. 4050 (August 25, 1972): 711.

20. *Vietnamese Intellectuals against U.S. Aggression* (Hanoi: Foreign Language, 1966), 11; Tồn Thất Tùng, "The Vietnam War and Man's Conscience," *Nhân Dân*, August 30, Hanoi Domestic Service translated from the Vietnamese, Science to Destroy Vietnamese Land, August 30, 1972, Folder 16, Box 02, Douglas Pike Collection: Unit 03—Technology, TTU Vietnam Archive.

21. Judith Colburn, "Losing Vietnam, Finding Vietnam," *Mother Jones* 8, no. 9 (1983): 22.

22. James H. Dwyer, "Summary of Proceedings of a Conference on Herbicide Exposure and Reproductive Epidemiology in Viet Nam Held April 10, 1982 at the Department of Medical Genetics, Mount Sinai School of Medicine," published August 1, 1982, 6, Box 33, Series III Subseries I, Young Collection, emphasis in the original.

23. Dwyer, "Summary of Proceedings," 6, Young Collection.

24. Dwyer, "Summary of Proceedings," 6, Young Collection; Cynthia R. Daniels, "Between Fathers and Fetuses: The Social Construction of Male Reproduction and the Politics of Fetal Harm," *Signs: Journal of Women in Culture and Society* 22, no. 3 (Spring 1997): 597–92.

25. Dwyer, "Summary of Proceedings," 14, Young Collection.

26. E. Cooperman and J. H. LeVan, "Scientific Cooperation with Vietnam," Letter to the Editor, *Science*, n.s., 205, no. 4406 (August 10, 1979): 204, 205; Kelly Moore, *Disrupting Science: Social Movements, American Scientists, and the Politics of the Military, 1945–1975* (Princeton: Princeton University Press, 2008), 180–81; "The Peking Visit: What Does It Mean for Science?" *Science News* 101, no. 10 (March 4, 1972): 148.

27. "International Symposium on Herbicides and Defoliants in War, the Long-Term

Effects on Man and Nature, Ho Chi Mihn City, 13–20 January 1983," Item 5595, Box 198, Series VIII Subseries II, Young Collection, https://specialcollections.nal.usda.gov/agentorange/agent-orange-item-id-5595.

28. "International Symposium on Herbicides and Defoliants in War, the Long-Term Effects on Man and Nature, Ho Chi Mihn City, 13–20 January 1983"; Tine Gammeltoft, *Haunting Images: A Cultural Account of Selective Reproduction in Vietnam* (Berkeley: University of California Press, 2014), 47. Gammeltoft notes that it was not until the mid-1990s that Vietnamese society began publicly acknowledging the effects of Agent Orange on its "victims."

29. "International Symposium on Herbicides and Defoliants in War, the Long-Term Effects on Man and Nature, Ho Chi Mihn City, 13–20 January 1983," Summary 3, 4, 5, Reproductive 1.

30. Elof Axel Carlson, "Commentary: International Symposium on Herbicides in the Vietnam War: An Appraisal," *BioScience* 33, no. 8 (September 1983): 512.

31. Hanoi Defoliation Study, "The Lasting Consequences of Chemical Warfare," Consequence of Chemical Warfare, 23 March 1983, Folder 02, Box 03, Douglas Pike Collection: Unit 03—Technology, TTU Vietnam Archive; Editorial Staff, *Herbicides and Defoliants in War: The Long-Term Effects on Man and Nature* (Honolulu: University Press of the Pacific, 2003).

32. Gammeltoft, *Haunting Images*, 36–38.

33. Cheri Pies, Malcolm Potts, and Bethany Young, "Quinacrine Pellets: An Examination of Nonsurgical Sterilization," *International Family Planning Perspectives* 20, no. 4 (December 1994): 137–41.

34. Diane Niblack Fox tells Mrs. Hà's story in several publications, most explicitly in her essay "Chemical Politics and the Hazards of Modern Warfare," in *Synthetic Planet: Chemical Politics and the Hazards of Modern Life*, ed. Monica J. Casper, 85–89 (New York: Routledge, 2003). Fox notes that she used a pseudonym in telling this woman's story.

35. Gammeltoft, *Haunting Images*, 2, 3, 36–38. Gammeltoft's interviews took place in the late 1990s and early 2000s.

36. Viet Thanh Nguyen, *Nothing Ever Dies: Vietnam and the Memory of War* (Cambridge: Harvard University Press, 2016), 258.

37. Sharon Seah Li-Lian, "Truth and Memory: Narrating Viet Nam," *Asian Journal of Social Science* 29, no. 3 (2001): 388.

38. Li-Lian, "Truth and Memory"; Patrick Hagopian, "Photography as a Locus of Memory," in *Locating Memory: Photographic Acts*, ed. Annette Kuhn and Kirsten McAllister, 201–3 (New York: Berghahn Books, 2008); and Vương Trí Nhàn, "The Diary of Đặng Thùy Trâm and the Postwar Vietnamese Mentality," *Journal of Vietnamese Studies* 3, no. 2 (Summer 2008): 181–86.

39. Patrick Hagopian, "Photography as a Locus of Memory."

40. Frances Starner, *Vietnamese Woman Holds Deformed Baby*, May 17, 1980, UPI, in Getty Images, http://www.gettyimages.com/license/514679672.

41. Philip Jones Griffiths, *Agent Orange: "Collateral Damage" in Vietnam* (London: Trolley Books, 2003).

42. Griffiths, *Agent Orange*, 38.

43. Christopher Hitchens, "The Seventies: The Vietnam Syndrome," *Vanity Fair*, March 26, 2007, https://www.vanityfair.com.

44. Hitchens, "Vietnam Syndrome."

45. Charles Waugh, "Preface," in *Family of Fallen Leaves: Stories of Agent Orange by Vietnamese Writers* (Athens: University of Georgia Press, 2010), xvii.

46. Suong Nyet Minh, "Thirteen Harbors," in *Family of Fallen Leaves*, 40.

47. Minh, "Thirteen Harbors," 48.

48. Minh Chuyen, "Dr. Le Cao Dai and the Agent Orange Sufferers," in *Family of Fallen Leaves*, 161.

49. Chuyen, "Dr. Le Cao Dai," 161.

50. Reagan, "'My Daughter Was Genetically Drafted with Me'"; Naomi Rogers, *Dirt and Disease: Polio before FDR* (New Brunswick, NJ: Rutgers University Press, 1992), 166, 172; Gretchen Krueger, "'For Jimmy and the Boys and Girls of America': Publicizing Childhood Cancers in Twentieth-Century America," *Bulletin of the History of Medicine* 81, no. 1 (2007): 70–93; Amy M. Hay, "Recipe for Disaster: Chemical Wastes, Community Activists, and Public Health at Love Canal, 1945–2000," PhD diss., Michigan State University, East Lansing, 2005; and Sharon Stephens, "Reflections on Environmental Justice: Children as Victims and Actors," *Social Justice* 23, no. 4 (2009): 72–76.

51. Fred Setterberg and Lonny Shavelson, *Toxic Nation: The Fight to Save Our Communities from Chemical Contamination* (New York: John Wiley & Sons, 1993), 7–10.

52. Kathryn M. Kuivila, Holly D. Barnett, and Jody L. Edmunds, "Herbicide Concentrations in the Sacramento–San Joaquin Delta, California," published in *U.S. Geological Survey Toxic Substances Hydrology Program—Proceedings of the Technical Meeting, Charleston, South Carolina, March 8–12, 1999*, vol. 2, *Contamination of Hydrologic Systems and Related Ecosystems* (Washington DC: U.S. Geological Survey Water—Resources Investigations Report, 1999), https://ca.water.usgs.gov; Molly Joel Coye, "The Health Effects of Agricultural Production: I. The Health of Agricultural Workers," *Journal of Public Health Policy* 6, no. 3 (September 1985): 351, 352; Agency for Toxic Substances and Disease Registry, "Health Consultation: Ambient Air and Indoor Dust, McFarland Study Area, McFarland, Kern County, California, EPA Facility ID: CA0001118603" (Atlanta, GA, March 22, 2006), 3, https://www.atsdr.cdc.gov.

53. Donald B. Taylor, "Girl, 9, Describes Ordeal of Poisoning at Pesticide Hearing for Farm Workers," *Los Angeles Times*, July 26, 1980, 30.

54. Laura Pulido, *Environmentalism and Economic Justice: Two Chicano Struggles in the Southwest* (Tucson: University of Arizona Press, 1996), 58; Linda Nash, *Inescapable Ecologies: A History of Environment, Disease, and Knowledge* (Berkeley: University of California Press, 2006), 136–40.

55. Nash, *Inescapable Ecologies*, 142.

56. Judith Walzer Leavitt, "'Typhoid Mary' Strikes Back: Bacteriological Theory and Practice in Early Twentieth Century Public Health," *Isis* 83, no. 4 (December 1992): 608–49; Coye, "Health Effects," 350, 351; Nash, *Inescapable Ecologies*, 142, 143; Frederick Rowe Davis, *Banned: A History of Pesticides and the Science of Toxicology* (New Haven: Yale University Press, 2015), 116–52; Sarah Vogel, *Is It Safe? BPA and the Struggle to Define the Safety of Chemicals* (Berkeley: University of California Press, 2012), 76–82; Nancy Langston, *Toxic Bodies: Hormone Disruptors and the Legacy of DES* (New Haven: Yale University Press, 2011), 78, 114–26. Langston's research shows that questions regarding low-dose exposure and toxicity arose as early as the 1940s and 1950s in the case of DES.

57. Nash, *Inescapable Ecologies*, 172–79. This paragraph summarizes Nash's excellent analysis of the geological and hydrological conditions that affected chemical migration in the San Joaquin Valley; Adam Tomkins, *Ghostworkers and Greens: The Cooperative Campaigns of Farmworkers and Environmentalists for Pesticide Reform* (Ithaca: IRL-Cornell University Press, 2016), 108–10.

58. Amy M. Hay, "Recipe for Disasters: Motherhood and Citizenship at Love Canal," *Journal of Women's History* 21, no. 1 (Spring 2009): 112–20; Sherilyn MacGregor, *Beyond Mothering Earth: Ecological Citizenship and the Politics of Care* (Vancouver: University of British Columbia Press, 2007), 48. Michelle Murphy provides an excellent discussion of the ways women activists were discredited with the label "hysterical housewives" in *Sick Building Syndrome and the Problem of Uncertainty: Environmental Politics, Technoscience, and Women Workers* (Durham, NC: Duke University Press, 2006), 103.

59. Michael Weisskopf, "Pesticides and Death amid Plenty," *Washington Post*, August 30, 1988.

60. Ronald B. Taylor, "'Cancer Clusters' Jolt Two Farming Towns," *Los Angeles Times*, September 19, 1985, 3, 32; Nash, *Inescapable Ecologies*, 182–84; Setterberg and Shavelson, *Toxic Nation*, 68–75.

61. "Cancer Testimony," *Press Democrat*, July 24, 1985, 4; Nash, *Inescapable Ecologies*, 182, 183; Taylor, "'Cancer Clusters' Jolt Two Farming Towns"; Setterberg and Shavelson, *Toxic Nation*, 18, 19; Tomkins, *Ghostworkers and Greens*, 111.

62. Pulido, *Environmentalism and Economic Justice*, 58.

63. Don Graff, "Pesticide Use Prompts Chavez Boycott," *Argus-Courier*, May 30, 1986, 4; Tomkins, *Ghostworkers and Greens*, 113.

64. Pat Hoffman, "UFW Fights Harvest of Poison," *The Witness*, July/August 1988, 9.

65. George McCrory, "Farm Worker Champion Speaks at Southwestern," *Vista Star-News*, May 7, 1987.

66. Graff, "Pesticide Use Prompts Chavez Boycott," 4; Tomkins, *Ghostworkers and Greens*, 108, 109.

67. "Nader, 15 Activist Groups Give Boost to Chavez-Led Boycott of California Grapes," *County Sun*, December 23, 1987, 10; "Nader Joins Boycott of California Grapes," *Los Angeles Times*, December 25, 1987, 50; "Nader and Friends Dump Grapes,"

Food and Justice 5, no. 2 (February 1988): 3; Lloyd G. Carter, "Cancer Stalks Farm Town's Children," *Press-Tribune*, July 15, 1988, 9.

68. *The Wrath of Grapes*, dir. Lorena Parlee and Lenny Bourin (Keene, CA: United Farm Workers, 1986); and Tomkins, *Ghostworkers and Greens*, 122.

69. Setterberg and Shavelson, *Toxic Nation*, 200.

70. "Cancer Fighter Turns on UFW," *Press Democrat*, November 23, 1987, 31; "Researcher Ties Pesticide to Town's Cancer Cluster," *Los Angeles Times*, December 1, 1987, 24; Setterberg and Shavelson, *Toxic Nation*, 138, 139.

71. Quoted in Setterberg and Shavelson, *Toxic Nation*, 205.

Conclusion

1. Interview with Ida Honorof, *Northcoast Now* (Eureka, CA: KBIQ, c. January 1994). The title of this chapter is taken from John Gwaltney's *The Dissenters: Voices from Contemporary America* (New York: Random House, 1986), which includes an interview with Maude DeVictor.

2. "Derwinski Approves Compensation for Soft Tissue Sarcomas; Secretary Acts in Response to Advisory Group Finding," *Agent Orange Review* 7, no. 3 (August 1990): 1, https://www.publichealth.va.gov.

3. "Service Connection Type II Diabetes and Agent Orange," https://www.va.gov/vetapp05/files1/0503754.txt; "Secretary Brown Announces VA Will Recognize Additional Conditions," *Agent Orange Review* 10, no. 2 (September 1993): 1, https://www.publichealth.va.gov; Adam Clymer, "Bill Passed to Aid Veterans Affected by Agent Orange," *New York Times*, January 31, 1991.

4. The Veterans Affairs website includes links to the recognized conditions along with the expanded chronological and geographic areas covered. See "Compensation," *US Department of Veterans Affairs*, https://www.benefits.va.gov/compensation/claims-postservice-agent_orange.asp.

5. Jeanne Mager Stellman, Steven D. Stellman, Richard Christian, Tracy Weber, and Carrie Tomasallo, "The Extent and Patterns of Usage of Agent Orange and Other Herbicides in Vietnam," *Nature* 422 (April 17, 2002): 681–87, http://www.vn-agentorange.org; Peter Sills, *Toxic War: The Story of Agent Orange* (Nashville, TN: Vanderbilt University Press, 2014), 195–202; "Hatfield Agent Orange Reports and Presentations (links)," *Hatfield Consultants*, https://www.hatfieldgroup.com; "Agent Orange—Agent Orange Defoliation Damage," Adrenoceptor (adrenoreceptor; adrenergic receptor) to Ambient, *Science Encyclopedia*, https://science.jrank.org/pages/122/Agent-Orange.html.

Bibliography

Archival Materials

The Alvin L. Young on Agent Orange Collection, Special Collections, National Agricultural Library, Beltsville, Maryland (physical and online)

The Dow Historical Collection, Science History Institute, Philadelphia, Pennsylvania

The Forest History Society, Durham, North Carolina

Ida Honorof Papers, personal collection of Faye Honorof

TTU Vietnam Archive and Archive, Texas Tech University, Lubbock, Texas (physical and online)

Newspapers and Newsletters

Agent Orange Review

Arizona Daily Sun [Flagstaff]

Arizona Republic

Bakersfield Californian

Brazosport Facts

Buffalo Veteran

Computerworld

Daily Chronicle [Centralia, CA]

Daily Missoulian [Missoula, MT]

Galveston Daily News

Humboldt Standard [Eureka, CA]

Independent Press-Telegram [Long Beach, CA]

Kokomo Tribune [IN]

Lincoln Evening Journal

Los Angeles Times

Milwaukee Journal

Mother Earth News

Nevada Evening Gazette [Reno]

New York Times

New Yorker

News Palladium [Benton Harbor, MI]

Northwestern Magazine

Pantagraph [Bloomington, IL]

Paris News [Texas]

Pasadena Independent Topics

Press-Tribune [Roseville, CA]

Red Bluff Daily News [CA]

Redlands Daily Facts [CA]

Report to the Consumer

San Antonio Express

Seattle Times

Sentinel [Santa Cruz, CA]

Southern Illinoisan [Carbondale]

Statesman Journal [Salem, OR]

Sun [San Bernardino County, CA]

Times Standard [Eureka, CA]

Tucson Daily Citizen

Ukiah Daily Journal [CA]

Van Nuys News [CA]

The Veteran

Village Voice [NY]

Washington Post

Weekly Oregon Statesman [Salem]

Primary Sources

"AAAS Officers, Committees, and Representatives for 1970." *Science*, n.s., 167, no. 3921 (February 20, 1970): 1154–57. doi:10.2307/1728702.

Afro-Asiatisches Solidaritätskomitee in der DDR. Vietnam-Ausschuss. *The Truth about U.S. Aggression in Vietnam: GDR Authors Unmask Imperialist Crimes*. Berlin: Vietnam Commission of the Afro-Asian Solidarity Committee of the German Democratic Republic, 1972.

Agency for Toxic Substances and Disease Registry. "Health Consultation: Ambient Air and Indoor Dust, McFarland Study Area, McFarland, Kern County, California, EPA Facility ID: CA0001118603." Atlanta, GA, March 22, 2006. https://www.atsdr.cdc.gov.

Anderson, Bruce. *The Mendocino Papers* Mendocino, CA: n.p., 2008.

Apocalypse Now. Directed by Francis Ford Coppola. Madrid: Universal, 2012.

Baughman, Robert, and Matthew Meselson. "An Analytical Method for Detecting TCDD (Dioxin): Levels of TCDD in Samples from Vietnam." *Environmental Health Perspectives* 5 (September 1, 1973): 27–35. doi:10.2307/3428110.

Beal, J. M. "Some Telemorphic Effects Induced in Sweet Pea by Application of 4-Chlorophenoxyacetic Acid." *Botanical Gazette* 105, no. 4 (1944): 471–74.

Berl, W. G. "The 1969 Meeting of the AAAS: A Brief Appraisal." *Science*, n.s., 167, no. 3921 (February 20, 1970): 1157–58. doi:10.2307/1728703.

Betts, R. R., and Frank H. Denton. *An Evaluation of Chemical Crop Destruction in Vietnam.* Santa Monica, CA: RAND 1975.

Bevan, William. "AAAS Council Meeting, 1970." *Science*, n.s., 171, no. 3972 (February 19, 1971): 709–11. doi:10.2307/1731481.

Biswell, Harold H. "Water Control by Rangeland Management." *Journal of Range Management* 22, no. 4 (1969): 227–30. doi:10.2307/3895922.

Blackman, G. E. "Plant-Growth Substances as Selective Weed-Killers: A Comparison of Certain Plant-Growth Substances with other Selective Herbicides." *Nature* 155, no. 3939 (1945): 500, 501.

Brandt, E. Ned. *Growth Company: Dow Chemical's First Century.* East Lansing: Michigan State University Press, 1997.

Brice, Anthony. "Six Degrees Interviews Maude DeVictor—African American Heroine to Military Veterans." *Blog Talk Radio.* http://www.blogtalkradio.com.

Brightman, Carol. "The 'Weed Killers'—A Final Word." *Viet-Report* 2, no. 7 (October 1966): 3.

Browning, Frank, and Dorothy Forman, eds. *The Wasted Nations: Report of the International Commission of Enquiry into US Crimes in Indochina, June 20–25, 1971.* New York: Harper & Row, 1972.

Browning, Frank, and Banning Garrett. "The New Opium War." *Ramparts*, May 1971, 32. https://www.unz.org.

Burk, Robert E. "What Chemistry Is Doing to Us and for Us." *Scientific Monthly* 49, no. 6 (December 1939): 491–593.

Carlson, Elof Axel. "Commentary: International Symposium on Herbicides in the Vietnam War: An Appraisal." *BioScience* 33, no. 8 (September 1983).

Carson, Rachel. *Silent Spring.* New York: Houghton Mifflin, 2002. Originally published 1962.

Carter, J. F., Charles Benbrook, James V. Parochetti, S. F. Cook, Richard Novick, John E. Donalds, George M. Happ, and M. D. Thorne. "CAST Profile Evokes Avid Responses." *BioScience* 29, no. 5 (1979): 276–80. doi:10.2307/1307817.

Constable, John D., Robert E. Cook, Matthew Meselson, and Arthur H. Westing. "AAAS and NAS Herbicide Reports." *Science*, n.s., 186, no. 4164 (November 15, 1974): 584–86. doi:10.2307/1739162.

Cooperman, E., and J. H. LeVan. "Scientific Cooperation with Vietnam." Letter to the Editor, *Science*, n.s., 205, no. 4406 (August 10, 1979): 204, 205.

Craig, Richard R. "La campaña permanente: Mexico's Antidrug Campaign." *Journal of Interamerican Studies and World Affairs* 20, no. 2 (May 1978).

———. "Operation Condor: Mexico's Antidrug Campaign Enters a New Era." *Journal of Interamerican Studies and World Affairs* 22, no. 3 (August 1980).

Cravens, Jay H. *A Well Worn Path*. Huntington, WV: University Editions, 1994.

"Currently in Print." *Bulletin of the Ecological Society of America* 53, no. 3 (September 1, 1972): 10–13. doi:10.2307/20165919.

D'Amato, Anthony A. "Book Reviews: *Against the Crime of Silence: Proceedings of the Russell International War Crime Tribunal* by John Duffett; *On Genocide* by Jean-Paul Sartre." *California Law Review* 57, no. 4 (October 1, 1969): 1033–38. doi:10.2307/3479579.

Dennis, W. W., and W. M. Dennis. "Chemicals: A Growth Industry." In special issue "Science and Security," *Analysts Journal* 8, no. 2 (March 1952): 61–65.

"Dioxin from Defoliation Found in Vietnam Fish." *Science News* 103, no. 18 (May 5, 1973): 287. doi:10.2307/4548258.

"Dioxin Toxicity Data Sent to Aid Italy." *Science News* 110, no. 23 (1976): 359. doi:10.2307/3961213.

"Dow Attacks Study Used to Ban 2,4,5-T." *Science News* 115, no. 11 (1979): 166. doi:10.2307/3964011.

Duffett, John, ed. *Against the Crime of Silence: Proceedings of the Russell International War Crimes Tribunal, Stockholm, Copenhagen*. New York: Bertrand Russell Peace Foundation, 1968.

Egler, Frank E. "Chemical Warfare in Southeast Asia." Review of *Harvest of Death*, by J. B. Neilands, G. H. Orians, E. W. Pfeiffer, Alje Vennema, and Arthur H. Westing. *Ecology* 53, no. 6 (1972): 1207–8. doi:10.2307/1935437.

Emmelin, Lars. "The Stockholm Conferences." *Ambio* 1, no. 4 (1972): 135–40.

Falk, Richard A. "Introduction." In *The Wasted Nations: Report of the International Commission of Enquiry into United States Crimes in Indochina, June 20–25, 1971*, edited by Frank Browning and Dorothy Forman, xv. New York: Harper Colophon Books, 1972.

Finn, James. "American Catholics and Social Movements." In *Contemporary Catholicism in the United States*, ed. Philip Gleason, 133–45. South Bend: University of Notre Dame, 1969.

Fishleder, Jack. *American Biology Teacher* 34, no. 2 (February 1, 1972): 103–4. doi:10.2307/4443833.

Front matter. *Science*, n.s., 179, no. 4080 (March 30, 1973).

Galston, Arthur. "Changing the Environment: Herbicides in Vietnam, II." In special issue "Chemical and Biological Warfare," *Scientist and Citizen* 9, no. 7 (1967): 122–28.

Galston, Arthur W., and Ethan Signer. "Education and Science in North Vietnam." *Science* 174, no. 4007 (1971): 379–85.

Garmon, L. "Pot-Smokers May Be Imperiled by Paraquat-Spraying Program." *Science News* 124, no. 4 (July 23, 1983): 55.

Geiger, Roger L. "Science, Universities, and National Defense, 1945–1970." *Osiris* 7 (1992): 26–48.

Gillette, Robert. "AAAS Council Meeting: Vietnam Resolutions; Bylaws Voted." *Science*, n.s., 179, no. 4070 (January 19, 1973): 258–62. doi:10.2307/1735663.

Green, Kass. *Economic, Herbicides and People: A Case Study of Phenoxy Herbicides in Western Oregon*. New York: Council on Economic Priorities, 1982.

Grümmer, Gerhard. Afro-Asiatisches Solidaritätskomitee in der DDR, and Vietnam-Ausschuss. *Accusation from the Jungle*. Berlin: Vietnam Commission of the Afro-Asian Solidarity Committee of the German Democratic Republic, 1972.

———. *Genocide with Herbicides Report, Analysis, Evidence*. Berlin: Vietnam Commission of the Afro-Asian Solidarity Committee of the German Democratic Republic, 1971.

———. *Giftküchen des Teufels*. Berlin: Militärverl. d. Dt. Demokrat. Republik, 1988.

———. *Herbicides in Vietnam*. Berlin: Vietnam Commission of the Afro-Asian Solidarity Committee of the German Democratic Republic, 1969.

Gunby, P. "Dispute over Some Herbicides Rages in Wake of Agent Orange." *Journal of the American Medical Association* 241, no. 14 (April 6, 1979): 1443–44.

Gwaltney, John Langston. *The Dissenters: Voices from Contemporary America*. New York: Random House, 1986.

Hamner, Charles L., and H. B. Tukey. "The Herbicidal Action of 2,4 Dichlorophenoxyacetic and 2,4,5 Trichlorophenoxyacetic Acid on Bindweed." *Science* 100, no. 2590 (1944): 154–55.

Hawkes, Nigel. "Human Environment Conference: Search for a Modus Vivendi." *Science*, n.s., 175, no. 4023 (February 18, 1972): 736–38. doi:10.2307/1733074.

Haynes, Eldridge. "Industrial Production and Manufacturing." *Journal of Marketing* 8, no. 1 (July 1943): 16.

Hearings before the Subcommittee on Energy, Natural Resources, and the Environment of the Committee on Commerce; United States Senate, Ninety-First Congress, Second Session on Effects of 2,4,5-T on Man and the Environment (Washington, DC, 1970).

Henig, Robin Marantz. "Agriculture's Strange Bedfellows: CAST-Industry Tie Raises Credibility Concerns." *BioScience* 29, no. 1 (1979): 9–59.

Herbicide Hearing Committee. "Report on the Use and Control of Simazin [*sic*] and Other Herbicides by the County of Los Angeles." County of Los Angeles Health Department, March 7, 1972. http://www.cdpr.ca.gov.

Hibbert, Alden R., Edwin A. Davis, and David G. Scholl. *Chaparral Conversion Potential in Arizona, Part I: Water Yield Response and Effects on Other Resources*. Series: USDA Forest Service Research Paper RM-126. Fort Collins, CO: Rocky Mountain Forest and Range Experiment Station, Forest Service, US Dept. of Agriculture, 1974.

Hill, Lawrence W., and Raymond M. Rice. "Converting from Brush to Grass Increases Water Yield in Southern California." *Journal of Range Management* 16, no. 6 (1963): 300–305. doi:10.2307/3895373.

Hoffman, Pat. "UFW Fights Harvest of Poison." *The Witness*, July/August (1988): 9.

Holden, Constance. "Vietnam Land Devastation Detailed." *Science*, n.s., 175, no. 4023 (February 18, 1972): 737. doi:10.2307/1733075.

Holmes, E. "The Role of Industrial Research and Development in Weed Control in Europe." *Weeds* 6, no. 3 (1958): 245–50. doi:10.2307/4040153.

Honorof, Ida. Interview, *Northcoast Now*. Eureka, CA: KBIQ, c. January 1994.

Hormay, August, Fred J. Alberico, and P. B. Lord. "Experiences with 2,4-D Spraying on the Lassen National Forest." *Journal of Range Management* 15, no. 6 (1962): 325–28. doi:10.2307/3894765.

Huynh, Quang Nhuong, and Mai Vo-Dinh. *The Land I Lost: Adventures of a Boy in Vietnam*. New York: Harper Trophy, 1982.

Judd, B. Ira. "Range Reseeding Success on the Tonto National Forest, Arizona." *Journal of Range Management* 19, no. 5 (1966): 296–301. doi:10.2307/3895724.

Kennedy, Jane, and Jeanne Weimann. "Interview: 'Dangerous Jane' Kennedy." *Off Our Backs* 5, no. 4 (April 1, 1975): 8–9. doi:10.2307/25783983.

Kuivila, Kathryn M., Holly D. Barnett, and Jody L. Edmunds. "Herbicide Concentrations in the Sacramento–San Joaquin Delta, California." Published in *U.S. Geological Survey Toxic Substances Hydrology Program—Proceedings of the Technical Meeting, Charleston, South Carolina, March 8–12, 1999*. Vol 2 of *Contamination of Hydrologic Systems and Related Ecosystems*. Washington, DC: U.S. Geological Survey Water—Resources Investigations Report, 1999. https://ca.water.usgs.gov.

Landrigan, Philip J., MD, MSc, Kenneth E. Powell, MD, MPH, Levy M. James, and Philip R. Taylor, MD. "Paraquat and Marijuana: Epidemiologic Risk Assessment." *American Journal of Public Health* 73, no. 7 (1983): 784–88.

Langer, Elinor. "Chemical and Biological Warfare (I): The Research Program." *Science*, n.s. 155, no. 3759 (January 13, 1967): 174.

———. "National Teach-In: Professors Debating Viet Nam, Question Role of Scholarship in Policy-Making." *Science* 148, no. 3673 (May 21, 1965): 1075–77.

"Law and Responsibility in Warfare: The Vietnam Experience." *Instant Research on Peace and Violence* 4, no. 1 (1974): 1–14.

Lederer, Edgar. "Report of the Sub-Committee on Chemical Warfare in Vietnam." In *Against the Crime of Silence: Proceedings of the Russell International War Crimes Tribunal*, edited by John Duffett, 338–66. New York: O'Hare Books, 1968.

Leopold, Aldo. "Grass. Brush, Timber, and Fire in Southern Arizona." *Journal of Forestry* 22, no. 1–10 (1924). https://www.nps.gov/seki/learn/nature/upload /leopold24.pdf.

———. *River of the Mother of God and Other Essays*. Edited by Susan L. Flader and J.

Baird Callicott. Madison: University of Wisconsin Press, 1993. https://archive
.org/details/riverofmotherofg00leop.

———. "The Virgin Southwest." In *The River of the Mother of God: And Other Essays*,
edited by Susan L. Flader and J. Baird Callicott, 173–80. Madison: University of
Wisconsin Press, 1992.

Lien, Huy, Charles Waugh, and Christopher Waldrep. *Family of Fallen Leaves: Stories of Agent Orange by Vietnamese Writers*. Athens: University of Georgia Press,
2010.

Linedecker, Clifford L., Michael Ryan, and Maureen Ryan. *Kerry, Agent Orange and
an American Family*. New York: Dell, 1983.

Luce, Don. "A Decade of Atrocity." In *Wasted Nations: Report of the International
Commission of Enquiry into United States Crimes in Indochina, June 20–25, 1971*,
edited by Frank Browning and Dorothy Forman. New York: Harper Colophon
Books, 1972.

"Major Congressional Action: Effects of Herbicides." *1970 CQ Almanac* 26 (February 1970).

Marcus, Alan. *Cancer from Beef: DES, Federal Food Regulation, and Consumer Confidence*. Baltimore: Johns Hopkins University Press, 1994.

Marth, Paul C., and John W. Mitchell. "2,4-Dichlorophenoxyacetic Acid as a Differential Herbicide." *Botanical Gazette* 106, no. 2 (1944): 224–32.

"Maude DeVictor: Fired for Helping Agent Orange Victims." *The Veteran* 15, no. 1
(Spring 1985): 3.

Mayr, Otto. "Revolution in Electrical Technology (1870–1900): AAAS Symposium,
December 27 1970, Chicago." *Science*, n.s., 170, no. 3964 (December 18, 1970):
1339–41. doi:10.2307/1730529.

McCarthy, Richard D. *The Ultimate Folly: War by Pestilence, Asphyxiation, and Defoliation*. New York: Alfred A. Knopf, 1969.

McCormick, Ronald J. *Plain Green Wrapper: A Forester's Story*. N.p.: R. J. McCormick, 2009.

McKell, Cyrus M., J. R. Goodin, and Cameron Duncan. "Chaparral Manipulation Affects Soil Moisture Depletion Patterns and Seedling Establishment." *Journal of
Range Management* 22, no. 3 (1969): 159–65. doi:10.2307/3896333.

McWilliams, Carey. "The Farmers Get Tough." *American Mercury* 33 (October 1934).
http://www.unz.org.

Meyer, Jean. "Starvation as a Weapon: Herbicides in Vietnam, I." In special issue
"Chemical and Biological Warfare," *Scientist and Citizen* 9, no. 7 (1967): 115–21.

Myrdal, Gunnar, Hans Göran Franck, and International Commission of Enquiry
into US Crimes in Indochina. *The Effects of Modern Weapons on the Human Environment in Indochina: Documents Presented at a Hearing Organized by the International Commission in Cooperation with the Stockholm Conference on Vietnam and*

the Swedish Committee for Vietnam, Stockholm June 2–4, 1972. Stockholm: International Commission of Enquiry into US Crimes in Indochina, 1972.

Neilands, J. B. "Vietnam: Progress of the Chemical War." *Asian Survey* 10, no. 3 (March 1, 1970): 209–29. doi:10.2307/2642575.

Neilands, J. B., G. H. Orians, E. W. Pfeiffer, Alje Vennema, and Arthur H. Westing. *Harvest of Death: Chemical Warfare in Vietnam and Cambodia*. New York: Free Press, 1972.

Nelson, Bryce. "Herbicides in Vietnam: AAAS Board Seeks Field Study." *Science* 163, no. 3862 (1969): 58–59.

Newton, Michael, and L. T. Burcham. "Defoliation Effects on Forest Ecology." *Science* 161, no. 3837 (1968): 109.

Newton, Michael, Logan A. Norris, and Arthur W. Galston. "Herbicide Usage." *Science* 168, no. 3939 (1970): 1606–7.

Nguyen, Viet Thanh. *Nothing Ever Dies: Vietnam and the Memory of War*. Cambridge: Harvard University Press, 2017.

Nutman, P. S., H. G. Thornton, and J. H. Quastel. "Plant-Growth Substances as Selective Weed-Killers: Inhibition of Plant Growth by 2:4-Dichlorophenoxyacetic Acid and Other Plant-Growth Substances." *Nature* 155, no. 3939 (1945): 498–500.

Oberle, Mark. "Forest Fires: Suppression Policy Has Its Ecological Drawbacks." *Science* 165, no. 3893 (1969): 568–71.

O'Brien, Tim. *The Things They Carried: A Work of Fiction*. Boston: Houghton Mifflin Harcourt, 2010.

Ochs, Jacki, and Betsy Sussler. "Secret Agent." *BOMB* 1, no. 2 (1982): 44.

"Paraquat Contamination of Marijuana—United States." *Morbidity and Mortality Weekly Report* 28, no. 8 (March 2, 1979).

"The Peking Visit: What Does It Mean for Science?" *Science News* 101, no. 10 (March 4, 1972): 148.

Perry, Chester A., Cyrus M. McKell, Joe R. Goodin, and Thomas M. Little. "Chemical Control of an Old Stand of Chaparral to Increase Range Productivity." *Journal of Range Management* 20, no. 3 (1967): 166–69. doi:10.2307/3895799.

"Plant Growth Regulators." *Science*, n.s., 103, no. 2677 (April 19, 1946).

Quastel, J. H. "2,4-Dichlorophenoxyacetic Acid (2,4-D) as a Selective Herbicide." *Agricultural Control Chemicals*. Collected Papers from the Symposia on Economic Poisons. Washington, DC: American Chemical Society, Division of Agricultural and Food Chemistry, 1950.

Range Seeding Equipment Committee. *Chemical Control of Range Weeds: Handbook*. Rev. ed. Washington, DC: U.S. Department of Agriculture, U.S. Department of the Interior, Range Seeding Equipment Committee, 1969.

Rose, Hilary A., and Steven P. R. Rose. "Chemical Spraying as Reported by Refugees from South Vietnam." *Science*, n.s., 177, no. 4050 (August 25, 1972): 710–12. doi:10.2307/1734766.

Rudner, Richard. "Report of the Delegate to the American Association for the Advancement of Science." *Proceedings and Addresses of the American Philosophical Association* 44 (January 1, 1970): 104–11. doi:10.2307/3129687.

Russell, E. John. *A History of Agricultural Science in Great Britain: 1620–1954*. London: George Allen & Unwin, 1966.

Russo, Anthony. "Inside the RAND Corporation and Out: My Story." *Ramparts* 10, no. 10 (April 1972): 45–55. http://www.unz.org.

———. *A Statistical Analysis of the U.S. Crop Spraying Program in South Vietnam*. Santa Monica, CA: RAND, 1967.

Sanchez, Thomas, and Lawrence Clark Powell. *Angels Burning: Native Notes from the Land of Earthquake and Fire*. Santa Barbara, CA: Capra Press, 1987.

"Science and Life in the World." *Science*, n.s., 103, no. 2683 (May 31, 1946).

"Science News." *Science—Supplement* 95, no. 2471 (May 8, 1942): 11.

Setterberg, Fred, and Lonny Shavelson. *Toxic Nation: Causes, Consequences, and the Fight to Save Our Communities*. New York: J. Wiley, 1993.

Shapley, Deborah. "Herbicides: Academy Finds Damage in Vietnam after a Fight of Its Own." *Science*, n.s., 183, no. 4130 (March 22, 1974): 1177–80. doi:10.2307/1737852.

Shoecraft, Billee. *Sue the Bastards!* Introduction by Frank E. Egler. Phoenix: Franklin Press, 1971.

Slade, R. E., W. G. Templeman, and W. A. Sexton. "Plant-Growth Substances as Selective Weed-Killers: Differential Effect of Plant-Growth Substances on Plant Species." *Nature*, 155, no. 3939 (1945): 497–99.

Smith, R. Jeffrey. "Dioxins Have Been Present since the Advent of Fire, Says Dow." *Science* 202, no. 4373 (1978): 1166–67.

Smith, R. Jeffrey. "EPA Halts Most Use of Herbicide 2,4,5-T." *Science* 203, no. 4385 (1979): 1090–91.

Snell, David. "Snake in the Crab Grass." *Life*, October 12, 1962.

Stanford Biology Study Group. *The Destruction of Indochina: A Legacy of Our Presence*. Stanford, CA: Stanford Biology Study Group, 1970.

Stewart, Kevin. *Tales of the Emerald Triangle: Memoirs of a Marijuana Grower*. Self-published, 2002.

Stone, Edward C. "Preserving Vegetation in Parks and Wilderness." *Science* 150, no. 3701 (1965): 1261–67.

Sutton, Mike. "Agent Orange Hearings: Studies while Nam Vets Suffer." *The Veteran* 12, no. 1 (February/March 1982): 3.

Tavares, Dennis. *How Mendocino County Went to Pot: Memories of Life in Mendocino Redwood Country in the Last Half of the 1900s*. N.p.: Trafford, 2014.

Thompson, H. E., Carl P. Swanson, and A. G. Norman. "New Growth-Regulating Compounds. I. Summary of Growth-Inhibitory Activities of Some Organic Compounds as Determined by Three Tests." *Botanical Gazette* 107, no. 4 (1946): 476–507.

Uhl, Michael, and Tod Ensign. *GI Guinea Pigs: How the Pentagon Exposed Our Troops to Dangers More Deadly than War*. New York: Playboy Press, 1980.

United States, and Range Seeding Equipment Committee. *Chemical Control of Range Weeds: [Handbook]*. Washington: US Dept. of Agriculture, US Dept. of the Interior, Range Seeding Equipment Committee, 1969.

"U.S. Attack Aims at Cutting Drug Traffic: Planes Patrol Mexican Border." *Progress Bulletin*, September 14, 1969, 1.

U.S. Crimes in Vietnam. Hanoi: Vietnam Courier, 1966. https://www.vietnam.ttu. edu/virtualarchive.

Van Strum, Carol. *A Bitter Fog: Herbicides and Human Rights*. San Francisco: Sierra Club Books, 1983.

Vietnam: The Secret Agent. Directed by Jacki Ochs. New York: Green Mountain Post Films; Human Arts Association; First-Run Features, 1983.

Vietnamese Intellectuals against U.S. Aggression. Hanoi: Foreign Languages, 1966.

"Volume Information." *Science*, n.s., 179, no. 4080 (March 30, 1973): i–xiv. doi:10.2307/1735046.

Westing, Arthur H. "AAAS Herbicide Assessment Commission." *Science*, n.s., 179, no. 4080 (March 30, 1973): 1278–79. doi:10.2307/1735048.

Whiteside, Thomas. *Defoliation*. New York: Ballantine Books, 1970.

———. *The Withering Rain: America's Herbicidal Folly*. New York: Dutton, 1971.

Whyte, William H. *The Organization Man*. Philadelphia: University of Pennsylvania Press, 2002.

Wilson, Sloan. *The Man in the Gray Flannel Suit*. London: Penguin, 2005.

Wolfle, Dael. "AAAS Council Meeting, 1969." *Science*, n.s., 167, no. 3921 (February 20, 1970): 1151–53. doi:10.2307/1728701.

World Federation of Scientific Workers. *Scientific World, Volumes 14–21* (London: World Federation of Scientific Workers, 1970).

The Wrath of Grapes. Directed by Lorena Parlee and Lenny Bourin. Keene, CA: United Farm Workers, 1986.

Zumwalt, Elmo R., Elmo Zumwalt, and John Pekkanen. *My Father, My Son*. New York: Dell, 1987.

Secondary Sources

Alexander, Michelle. "The War on Drugs and the New Jim Crow." *Race, Poverty & the Environment* 17, no. 1 (Spring 2010): 75–77.

Anderson, J. L. *Industrializing the Corn Belt: Agriculture, Technology, and Environment, 1945–1972*. DeKalb: Northern Illinois University Press, 2009.

———. "War on Weeds: Iowa Farmers and Growth-Regulator Herbicides." *Technology and Culture* 46, no. 4 (2005): 719–44.

Anderson, M. Kat, Michael G. Barbour, and Valerie Whitworth. "A World of Balance and Plenty: Land, Plants, Animals, and Humans in a Pre-European California." *California History* 76, no. 2/3 (1997): 12–47. doi:10.2307/25161661.

Arnold, David John. *The Problem of Nature: Environment, Culture and European Expansion*. Oxford: Blackwell, 1996.

————. *The Tropics and the Traveling Gaze: India, Landscape, and Science, 1800–1856*. Seattle: University of Washington Press, 2006.

Aso, Mitchitake, and Annick Guénel. "The Itinerary of a North Vietnamese Surgeon: Medical Science and Politics during the Cold War." *Science, Technology & Society* 18, no. 3 (2013): 291–306.

Bailey, Beth L. *Sex in the Heartland*. Cambridge: Harvard University Press, 1999.

Bailey, Beth L., and David Farber. *America in the Seventies*. Lawrence: University Press of Kansas, 2004.

Bankoff, Greg. "A Question of Breeding: Zootechny and Colonial Attitudes toward the Tropical Environment in the Late Nineteenth-Century Philippines." *Journal of Asian Studies* 60, no. 2 (2001): 413–37. doi:10.2307/2659699.

Bartilow, Horace A. "Drug Wars Collateral Damage." *Latin American Research Review* 49, no. 2 (April 2014): 24–46.

Baum, Dan. "Legalize It All: How to Win the War on Drugs." *Harper's Magazine*, March 22, 2016. https://harpers.org.

Beck, Ulrich. *Risk Society towards a New Modernity*. Translated by Mark Ritter. London: Sage, 1992.

Bewley-Taylor, David R. "The Cost of Containment: The Cold War and US International Drug Control at the UN, 1950–58." *Diplomacy & Statecraft* 10, no. 1 (March 1999): 147–71. America: History and Life with Full Text, EBSCOhost.

Biggs, David A. "Managing a Rebel Landscape: Conservation, Pioneers, and the Revolutionary Past in the U Minh Forest, Vietnam." *Environmental History* 10, no. 3 (2005): 448–76.

————. *Quagmire: Nation-Building and Nature in the Mekong Delta*. Seattle: University of Washington Press, 2011.

Björk, Tord. "Challenging Western Environmentalism at the United Nations Conference on Human Environment at Stockholm 1972." RIO + 20 / STH + 40, Paper II, June 2012, 2:21. http://www.aktivism.info/rapporter/ChallengingUN72.pdf.

Blaser, Arthur W. "How to Advance Human Rights without Really Trying: An Analysis of Nongovernmental Tribunals." *Human Rights Quarterly* 14, no. 3 (1992): 339–70. doi:10.2307/762370.

Bonaccorsi, Aurora, Roberto Fanelli, and Gianni Tognoni. "In the Wake of Seveso." *Ambio* 7, no. 5/6 (1978): 234–39.

Bovey, Rodney W. *Woody Plants and Woody Plant Management: Ecology, Safety, and Environmental Impact*. New York: M. Dekker, 2001.

Bradley, Mark, and Marilyn Blatt Young. *Making Sense of the Vietnam Wars: Local, National, and Transnational Perspectives*. Oxford: Oxford University Press, 2008.

Brady, Emily. *Humboldt: Life on America's Marijuana Frontier*. Boston: Grand Central, 2013.

Brailsford, Ian. "Madison Avenue Takes on the 'Teenage Hop Heads': The Advertising Council's Campaign against Drugs in the Nixon Era." *Australasian Journal of American Studies* 18, no. 2 (1999): 43–59.

Braunstein, Peter, and Michael William Doyle. *Imagine Nation: The American Counterculture of the 1960s and '70s.* New York: Routledge, 2002.

Brocheux, Pierre, Daniel Hémery, and Ly Lan Dill-Klein. *Indochina: An Ambiguous Colonization, 1858–1954.* Berkeley: University of California Press, 2009.

Brock, Emily K. *Money Trees: The Douglas Fir and American Forestry, 1900–1944.* Corvallis: Oregon State University Press, 2015.

Brown, Beverly A. *In Timber Country: Working People's Stories of Environmental Conflict and Urban Flight.* Philadelphia: Temple University Press, 1995.

Buckingham, William A. *Operation Ranch Hand: The Air Force and Herbicides in Southeast Asia, 1961–1971.* Washington, DC: Office of Air Force History, United States Air Force, 1981.

Buhs, Joshua Blu. *The Fire Ant Wars: Nature, Science, and Public Policy in Twentieth-Century America.* Chicago: University of Chicago Press, 2004.

Bui, Thi Phuong-Lan. "When the Forest Became the Enemy and the Legacy of American Herbicidal Warfare in Vietnam." PhD dissertation, Harvard University, 2003.

Buzzanco, Robert. *Masters of War: Military Dissent and Politics in the Vietnam Era.* Cambridge: Cambridge University Press, 2005.

Campos, Isaac. *Home Grown: Marijuana and the Origins of Mexico's War on Drugs.* Chapel Hill: University of North Carolina Press, 2014.

Cannon, Brian Q. "Water and Economic Opportunity: Homesteaders, Speculators, and the U.S. Reclamation Service, 1904–1924." *Agricultural History* 76, no. 2 (April 1, 2002): 188–207. doi:10.2307/3744998.

Carle, David. *Burning Questions: America's Fight with Nature's Fire.* Westport, CT: Praeger, 2002.

Cecil, Paul F. *Herbicidal Warfare: The Ranch Hand Project in Vietnam.* New York: Praeger, 1986.

Chandler, Alfred D. *Shaping the Industrial Century: The Remarkable Story of the Evolution of the Modern Chemical and Pharmaceutical Industries.* Cambridge: Harvard University Press, 2005.

Coye, Molly Joel. "The Health Effects of Agricultural Production: I. The Health of Agricultural Workers." *Journal of Public Health Policy* 6, no. 3 (1985): 349–70. doi:10.2307/3342402.

———. "The Health Effects of Agricultural Production: II. The Health of the Community." *Journal of Public Health Policy* 7, no. 3 (1986): 340–54. doi:10.2307/3342461.

Crafts, A. S. "Selectivity of Herbicides." *Plant Physiology* 21, no. 3 (1946): 345–61.

Crane, Jeff, and Michael Egan. *Natural Protest: Essays on the History of American Environmentalism.* New York: Routledge, 2009.

Crummett, Warren B. *Decades of Dioxin: Limelight on a Molecule.* Philadelphia: Xlibris, 2002.

Daniel, Pete. *Toxic Drift Pesticides and Health in the Post–World War II South.* Baton Rouge: Louisiana State University Press, 2007.

Daum, Andreas W., Lloyd C. Gardner, and Wilfried Mausbach, eds. *America, the Vietnam War, and the World: Comparative and International Perspectives.* Washington, DC: German Historical Institute, 2003.

Davidson, Phillip B. *Vietnam at War: The History: 1946–1975.* New York: Oxford University Press, 1991.

Davis, Frederick Rowe. *Banned: A History of Pesticides and the Science of Toxicology.* New Haven: Yale University Press, 2015.

Davis, Joshua Clark. *From Head Shops to Whole Foods: The Rise and Fall of Activist Entrepreneurs.* New York: Columbia University Press, 2017.

Davis, Mike. *Ecology of Fear: Los Angeles and the Imagination of Disaster.* London: Picador, 2000.

DeBenedetti, Charles. *An American Ordeal: The Antiwar Movement of the Vietnam Era.* Syracuse: Syracuse University Press, 1990.

De Groot, Gerard J. "Ronald Reagan and Student Unrest in California, 1966–1970." *Pacific Historical Review* 65, no. 1 (February 1996).

Dennis, Michael Aaron. "'Our First Line of Defense': Two University Laboratories in the Postwar American State." *Isis* 85, no. 3 (1994): 427–55.

Dittmer, John. *The Good Doctors: The Medical Committee for Human Rights and the Struggle for Social Justice in Health Care.* New York: Bloomsbury Press, 2009.

Doyle, Jack. *Trespass against Us: Dow Chemical & the Toxic Century.* Monroe, ME: Common Courage Press, 2004.

Dreyfuss, Robert. "Apocalypse Still." *Mother Jones*, January 1, 2000. http://www.motherjones.com.

Dunlap, Thomas. *DDT: Scientists, Citizens, and Public Policy.* Princeton: Princeton University Press, 1981.

———. *DDT, "Silent Spring," and the Rise of Environmentalism: Classic Texts* (Seattle: University of Washington Press, 2008).

Dyck, Erika. *Psychedelic Psychiatry: LSD from Clinic to Campus.* Baltimore: Johns Hopkins University Press, 2008.

Elliott, David W. P. *The Vietnamese War: Revolution and Social Change in the Mekong Delta 1930–1975.* Armonk, NY: M. E. Sharpe, 2006.

Elliott, Duong Van Mai. "RAND in Southeast Asia: A History of the Vietnam War Era." (RAND, 2010). http://rand.org.

Engber, Daniel. "Test-Tube Piggies: How Did the Guinea Pig Become a Symbol of Science?" *Slate*, June 18, 2012. http://www.slate.com.

Espeland, Wendy Nelson. *The Struggle for Water: Politics, Rationality, and Identity in the American Southwest.* Chicago: University of Chicago Press, 1998.

Falck, Zachary J. S. *Weeds: An Environmental History of Metropolitan America*. Pittsburgh: University of Pittsburgh Press, 2010.

Ferguson, Cody. *This Is Our Land: Grassroots Environmentalism in the Late Twentieth Century*. New Brunswick, NJ: Rutgers University Press, 2015.

Fermaglich, Kirsten. *American Dreams and Nazi Nightmares: Early Holocaust Consciousness and Liberal America, 1957–1965*. Waltham, MA: Brandeis University Press, 2007.

Ffolliott, Peter F., Malchus B. Baker, and Leonard F. DeBano. "Arizona Watershed Management Program." *Journal of the Arizona-Nevada Academy of Science* 35, no. 1 (2003): 5–10.

The Fog of War: Eleven Lessons from the Life of Robert S. McNamara. Directed by Errol Morris. New York: Sony Pictures Classics, 2004.

Fonda, Jane. *My Life So Far*. New York: Random House, 2005.

Forrestal, Dan J. *Faith, Hope, and $5000: The Story of Monsanto: The Trials and Triumphs of the First 75 Years*. New York: Simon and Schuster, 1977.

Fox, Diane Niblack. "Agent Orange, Vietnam, and the United States: Blurring the Boundaries." In *Vietnam and the West: New Approaches*, edited by Wynn Wilcox, 140–55. Ithaca: Cornell University, Southeast Asia Program, 2010.

———. "Chemical Politics and the Hazards of Modern Warfare: Agent Orange." In *Synthetic Planet: Chemical Politics and the Hazards of Modern Life*, edited by Monica J. Casper, 73–90. New York: Routledge, 2003.

Freeman, Michael. "Speaking about the Unspeakable: Genocide and Philosophy." *Journal of Applied Philosophy* 8, no. 1 (1991): 3–17.

French, Wesley Tyler. "Drug War Zone: Frontline Dispatches from the Streets of EL Paso and Juárez." *Oral History Review* 38, no. 2 (June 2011): 378–80. EBSCOhost.

Frenkel, Stephen. "Jungle Stories: North American Representations of Tropical Panama." *Geographical Review* 86, no. 3 (1996): 317–33. doi:10.2307/215497.

Frost, Jennifer. *An Interracial Movement of the Poor: Community Organizing and the New Left in the 1960s*. New York: New York University Press, 2005.

Frydl, Kathleen. *The Drug Wars in America, 1940–1973*. Cambridge: Cambridge University Press, 2013.

Gammeltoft, Tine. *Haunting Images: A Cultural Account of Selective Reproduction in Vietnam*. Berkeley: University of California Press, 2014.

Giddens, Anthony. *The Consequences of Modernity*. Stanford: Stanford University Press, 1990.

Gleason, Philip. *Contemporary Catholicism in the United States*. Notre Dame, IN: University of Notre Dame, 1969.

Gogman, Lars. "Vietnam in the Collections." Labor Movement and Archive. http://www.arbark.se/wib/vietnam-in-the-collections.pdf.

Goldstein, Jonathan. "Agent Orange on Campus: The Summit-Spicerack Controversy at the University of Pennsylvania, 1965–1967." In *Sights on the Sixties*, edited by Barbara Tischler, 43–62. New Brunswick, NJ: Rutgers University Press, 1992.

Gough, Michael. *Dioxin, Agent Orange: The Facts.* New York: Springer, 1986.

Griffiths, Philip Jones. *Agent Orange: Collateral Damage in Viet Nam.* London: Trolley, 2003.

Gross, Harriet. "Jane Kennedy: Making History through Moral Protest." *Frontiers: A Journal of Women Studies* 2, no. 2 (July 1, 1977): 73–81. doi:10.2307/3346016.

Haak, Wilbur A., Lynn F. Haak, and Gila County Historical Museum Archive. *Globe.* Charleston, SC: Arcadia, 2008.

Hall, Mitchell K. *Because of Their Faith: CALCAV and Religious Opposition to the Vietnam War.* New York: Columbia University Press, 1990.

Halliwell, Martin. "Cold War Ground Zero: Medicine, Psyops and the Bomb." *Journal of American Studies* 44, no. 2 (2010): 313–31.

Hanson, Hilary. "Nixon Aides Suggest Colleague Was Kidding about Drug War Being Designed to Target Black People." *Huffington Post.* http://www.huffingtonpost.com.

Harrington, Jerry. "The Midwest Agricultural Chemical Association: A Regional Study of an Industry on the Defensive." *Agricultural History* 70, no. 2 (1996): 415–38.

Harvey, David. *The Condition of Post Modernity: An Enquiry into the Origins of Cultural Change* (Cambridge, MA: B. Blackwell, 1980).

Hay, Amy. "Recipe for Disaster: Chemical Wastes, Community Activists, and Public Health at Love Canal, 1945–2000." PhD dissertation, Michigan State University, 2005.

———. "Recipe for Disaster: Motherhood and Citizenship at Love Canal." *Journal of Women's History* 21, no. 1 (Spring 2009): 111–34.

Hazlett, Maril. "Voices from the Spring: *Silent Spring* and the Ecological Turn in American Health." In *Seeing Nature through Gender,* edited by Virginia Scharff, 103–28. Lawrence: University Press of Kansas, 2003.

Helling, Thomas S., and Daniel Azoulay. "Ton That Tung's Livers." *Annals of Surgery* 259, no. 6 (2014): 1245–52.

Herbicides and Defoliants in War: The Long-Term Effects on Man and Nature. Honolulu: University Press of the Pacific, 2003.

Herzog, Tobey C. *Vietnam War Stories: Innocence Lost.* London: Routledge, 1992.

———. *Writing Vietnam, Writing Life: Caputo, Heinemann, O'Brien, Butler.* Iowa City: University of Iowa Press, 2008.

Hickey, Gerald. *Window on a War: An Anthropologist in the Vietnam Conflict.* Lubbock: Texas Tech University Press, 2002.

Hitchens, Christopher. "The Seventies: The Vietnam Syndrome." Photographs by James Nachtwey. *Vanity Fair,* March 26, 2007. https://www.vanityfair.com.

Hornblum, Allen. "Subjected to Medical Experimentation: Pennsylvania's

Contribution to 'Science' in Prisons." *Pennsylvania History* 67, no. 3 (2000): 415–26.

Hunt, Andrew E. *The Turning: A History of Vietnam Veterans Against the War*. New York: New York University Press, 1999.

Jackson, Kenneth T. *Crabgrass Frontier: The Suburbanization of the United States*. New York: Oxford University Press, 1987.

Jenkins, Virginia Scott. *The Lawn: A History of an American Obsession*. Washington, DC: Smithsonian Institution Press, 1994.

Johnson, Thomas J., and Wayne Wanta. "Influence Dealers: A Path Analysis Model of Agenda Building during Richard Nixon's War on Drugs." *Journalism & Mass Communication Quarterly* 73, no. 1 (1996): 181–94. EBSCOhost.

Jurca, Catherine. "The Sanctimonious Suburbanite: Sloan Wilson's *The Man in the Gray Flannel Suit*." *American Literary History* 11, no. 1 (1999): 82–106.

Kennedy, Patrick D. "Reactions against the Vietnam War and Military-Related Targets on Campus: The University of Illinois as a Case Study, 1965–1972." *Illinois Historical Journal* 84, no. 2 (July 1, 1991): 101–18.

Kinkela, David. *DDT and the American Century Global Health, Environmental Politics, and the Pesticide That Changed the World*. Chapel Hill: University of North Carolina Press, 2011.

Klejment, Anne. "The Spirituality of Dorothy Day's Pacifism." *U.S. Catholic Historian* 27, no. 2 (April 1, 2009): 1–24. doi:10.2307/40468572.

Klinghoffer, Arthur Jay, and Judith Apter Klinghoffer. *International Citizens' Tribunals: Mobilizing Public Opinion to Advance Human Rights*. New York: Palgrave, 2002.

Kopp, James J. *Eden within Eden: Oregon's Utopian Heritage*. Corvallis: Oregon State University Press, 2009.

Krache Morris, Evelyn Frances. "Into the Wind: The Kennedy Administration and the Use of Herbicides in South Vietnam." PhD dissertation, Georgetown University, 2012. ProQuest.

Krech, Shepard III. *The Ecological Indian: Myth and History*. New York: W. W. Norton, 2000.

Krueger, Gretchen Marie. "'For Jimmy and the Boys and Girls of America': Publicizing Childhood Cancers in Twentieth-Century America." *Bulletin of the History of Medicine* 81, no. 1 (2007): 70–93.

Kuhn, Annette, and Kirsten Emiko McAllister. *Locating Memory: Photographic Acts*. New York: Berghahn Books, 2008.

Kuzmarov, Jeremy. "From Counter-Insurgency to Narco-Insurgency: Vietnam and the International War on Drugs." *Journal of Policy History* 20, no. 3 (July 2008): 344–78. EBSCOhost.

———. *The Myth of the Addicted Army: Vietnam and the Modern War on Drugs*. Amherst: University of Massachusetts Press, 2009.

Langston, Nancy. *Forest Dreams, Forest Nightmares: The Paradox of Old Growth in the Inland West.* Seattle: University of Washington Press, 1995.

———. *Toxic Bodies: Hormone Disruptors and the Legacy of DES.* New Haven: Yale University Press, 2011.

Latham, Michael E. "Knowledge at War: American Social Science and Vietnam." In *A Companion to the Vietnam War,* edited by Marilyn B. Young and Robert Buzzanco, 434–49. Malden, MA: Blackwell, 2006.

Layne, Linda L. "In Search of Community: Tales of Pregnancy Loss in Three Toxically Assaulted U.S. Communities." In special issue "Earthwork: Women and Environments," *Women's Studies Quarterly* 29, no. 1/2 (2001): 25–50.

Lear, Linda J. "Bombshell in Beltsville: The USDA and the Challenge of 'Silent Spring.'" *Agricultural History* 66, no. 2 (1992): 151–70.

———. *Rachel Carson: Witness for Nature.* New York: Penguin, 2009. Originally published 1997.

Leavitt, Judith W. "'Typhoid Mary' Strikes Back: Bacteriological Theory and Practice in Early Twentieth-Century Public Health." *Isis* 83 (1992).

Li-Lian, Sharon Seah. "Truth and Memory: Narrating Viet Nam." *Asian Journal of Social Science* 29, no. 3 (2001): 381–400.

Lichtman, Sarah A. "Do-It-Yourself Security: Safety, Gender, and the Home Fallout Shelter in Cold War America." *Journal of Design History* 19, no. 1 (2006): 39–55.

Lieberman, Marvin B. "Magnesium Industry in Transition." *Journal of Industrial Organization* June 19, 2001, 3.

Lind, Michael. *Vietnam, the Necessary War: A Reinterpretation of America's Most Disastrous Military Conflict.* New York: Free Press, 1999.

Logevall, Fredrik. "America Isolated: The Western Powers and the Escalation of the War." In *America, the Vietnam War, and the World: Comparative and International Perspectives,* edited by Andreas W. Daum, Lloyd C. Gardner, and Wilfried Mausbach, 175–96. New York: German Historical Institute; Cambridge University Press, 2003.

———. *Choosing War: The Lost Chance for Peace and the Escalation of War in Vietnam.* Berkeley: University of California Press, 1999.

Lytle, Mark H. *The Gentle Subversive: Rachel Carson, "Silent Spring," and the Rise of the Environmental Movement.* New York: Oxford University Press, 2007.

MacGregor, Sherilyn. *Beyond Mothering Earth: Ecological Citizenship and the Politics of Care.* Vancouver: University of British Columbia Press, 2006.

MacQueen, Graeme, and Joanna Santa-Barbara. "Peace Building through Health Initiatives." *BMJ: British Medical Journal* 321, no. 7256 (July 29, 2000): 293–96.

Maraniss, David. *They Marched into Sunlight: War and Peace, Vietnam and America, October 1967.* New York: Simon & Schuster, 2003.

Marcus, Alan I. *Cancer from Beef: DES, Federal Food Regulation, and Consumer Confidence.* Baltimore: Johns Hopkins University Press, 1994.

Markowitz, Gerald, and David Rosner. *Deceit and Denial: The Deadly Politics of Industrial Pollution.* Berkeley: University of California Press, 2002.

Mart, Michelle. *Pesticides, A Love Story: America's Enduring Embrace of Dangerous Chemicals.* Lawrence: University of Kansas Press, 2015.

Martin, Emily. *Flexible Bodies.* New York: Beacon Press, 1995.

Martini, Edwin A. *Agent Orange: History, Science, and the Politics of Uncertainty.* Amherst: University of Massachusetts Press, 2012.

Martucci, Jessica. *Back to the Breast: Natural Motherhood and Breastfeeding in America.* Chicago: University of Chicago Press, 2015.

Mason, Bobbie Ann. *In Country.* New York: Harper & Row, 1985.

May, Elaine Tyler. *Homeward Bound: American Families in the Cold War Era.* New York: Basic Books, 2008.

McElewee, Pamela D. *Forests Are Gold: Trees, People, and Environmental Rule in Vietnam.* Seattle: University of Seattle Press, 2016.

McMillian, John Campbell. *Smoking Typewriters: The Sixties Underground Press and the Rise of Alternative Media in America.* New York: Oxford University Press, 2014.

McWilliams, John C. "Unsung Partner against Crime: Harry J. Anslinger and the Federal Bureau of Narcotics, 1930–1962." *Pennsylvania Magazine of History and Biography* 113, no. 2 (1989): 207–36.

Meconis, Charles A. *With Clumsy Grace: The American Catholic Left, 1961–1975.* New York: Seabury Press, 1979.

Meikle, Jeffrey. *American Plastic: A Cultural History.* New Brunswick, NJ: Rutgers University Press, 1995.

Merton, Thomas. *Faith and Violence: Christian Teaching and Christian Practice.* Notre Dame, IN: University of Notre Dame Press, 1994.

Meyer, Alan D., and David P. Anderson. "The Air National Guard and the War on Drugs: Non-State Actors before 9/11." *Air Power History* 55, no. 3 (2008): 12–29. EBSCOhost.

Meyer, David S., and Deana A. Rohlinger. "Big Books and Social Movements: A Myth of Ideas and Social Change." *Social Problems* 59, no. 1 (2012): 136–53. doi:10.1525/sp.2012.59.1.136.

Miller, Edward. *Misalliance: Ngo Dinh Diem, the United States, and the Fate of South Vietnam.* Cambridge: Harvard University Press, 2013.

———. "War Stories: The Taylor-Buzzanco Debate and How We Think about the Vietnam War." *Journal of Vietnamese Studies* 1, no. 1–2 (2006): 453–84. doi: 10.1525/vs.2006.1.1-2.453.

Milne, David. *America's Rasputin: Walt Rostow and the Vietnam War.* New York: Hill and Wang, 2008.

Mittman, Gregg. *Breathing Space: How Allergies Shape Our Lives and Landscapes.* New Haven: Yale University Press, 2007.

Mollin, Marian. "Communities of Resistance: Women and the Catholic Left of the

Late 1960s." *Oral History Review* 31, no. 2 (2004): 29–51.

Moon, Penelope Adams. "'Peace on Earth: Peace in Vietnam': The Catholic Peace Fellowship and Antiwar Witness, 1964–1976." *Journal of Social History* 36, no. 4 (July 1, 2003): 1033–57. doi:10.2307/3790362.

Moore, Kelly. *Disrupting Science: Social Movements, American Scientists, and the Politics of the Military, 1945–1975*. Princeton: Princeton University Press, 2013.

Morgan, Joseph G. "A Change of Course: American Catholics, Anticommunism, and the Vietnam War." *U.S. Catholic Historian* 22, no. 4 (October 1, 2004): 117–30. doi:10.2307/25154936.

Murphy, Michelle. *Sick Building Syndrome and the Problem of Uncertainty: Environmental Politics, Technoscience, and Women Workers*. Durham, NC: Duke University Press, 2006.

Musto, David. The History of the Marihuana Tax Act of 1937." Schaffer Library of Drug Policy. https://druglibrary.net. The article first appeared in the *Archives of General Psychiatry* 26 (February 1972).

Nash, Linda Lorraine. *Inescapable Ecologies: A History of Environment, Disease, and Knowledge*. Berkeley: University of California Press, 2006.

Neer, Robert M. *Napalm: An American Biography*. Cambridge: Harvard University Press, 2015.

Nguyen, Nathalie Huynh Chau. "Military Doctors in South Vietnam: Wartime and Post-war Lives." *Oral History* 43, no. 1 (2015): 85–96.

Nhàn, Vương Trí. "The Diary of Đặng Thùy Trâm and the Postwar Vietnamese Mentality." *Journal of Vietnamese Studies* 3, no. 2 (2008): 180–95. doi:10.1525/vs.2008.3.2.180.

Nixon, Rob. *Slow Violence and the Environmentalism of the Poor*. Cambridge: Harvard University Press, 2011.

Olmsted, Kathryn S. *Right out of California: The 1930s and the Big Business Roots of Modern Conservatism*. New York: New Press, 2015.

Paterson, Thomas. *Kennedy's Quest for Victory: American Foreign Policy, 1961–1963*. New York: Oxford University Press, 1989.

Peluso, Nancy Lee, and Peter Vandergeest. "Political Ecologies of War and Forests: Counterinsurgencies and the Making of National Natures." *Annals of the Association of American Geographers* 101, no. 3 (2011): 587–608.

Pembleton, Matthew R. "The Voice of the Bureau: How Frederic Sondern and the Bureau of Narcotics Crafted a Drug War and Shaped Popular Understanding of Drugs, Addiction, and Organized Crime in the 1950s." *Journal of American Culture* 38, no. 2 (June 2015): 113–29. EBSCOhost.

Peters, Shawn Francis. *The Catonsville Nine: A Story of Faith and Resistance in the Vietnam Era*. Oxford: Oxford University Press.

Peterson, Gale E. "The Discovery and Development of 2,4-D." *Agricultural History* 41, no. 3 (1967): 243–54.

Pies, Cheri, Malcolm Potts, and Bethany Young. "Quinacrine Pellets: An Examination of Nonsurgical Sterilization." *International Family Planning Perspectives* 20, no. 4 (December 1994): 137–41.

Pike, Douglas. *Viet Cong: The Organization and Techniques of the National Liberation Front of South Vietnam.* Cambridge, MA: MIT Press, 1966.

Pulido, Laura. *Environmentalism and Economic Justice: Two Chicano Struggles in the Southwest.* Tucson: University of Arizona Press, 1996.

Pyne, Stephen J. *California: A Fire Survey.* Tucson: University of Arizona Press, 2016.

———. *Fire in America: A Cultural History of Wildland and Rural Fire.* Princeton: Princeton University Press, 1982.

———. "Fire Policy and Fire Research in the U.S. Forest Service." *Journal of Forest History* 25, no. 2 (1981): 64–77. doi:10.2307/4004547.

———. *The Southwest: A Fire Survey.* Tucson: University of Arizona Press, 2016.

Rambo: First Blood. Directed by Ted Kotcheff. Hollywood, CA: Orion Pictures [MGM], 1982.

Rasmussen, Karen, and Sharon D. Downey. "Dialectical Disorientation in Vietnam War Films: Subversion of the Mythology of War." *Quarterly Journal of Speech* 77, no. 2 (May 1, 1991): 176–95. doi:10.1080/00335639109383951.

Rasmussen, Nicolas. "Plant Hormones in War and Peace: Science, Industry, and Government in the Development of Herbicides in 1940s America." *Isis* 92, no. 2 (2001): 291–316.

Reagan, Leslie J. "From Hazard to Blessing to Tragedy: Representations of Miscarriage in Twentieth-Century America." *Feminist Studies* 29, no. 2 (2003): 356–78.

———. "'My Daughter Was Genetically Drafted with Me': US-Vietnam War Veterans, Disabilities and Gender." *Gender & History* 28, no. 3 (2016): 833–53.

———. "Representations and Reproductive Hazards of Agent Orange." *Journal of Law, Medicine, and Ethics* 39, no. 1 (2011): 54–61. doi:10.1111/j.1748-720X .2011.00549.x.

Rinner, Susanne. *The German Student Movement and the Literary Imagination: Transnational Memories of Protest and Dissent.* New York: Berghahn Books, 2013.

Robbins, Paul. *Lawn People: How Grasses, Weeds, and Chemicals Make Us Who We Are.* Philadelphia, PA: Temple University Press, 2007.

Robbins, William G. "Cornucopian Dreams: Remaking Nature in Postwar Oregon." *Agricultural History* 76, no. 2 (2002): 208–19.

———. *Landscapes of Conflict: The Oregon Story: 1940–2000.* Seattle: University of Washington Press, 2014.

———. *Landscapes of Promise: The Oregon Story, 1800–1940.* Seattle: University of Washington Press, 2009.

———. *Lumberjacks and Legislators: Political Economy of the U.S. Lumber Industry, 1890–1941.* College Station: Texas A&M University Press, 1982.

———. "The Social Context of Forestry: The Pacific Northwest in the

Twentieth Century." *Western Historical Quarterly* 16, no. 4 (1985): 413–27. doi:10.2307/968606.

Rogers, Naomi. *Dirt and Disease: Polio before FDR.* New Brunswick, NJ: Rutgers University Press, 1996.

Rome, Adam Ward. *The Bulldozer in the Countryside: Suburban Sprawl and the Rise of American Environmentalism.* Cambridge: Cambridge University Press, 2001.

———. "'Give Earth a Chance': The Environmental Movement and the Sixties." *Journal of American History* 90, no. 2 (2003): 525–54. doi:10.2307/3659443.

Rossinow, Doug. "The New Left in the Counterculture: Hypotheses and Evidence." *Radical History Review* no. 67 (Winter 1997): 79–120. doi:10.1215/01636545-1997-67-79.

Russell, Edmund. *War and Nature: Fighting Humans and Insects with Chemicals from World War I to "Silent Spring."* Cambridge: Cambridge Univ. Press, 2001.

Said, Edward W. *Culture and Imperialism.* London: Vintage, 2007.

Schama, Simon. *Landscape and Memory.* London: Harper Perennial, 2004.

Schlesinger, Arthur M. *The Vital Center: The Politics of Freedom.* Cambridge, MA: Riverside, 1949.

Schlich, Thomas. "Linking Cause and Disease in the Laboratory: Robert Koch's Method of Superimposing Visual and 'Functional' Representations of Bacteria." *History and Philosophy of the Life Sciences* 22, no. 1 (2000): 43–58.

Schulman, Bruce. *The Seventies: The Great Shift in American Culture, Society, and Politics.* New York: Free Press, 2001.

Schulte, Terrianne K. "Citizen Experts: The League of Women Voters and Environmental Conservation." *Frontiers: A Journal of Women Studies* 30, no. 3 (2009): 1–29.

Scott, James C. *The Art of Not Being Governed: An Anarchist History of Upland Southeast Asia.* New Haven: Yale University Press, 2010.

Scott, Wilbur J. "Competing Paradigms in the Assessment of Latent Disorders: The Case of Agent Orange." *Social Problems* 35, no. 2 (1988): 145–61. doi:10.2307/800737.

———. *The Politics of Readjustment: Vietnam Veterans since the War.* New York: Aldine de Gruyter, 1993.

Schuck, Peter H. *Agent Orange on Trial: Mass Toxic Disasters in the Courts.* Cambridge: Belknap Press of Harvard University Press, 1987.

Schwenkel, Christina. "Exhibiting War, Reconciling Pasts: Photographic Representation and Transnational Commemoration in Contemporary Vietnam." *Journal of Vietnamese Studies* 3, no. 1 (2008): 36–77. doi:10.1525/vs.2008.3.1.36.

Sellers, Christopher C. "Body, Place and the State: The Makings of an 'Environmentalist' Imaginary in the Post–World War II U.S. *Radical History Review* 74 (Spring 1999): 31–64. doi:10.1215/01636545-1999-74-31.

———. *Crabgrass Crucible: Suburban Nature and the Rise of Environmentalism in Twentieth-Century America.* Chapel Hill: University of North Carolina Press, 2012.

Shaw, Randy. *Beyond the Fields: Cesar Chavez, the UFW, and the Struggle for Justice in the 21st Century*. Berkeley: University of California Press, 2011.

Sills, Peter. *Toxic War: The Story of Agent Orange*. Nashville: Vanderbilt University Press, 2014.

Smith, Karen L. "The Campaign for Water in Central Arizona, 1890–1903." *Arizona and the West* 23, no. 2 (1981): 127–48.

———. *The Magnificent Experiment: Building the Salt River Reclamation Project, 1890–1917*. Tucson: University of Arizona Press, 1986.

Smith, Susan Livingston. *Toxic Exposures: Mustard Gas and the Health Consequences of World War II in the United States*. New Brunswick, NJ: Rutgers University Press, 2017.

Solovey, Mark. "Introduction: Science and the State during the Cold War: Blurred Boundaries and a Contested Legacy." *Social Studies of Science* 31, no. 2 (April 1, 2001): 165–70. doi:10.2307/3183110.

Sowards, Adam M. "Administrative Trials, Environmental Consequences, and the Use of History in Arizona's Tonto National Forest, 1926–1996." *The Western Historical Quarterly*. 31, no. 4 (2000): 189.

———. "Reclamation, Ranching, and Reservation: Environmental, Cultural, and Governmental Rivalries in Transitional Arizona." *Journal of the Southwest* 40, no. 3 (1998): 333–61.

Speece, Darren Frederick. *Defending Giants: The Redwood Wars and the Transformation of American Environmental Politics*. Seattle: University of Washington Press, 2017.

Steinberg, Ted. *American Green: The Obsessive Quest for the Perfect Lawn*. New York: W. W. Norton, 2006.

———. "Lawn and Landscape in World Context, 1945–2000." *OAH Magazine of History* 19, no. 6 (2005): 62–68.

Stellman, Jeanne Mager, Steven D. Stellman, Richard Christian, Tracy Weber, and Carrie Tomasallo. "The Extent and Patterns of Usage of Agent Orange and Other Herbicides in Vietnam." *Nature* 422 (April 17, 2002): 681–87. http://www.vn-agentorange.org.

Stephens, Sharon. "Reflections on Environmental Justice: Children as Victims and Actors." In *Environmental Crime: A Reader Environmental Crime*, edited by R. D. White, 222–43. London: Routledge, 2009.

Stoll, Steven. *The Fruits of Natural Advantage: Making the Industrial Countryside in California*. Berkeley: University of California Press, 1998.

Tesh, Sylvia Noble. *Uncertain Hazards: Environmental Activists and Scientific Proof*. Ithaca: Cornell University Press, 2000.

Thorpe, Charles. *Oppenheimer: The Tragic Intellect*. Chicago: University of Chicago Press, 2008.

Tischler, Barbara L., ed. *Sights on the Sixties*. New Brunswick, NJ: Rutgers University Press, 1992.

Tomes, Nancy. "The Great American Medicine Show Revisited." *Bulletin of the History of Medicine* 79 (2005): 627–63. doi:10.2307/2675104.

———. "Merchants of Health: Medicine and Consumer Culture in the United States, 1900–1940." *Journal of American History* 88, no. 2 (2001): 519–47.

Tompkins, Adam. *Ghostworkers and Greens: The Cooperative Campaigns of Farmworkers and Environmentalists for Pesticide Reform*. Ithaca: IRL, imprint of Cornell University Press, 2016.

Trost, Cathy. *Elements of Risk: The Chemical Industry and Its Threat to America*. New York: Times Books, 1984.

Vail, David. *Chemical Lands: Pesticides, Aerial Spraying, and Health in North America's Grasslands since 1945*. Tuscaloosa: University of Alabama Press, 2018.

———. "'Kill That Thistle': Rogue Sprayers, Bootlegged Chemicals, Wicked Weeds, and the Kansas Chemical Laws, 1945–1980." *Kansas History* 32, no. 2 (2012).

Vaught, David. *Cultivating California: Growers, Specialty Crops, and Labor, 1875–1920*. Baltimore: Johns Hopkins University Press, 1999.

Vogel, Sarah A. *Is It Safe? BPA and the Struggle to Define the Safety of Chemicals*. Berkeley: University of California Press, 2013.

Wang, Jessica. *American Science in an Age of Anxiety: Scientists, Anticommunism, and the Cold War*. Chapel Hill: University of North Carolina Press, 1999.

Weimer, Daniel. *Seeing Drugs: Modernization, Counterinsurgency, and U.S. Narcotics Control in the Third World, 1969–1976*. Kent, OH: Kent State University Press, 2011.

Wells, H. G. *Men like Gods*. New York: Ferris, 1923.

Wesley, Marilyn. "Truth and Fiction in Tim O'Brien's 'If I Die in a Combat Zone' and 'The Things They Carried.'" *College Literature* 29, no. 2 (2002): 1–18.

Whorton, James. *Before "Silent Spring": Pesticides and Public Health in Pre-DDT America*. Princeton: Princeton University Press, 1974.

Wilcox, Fred A. *Waiting for an Army to Die: The Tragedy of Agent Orange*. Cabin John, MD: Seven Locks Press, 1989.

Wishnia, Steven. "Debunking the Hemp Conspiracy Myth." *Alternet*, February 20, 2008. http://www.alternet.org

Wurster, Charles F. *DDT Wars: Rescuing Our National Bird, Preventing Cancer, and Creating the Environmental Defense Fund*. New York: Oxford University Press, 2015.

Young, Marilyn Blatt, and Robert Buzzanco. *A Companion to the Vietnam War*. Malden, MA: Blackwell, 2002.

Zierler, David. *The Invention of Ecocide: Agent Orange, Vietnam, and the Scientists Who Changed the Way We Think about the Environment*. Athens: University of Georgia Press, 2011.

Index

Page numbers in italics refer to figures.